Reservoir Engineering Models

Reservoir Engineering Models

Analytical and Numerical Approaches

Turgay Ertekin and
Luis F. Ayala

McGraw Hill Education

New York Chicago San Francisco
Athens London Madrid
Mexico City Milan New Delhi
Singapore Sydney Toronto

Library of Congress Control Number: 2018945924

Reservoir Engineering Models: Analytical and Numerical Approaches

1 2 3 4 5 6 7 8 9 QVS 22 21 20 19 18

ISBN 978-1-259-58563-0
MHID 1-259-58563-8

Sponsoring Editor
Summers Scholl

Editorial Supervisor
Donna M. Martone

Acquisitions Coordinator
Elizabeth Houde

Project Manager
Poonam Bisht

Copy Editor
Mohammad Taiyab Khan

Proofreader
A. Nayyer Shamsi

Indexer
Dianna Haught

Production Supervisor
Pamela A. Pelton

Composition

Art Director, Cover
Jeff Weeks

Contents

Preface

For the last 100 years mathematical models of hydrocarbon reservoirs have become indisputable reservoir engineering and reservoir management tools. Albeit, some of these mathematical models as reservoir analysis tools were earlier treated with jaundiced eyes, but nowadays they are much more widely accepted and used. The electronic explosion we have been observing during the last 50 years and the associated developments in scientific computing and computer technologies have been instrumental in the change of winds toward the acceptance and use of mathematical models of hydrocarbon reservoirs.

From a scientific perspective, developing and using a mathematical model for a process require a certain degree of maturity in mathematics, physics, chemistry, thermodynamics, and transport phenomena, and a good degree of sophistication and experience on different computational platforms. There is no doubt that today's petroleum engineers and scientists are facing much more challenging reservoirs that need to be understood and analyzed. The new class of reservoirs of today will require more engineering and science so that another barrel of oil or another cubic foot of gas is produced. The presence of much more pronounced nonlinearities in the recovery processes and, in turn, mathematical expressions add even more complexity to the solution protocols.

The principal aim of this book is to describe how such different coupled physical, chemical, and thermodynamic phenomena can be assembled to capture and represent the mostly nonlinear phenomena involved. We believe that once this stage is effectively accomplished the mathematical models used by the engineers will be much more understandable and the friendly face of mathematical models of petroleum reservoirs will surface. Here by the word "understandable" we mean that the mathematical models should not be only perceived as a bunch of mathematical expressions that are put together but the physical, chemical, and thermodynamic effects corresponding to each group of terms are also clearly recognized. With this goal in mind, here we present a summary of the topical items covered in each chapter.

Chapter 1 briefly presents the overall framework of mathematical modeling of hydrocarbon reservoirs and summarizes the fundamental equations that are needed to be brought together in developing a hydrocarbon reservoir transport model. In *Chapter 2,* we attempt to give a concise mathematical development and representation of the generalized transport equation from the Lagrangian point of view as well as a discussion of Reynold's transport theorem. The chapter concludes with a review of equations of flow in porous media and a discussion of some useful coordinate transformations. Basic reservoir engineering principles are reviewed in *Chapter 3.* The discussion includes a review of basic rock and fluid properties, types of reservoir fluids, as well as some important concepts such as fluid flow potential, relative permeability theory, steady-state and unsteady-state flow regimes, and homogeneous-heterogeneous-isotropic-anisotropic reservoir flow domains. The hydrocarbon fluid models and the relevant solution thermodynamics are the subjects of *Chapter 4*. The various treatments of reservoir flow such as black oil representation, compositional formalisms, phase-equilibria equations, and how each of them are introduced into the reservoir flow equations are discussed in detail. *Chapter 5* focuses on how reservoir engineering transport equations are assembled. After developing the most comprehensive form of the compositional transport equations, then, in a systematic manner the generalized compositional flow equation is collapsed into its subsets. These subsets include different single-phase flow, two-phase flow, and three-phase flow (black-oil representation). By the end of Chapter 5, the reader is expected to develop a good understanding of how the flow equations of porous media are structured. The next step that will be of interest is how these equations are solved so that they become much more useful in practice. *Chapter 6* provides one of the most widely used analytical solution that is adapted to be used as an inverse solution tool for well test analysis applications. The entire process is considered as a systems analysis technique. The weaknesses and strengths of the approach are discussed in view of the simplifying assumptions that had to be incorporated to obtain a continuous solution. In this chapter, we also discuss some mathematical principles such as superposition principle and reciprocity principle, which are effective in removing some of the simplifying assumptions. The chapter concludes with a concise description of conformal mapping technique as applied to steady-state problems. *Chapter 7* is devoted to numerical reservoir engineering solutions which constitute the backbone of reservoir simulators. The finite-difference and finite-volume representation of flow equations are discussed. Analog representations for partial derivatives are developed using implicit and explicit representations. The implicit-pressure-explicit-saturation and simultaneous-solution methods are discussed for simultaneous solution of a system of partial differential equations. The Generalized Newton–Raphson protocol to solve systems of nonlinear algebraic equations is also introduced in this chapter. *Chapter 8* provides an introduction to workflows that have been developed more recently for surrogate modeling of hydrocarbon reservoirs. These knowledge-based models are considered as intelligent systems and in contrast to deterministic representations they capitalize on the inherent uncertainties of hydrocarbon reservoirs. What makes this type of applications attractive is when uncertainties encompass not only the distribution of the reservoir properties but also as they spread out to the description of processes and process controlling mechanisms.

We have coauthored this book in the hopes of sharing the excitement found in the assemblage of physics, chemistry, mathematics, and computing with the goal of developing and/or becoming a more prolific practitioner of mathematical models of hydrocarbon reservoirs. Please note that our goal is not to teach you advanced mathematics, physics, chemistry, or thermodynamics, rather to bring perspectives about how these fundamentally different scientific areas can be put to use together. Your engineering and scientific background and understanding of the fundamentals will provide the much needed glue that will keep everything in place.

Turgay Ertekin and Luis F. Ayala
University Park, PA
August 2018

Acknowledgments

We thank our families for being our source of strength, purpose, joy, and happiness. Their support and patience are the key reasons behind what we have accomplished in life. We dedicate this book to our families.

We also thank to our colleagues Dr. Qian Sun, Dr. Miao Zhang, and Dr. Venkataramana Putcha for assisting us with the typing of certain sections of the original manuscript, preparing some of the figures, and helping us with the subsequent revisions.

Nomenclature

A	=	cross-sectional area, L^2, ft^3; also a constant in Eq. 6.185
A_x	=	cross-sectional area along x-direction, L^2, ft^2
A_y	=	cross-sectional area along y-direction, L^2, ft^2
A_z	=	cross-sectional area along z-direction, L^2, ft^2
$A_{i,j,k}$	=	transmissibility coefficient using matrix notation between blocks (i, j, k) and $(i, j, k + 1)$, L^4t/m, STB/(Day-psia) for liquid, SCF/(Day-psia) for gas
B	=	an arbitrary vector field in Eq. 2.8; also formation volume factor, RB/STB, dimensionless
$B_{i,j,k}$	=	transmissibility coefficient using matrix notation between blocks (i, j, k) and $(i, j, k - 1)$, L^4t/m, STB/(Day-psia) for liquid, SCF/(Day-psia) for gas
B_g	=	gas formation volume factor, L^3/L^3, RB/SCF
B_o	=	oil formation volume factor, L^3/L^3, RB/STB
B_w	=	water formation volume factor, L^3/L^3, RB/STB
c	=	compressibility, Lt^2/m, psi^{-1}
c_f	=	fluid compressibility, Lt^2/m, psi^{-1}
c_g	=	gas compressibility, Lt^2/m, psi^{-1}
c_I	=	overall molar composition of species 'I' in thermodynamic system, fraction
c_o	=	oil compressibility, Lt^2/m, psi^{-1}
c_R	=	formation rock compressibility, Lt^2/m, psi^{-1}
c_t	=	total compressibility, Lt^2/m, psi^{-1}
c_w	=	water compressibility, Lt^2/m, psi^{-1}
c_ϕ	=	pore compressibility, Lt^2/m, psi^{-1}
c_1	=	integration constant used in Eq. 6.126, dimensionless.
C_A	=	shape factor, dimensionless
$C_{i,j,k}$	=	transmissibility coefficient using matrix notation of central block (i, j, k), L^4t/m, STB/(Day-psia) for liquid, SCF/(Day-psia) for gas
C_{Ia}	=	molar concentration of component "I" in the a-phase, n/L^3, moles/ft^3
C_{Il}	=	molar concentration of component "I" in the l-phase, n/L^3, moles/ft^3
C_{Iv}	=	molar concentration of component "I" in the v-phase, n/L^3, moles/ft^3

C_s	=	wellbore storage constant, L^4t^2/m, bbl/psi
C_{SD}	=	dimensionless wellbore storage constant, dimensionless
CV	=	control volume
d_n	=	distance from the actual well to the nth image well, L, ft
D	=	Inertial-turbulence (*IT*) flow factor for the system in Chap. 6, dimensionless; also flow domain characteristics in Chap.8
E	=	error encountered in the middle layers when training a recurrent ANN
$E_{i,j,k}$	=	transmissibility coefficient using matrix notation between blocks (i, j, k) and $(i + 1, j, k)$, L^4t/m, STB/(Day-psia) for liquid, SCF/(Day-psia) for gas
$f_{I,J}$	=	fugacity of *I*-th species in *J*-th phase, m/Lt^2, psia
F	=	Gibbs phase rule's degrees of freedom, dimensionless, and arbitrary function in Chap.6
g	=	local gravitational acceleration, L/t^2, ft/s^2
g_c	=	unit conversion factor, 32.174 lb-ft/(lbf-s^2)
G	=	elevation with respect to datum, L, ft
GOR	=	producing gas oil ratio, L^3/L^3, SCF/STB
WOR	=	producing water oil ratio, L^3/L^3, STB/STB
h	=	thickness, L, ft
i	=	imaginary number
I	=	input layer of an ANN model
J	=	quantity flux vector
K_I	=	mass composition-based equilibrium ratio for component "*I*" in a liquid-vapor system, dimensionless
K_G	=	mass composition-based equilibrium ratio for surface-gas component in a reservoir oil-gas system, dimensionless
K_O	=	mass composition-based equilibrium ratio for surface-oil component in a reservoir oil-gas system, dimensionless
K_I^*	=	molar composition-based equilibrium ratio for component "*I*" in a liquid-vapor system, dimensionless
$K_{I,J}$	=	mass composition-based equilibrium ratio for component "*I*" in phase "*J*" with respect to a reference phase in a multiphase system, dimensionless
k	=	permeability, L^2, darcy
$\overline{k}$	=	permeability tensor, L^2, darcy
$\overline{k}_{i,j,k}$	=	the geometric mean of the directional permeabilities in the plane that the well is orthogonal, L^2, darcy
k_{rg}	=	relative permeability to gas, dimensionless
k_{ro}	=	relative permeability to oil, dimensionless
k_{rw}	=	relative permeability to water, dimensionless
k_{rocw}	=	relative permeability to gas at irreducible water saturation, dimensionless
k_{rog}	=	relative permeability to oil in oil/gas system, dimensionless
k_{row}	=	relative permeability to oil in oil/water system, dimensionless
k_x	=	permeability along x direction, L^2, darcy
k_y	=	permeability along y direction, L^2, darcy

k_z	=	permeability along z direction, L^2, darcy
L_f	=	Distance to the fault, L, ft
m	=	degree of implicitness/explicitness in a numerical scheme (n, $n + 1/2$, or $n + 1$)
$\dot{m}_{IJ}$	=	mass flow of *I-th* component in *J-th* phase, m/t, lb/Day
M_J	=	total mass of *J-th* phase, m, lb
M_o	=	total mass of the reservoir oil phase, m, lb
M_g	=	total mass of the reservoir gas phase, m, lb
M_{IJ}	=	mass of *I-th* component in *J-th* phase, m, lb
M_{Oo}	=	mass of surface-oil in reservoir oil phase, m, lb
M_{G_o}	=	mass of surface-gas in reservoir oil phase, m, lb
M_{G_g}	=	mass of surface-gas in reservoir gas phase, m, lb
M_{o_g}	=	mass of surface-oil in reservoir gas phase, m, lb
MW	=	molecular weight, m/n, lb/lbmol
MW_{air}	=	molecular weight of air, m/n, 28.97 lb/lbmol
MW_{ISC}	=	molecular weight of *I*-th surface component at standard conditions, m/n, lb/lbmol
MW_{OSC}	=	molecular weight of surface-oil at standard conditions (stock-tank), m/n, lb/lbmol
MW_{gsc}	=	molecular weight of surface-gas at standard conditions, m/n, lb/lbmol
MW_J	=	molecular weight of the *J*-th phase, m/n, lb/lbmol
n	=	time-step level in a numerical formulation, symbol
n_g	=	relative permeability exponent of gas phase
n_{Ilx}	=	molar flux of component "*I*" of phase "*l*" along x-direction, n/L^2-t, lbmole/ft^2-day
n_o	=	relative permeability exponent of oil phase
n_w	=	relative permeability exponent of water phase
N	=	original oil in place, L^3, STB
$N_{i,j,k}$	=	transmissibility coefficient using matrix notation between blocks (i, j, k) and (i, $j + 1$, k), L^4t/m, STB/(Day-psia) for liquid, SCF/(Day-psia) for gas
N_{Ilx}	=	molar rate of component "*I*" of phase "*l*" along x-direction, n/t, lbmol/day
N_C	=	number of components, dimensionless
N_P	=	number of phases, dimensionless
N_{IJ}	=	moles of component "*I*" in phase "*J*", n, lbmol
N_J	=	total moles of phase "*J*", n, lbmol
o	=	output layer of an ANN model
p	=	pressure, m/Lt^2, psia
p_c	=	critical pressure, m/Lt^2, psia
p_{cgo}	=	gas/oil capillary pressure, m/Lt^2, psia
p_{cow}	=	oil/water capillary pressure, m/Lt^2, psia
p_e	=	reservoir external boundary pressure, m/Lt^2, psia
p_g	=	gas pressure, m/Lt^2, psia
p_i	=	initial pressure, m/Lt^2, psia
p_o	=	oil pressure, m/Lt^2, psia

p_{pr}	=	pseudoreduced pressure, dimensionless
p_R	=	average reservoir pressure, m/Lt^2, psia
p_{sc}	=	standard condition pressure, m/Lt^2, psia
p_t	=	dimensionless pressure drop at the wellbore, dimensionless
p_{sf}	=	sandface pressure, m/Lt^2, psia
p_w	=	water pressure, m/Lt^2, psia
p_{wf}	=	well bottomhole flowing pressure, m/Lt^2, psia
p_{ws}	=	shut-in well pressure, m/Lt^2, psia
p_δ	=	pressure at $\Delta t = \delta$, m/Lt^2, psia
Δp	=	pressure drop, m/Lt^2, psia
Δp_D	=	dimensionless pressure drop, dimensionless
$\Delta p'_D$	=	temporary dimensionless pressure drop, dimensionless
Δp^*_D	=	temporary variable used to reduce the order of the ODE in Eq. 6.117
Δp_{IT}	=	additional pressure drop due to inertial-turbulence effects
q	=	production or flow rate, L^3/t, RB/Day
q_{gsc}	=	gas production or flow rate, L^3/t, SCF/Day
q_{osc}	=	oil production or flow rate, L^3/t, STB/Day
q_{wsc}	=	water production or flow rate, L^3/t, STB/Day
q_D	=	dimensionless flow rate, dimensionless
Q^*_I	=	external injection or production of component "I", n/t, moles/day
$Q_{i,j,k}$	=	right-hand-side coefficient of governing flow equation using matrix notation for block (i, j, k) L^3/t, STB/Day for liquid, SCF/Day for gas
r	=	radial distance in both cylindrical and spherical coordinator system, L, ft
r_d	=	effective drainage radius, L, ft
r_D	=	dimensionless radius, dimensionless
r_e	=	Peaceman's equivalent well block radius, L, ft; and external reservoir radius in Chap. 6, L, ft
r_{eD}	=	dimensionless external radius, dimensionless
r_{inv}	=	radius of investigation, L, ft
r_w	=	well block radius, L, ft
R	=	universal gas constant; also, production mechanism and/or recovery process in Chap. 8
$R_{i,j}$	=	residual, L^3/t, STB/day for liquid, SCF/day for gas
R_s	=	solution surface-gas to surface-oil ratio in reservoir oil phase, L^3/L^3, SCF/STB
R_v	=	solution surface-oil to surface-gas ratio in reservoir gas phase, L^3/L^3, STB/SCF
s	=	skin factor, dimensionless
s_a	=	apparent skin due to restricted entry
S	=	boundary surface
S_c	=	sealing capacity of the fault, dimensionless
$S_{i,j,k}$	=	transmissibility coefficient using matrix notation between blocks (i, j, k) and $(i, j-1, k)$, L^{4}t/m, STB/(Day-psia) for liquid, SCF/(Day-psia) for gas

S_g = gas saturation, fraction
S_{gc} = critical gas saturation, fraction
S_{gi} = initial gas saturation, fraction
S_{gn} = normalized gas saturation, fraction
S_{iw} = irreducible water saturation, fraction
s_{IT} = skin due to inertial-turbulence flow effects, dimensionless
S_J = saturation of J-phase, fraction
S_o = oil saturation, fraction
S_{oi} = initial oil saturation, fraction
S_{on} = normalized oil saturation, fraction
S_{or} = residual oil saturation, fraction
S_{org} = residual oil saturation in oil/gas system, fraction
S_{orw} = residual oil saturation in oil/water system, fraction
S_w = water saturation, fraction
S_w^* = normalized water saturation, dimensionless
S_{wi} = initial water saturation, fraction
S_{wir} = irreducible water saturation, fraction
t = time, t, days
t_D = dimensionless times, dimensionless
t_p = total production time, t, days
t_s = stabilization time, t, days
t_{wsD} = dimensionless time when wellbore storage effects disappear, dimensionless
t_{ws} = time when wellbore storage effects disappear, t, days
Δt_e = where *pwf* is the measured bottomhole pressure just before the well is shut in, t, days
T = absolute temperature, T, °R
T^* = boundary transmissibility between adjacent blocks, L^4t/m, STB/(Day-psia)
T_G = surface-gas transmissibility factor between adjacent blocks, L^4t/m, SCF/(Day-psia)
T_O = surface-oil transmissibility factor between adjacent blocks, L^4t/m, STB/(Day-psia)
T_W = surface-water transmissibility factor between adjacent blocks, L^4t/m, STB/(Day-psia)
T_{IJ} = compositional transmissibility factor for the flow of *I*-th component in *J*-th phase between adjacent blocks, L^4t/m, STB/(Day-psia)
T_{sc} = standard condition temperature, T, °R
v = superficial velocity, L/t, RB/(Day-ft^2)
v_r = superficial velocity component along *r* direction, L/t, RB/(Day-ft^2)
v_x = superficial velocity component along *x* direction, L/t, RB/(Day-ft^2)
v_y = superficial velocity component along y direction, L/t, RB/(Day-ft^2)
v_z = superficial velocity component along *z* direction, L/t, RB/(Day-ft^2)
V = volume, L^3, ft^3; also, a weight matrix between input layer and an adjacent hidden layer in Chap. 8

V_b	=	bulk volume, L^3, ft^3
V_p	=	pore volume, L^3, ft^3
V_o	=	volume of reservoir oil, L^3, ft^3
V_g	=	volume of reservoir gas, L^3, ft^3
V_w	=	volume of reservoir water, L^3, ft^3
V_{ws}	=	total wellbore volume, L^3, ft
V_t	=	total fluid volume, L^3, ft^3
$(V_{Osc})_o$	=	volume of surface-oil in reservoir oil produced surface conditions, L^3, STB
$(V_{Osc})_g$	=	volume of surface-oil in reservoir gas produced surface conditions, L^3, STB
$(V_{Gsc})_g$	=	volume of surface-gas in reservoir gas produced surface conditions, L^3, SCF
$(V_{Gsc})_o$	=	volume of surface-gas in reservoir oil produced surface conditions, L^3, SCF
w	=	transform function in Chap. 2; work/kinetic energy in Chap. 3, m/Lt^2, psia
W	=	weight matrices between hidden layers
$W_{i,j,k}$	=	transmissibility coefficient using matrix notation between blocks (i, j, k) and $(i-1, j, k)$, L^4t/m, STB/(Day-psia) for liquid, SCF/(Day-psia) for gas
x	=	distance along x-direction in the Cartesian coordinate system, L, ft
x_e	=	location of the east boundary in a finite-control volume, L, ft
x_w	=	location of the west boundary in a finite-control volume, L, ft
Δx	=	control volume dimension along the x direction, L, ft
x_{IJ}	=	molar composition of I-th component in J-th phase, fraction
x_{O_o}	=	molar composition of surface-oil in reservoir oil phase, fraction
x_{G_o}	=	molar composition of surface-gas in reservoir oil phase, fraction
x_{G_g}	=	molar composition of surface-gas in reservoir gas phase, fraction
x_{O_g}	=	molar composition of surface-oil in reservoir gas phase, fraction
X	=	spatial transformation variable along x-direction in Chap. 2, L^{-1}; also temporary variable used on Boltzman transformation in Chap. 6, dimensionless.
Y	=	spatial transformation variable along y-direction in Chap. 2, L^{-1}
y	=	distance in y-direction in the Cartesian coordinate system
Δy	=	control volume dimension along the y-direction, L, ft
y_n	=	location of the north boundary in a finite-control volume, L, ft
y_s	=	location of the south boundary in a finite-control volume, L, ft
z	=	distance along z-direction in the Cartesian coordinate system
z_a	=	location of the above boundary in a finite-control volume, L, ft
z_b	=	location of the below boundary in a finite-control volume, L, ft
Δz	=	control volume dimension along the z-direction, L, ft
z_I	=	overall mass composition of species "I" in the thermodynamic system, fraction
z_G	=	overall mass composition of surface-gas in the thermodynamic system, fraction
z_O	=	overall mass composition of surface-oil in the thermodynamic system, fraction
Z	=	spatial transformation variable along z-direction in Chap. 2, L^{-1}; gas compressibility factor, dimensionless

Greek letters

α_c	=	volumetric conversion factor, 5.615 SCF/STB
α_J	=	fractional molar amount of *J*-th phase with respect to total system moles, fraction
β_J	=	fractional mass amount of *J*-th phase with respect to total system mass, fraction
β_g	=	fractional mass amount of gas phase with respect to total system mass, fraction
β_o	=	fractional mass amount of oil phase with respect to total system mass, fraction
β_n	=	root of the product function in Eq. 6.156
∇	=	divergence operator
θ	=	angle in θ direction in both cylindrical and spherical coordinator system, rad; or time-weighting factor in time-dependent numerical schemes ($\theta = 0$, ½, or 1), dimensionless
δ	=	time interval during a pressure buildup test, t, days in Chap. 6; also, errors encountered in each training epoch of an ANN model in Chap. 8
ρ	=	mass density, m/L^3, lb/ft^3
ρ_J	=	mass density of the *J*-th phase, m/L^3, lb/ft^3
ρ_o	=	mass density of reservoir oil phase, m/L^3, lb/ft^3
ρ_g	=	mass density of reservoir gas phase, m/L^3, lb/ft^3
ρ_w	=	mass density of reservoir water phase, m/L^3, lb/ft^3
ρ_{osc}	=	mass density of surface-oil at surface conditions, m/L^3, lb/ft^3
ρ_{gsc}	=	mass density of surface-gas at surface conditions, m/L^3, lb/ft^3
ρ_{wsc}	=	mass density of surface-water at surface conditions, m/L^3, lb/ft^3
$\hat{\rho}$	=	molar density, m/L^3, lbmol/ft^3
$\hat{\rho}_o$	=	molar density of reservoir oil phase, m/L^3, lbmol/ft^3
$\hat{\rho}_g$	=	molar density of reservoir gas phase, m/L^3, lbmol/ft^3
$\hat{\rho}_{osc}$	=	molar density of surface-oil at surface conditions, m/L^3, lbmol/ft^3
$\hat{\rho}_{Gsc}$	=	molar density of surface-gas at surface conditions, m/L^3, lbmol/ft^3
γ	=	unit conversion factor
γ_g	=	gas gravity, m/L^{2}t^2, psi/ft
γ_o	=	oil gravity, m/L^{2}t^2, psi/ft
γ_w	=	water gravity, m/L^{2}t^2, psi/ft
γ_{gsc}	=	surface-gas specific gravity, dimensionless
γ_{osc}	=	surface-oil specific gravity, dimensionless
$\xi_{i,j,k}$	=	accumulation coefficient in time-dependent numerical simulation equations, L^3/t, ft^3/day
Γ	=	generation term, m/L^3
Γ_{IJ}	=	rate of mass transfer of species "*I*" into phase "*J*" per unit time and bulk volume, m/t/L^3, lbm/day/ft^3
ϕ	=	porosity, fraction
Φ_g	=	gas flow potential, m/Lt^2, psia
Φ_o	=	oil flow potential, m/Lt^2, psia
Φ_w	=	water flow potential, m/Lt^2, psia

ψ	=	real gas potential, m/L t^3, psia^2/cp
ω_{IJ}	=	mass composition of *I*-th component in *J*-th phase, fraction
ω_{Oo}	=	mass composition of surface-oil in reservoir oil phase, fraction
ω_{Go}	=	mass composition of surface-gas in reservoir oil phase, fraction
ω_{Wo}	=	mass composition of surface-water in reservoir oil phase, fraction
ω_{Gg}	=	mass composition of surface-gas in reservoir gas phase, fraction
ω_{Og}	=	mass composition of surface-oil in reservoir gas phase, fraction
ω_{wg}	=	mass composition of surface-water in reservoir gas phase, fraction
ω_{Gw}	=	mass composition of surface-gas in reservoir water phase, fraction
ω_{Ow}	=	mass composition of surface-oil in reservoir water phase, fraction
ω_{Ww}	=	mass composition of surface-water in reservoir water phase, fraction
Ω	=	well productivity/injectivity index, L^4t/m, STB/(Day-psia) for liquid, SCF/(Day-psia) for gas; mathematical flow model with relevant physics and thermodynamics in Chap. 8
η	=	hydraulic diffusivity, L^2/t
ε	=	tolerance
λ	=	relative permeability coefficient in Chap. 3; also unit conversion factor in Chap. 6
λ_{Gg}	=	surface-oil mobility in reservoir gas flow, L^3t/m, SCF/(Day-psi-ft)
λ_{Go}	=	surface-oil mobility in reservoir oil flow, L^3t/m, SCF/(Day-psi-ft)
λ_{Gsc}	=	total surface-gas mobility in surface-volume units, L^3t/m, SCF/(Day-psi-ft)
λ_{Oo}	=	surface-oil mobility in reservoir oil flow, L^3t/m, STB/(Day-psi-ft)
λ_{Og}	=	surface-oil mobility in reservoir gas flow, L^3t/m, STB/(Day-psi-ft)
λ_{Osc}	=	total surface-oil mobility in surface-volume units, L^3t/m, STB/(Day-psi-ft)
λ_{Ww}	=	surface-water mobility in reservoir water flow, L^3t/m, STB/(Day-psi-ft)
λ_{Wsc}	=	total surface-water mobility in surface-volume units, L^3t/m, STB/(Day-psi-ft)
μ_g	=	gas viscosity, m/Lt, cp
μ_o	=	oil viscosity, m/Lt, cp
μ_w	=	water viscosity, m/Lt, cp

CHAPTER 1
Introduction

In working with petroleum hydrocarbon reservoirs, we typically have a very small window to view the changes that are taking place in the system as a function of time, space, and geometry. At the same time, we are bound to analyze the performance of the system for optimization and forecasting purposes within a reasonable degree of accuracy. At the end, most of our analyses will be geared toward establishing optimized field management strategies. In building and/or using a mathematical model it is imperative to know and understand the underlying laws and principles, and relations that are used. Furthermore, a good degree of understanding of the implementation of certain assumptions and their net effects and implications on the results generated is a must. Finally, having insights on the suitable mathematical rearrangements of the relevant equations will only make the user more powerful. Obviously, without a good understanding of the physics and mathematics involved, it will not be possible to construct either the mathematical or the numerical model. In this chapter, we particularly address the following:

- What are the fundamental laws and relations that are used in the construction of a model?
- How fundamental laws and relations are expressed in a certain set of mathematical formalisms?
- What are the simplifying assumptions that are built into the formulation of the problem?
- What type of solution and information can be generated from the model?
- What are the practical implications of the solutions that are generated by the model?

1.1 Need for Reservoir Modeling

Petroleum engineering operations are known to be heavily capital intensive. For example, drilling a single exploratory well in ultra-deep waters may cost hundreds of millions of dollars. Implementing a full-scale field development project will result in expending hundreds of millions of dollars. Accordingly, before starting with the implementation of any project it will be extremely important to reduce the risk factors to acceptable levels so that a *"go ahead"* decision from the management can be obtained. During the planning and feasibility studies of any project in any engineering area it will be necessary to perform an in-depth analysis of processes that will be taking place in the system. Although similar analysis procedures are applicable, there is one important aspect of petroleum engineering analyses as the reservoir systems and their controlling parameters under investigation are not well defined. In mathematical words, almost invariably, the number of unknowns by far exceeds the number of equations available. Therefore, it will always be necessary to bring in our scientific and engineering understanding, creativity, and intuition together. Within this context, reservoir models become important means that enable us studying a large number of scenarios as we test the sensitivity of the results to some of the parameters that are not known. In petroleum engineering applications there are many external and internal factors that complicate the reservoir behavior. The presence of heterogeneity and anisotropy together with the presence of nonlinear intricacies of relationships between reservoir fluids and reservoir rock make the needed analysis even more challenging. At this juncture, reservoir models help us in establishing a good degree of formalism in reservoir analysis.

1.2 Purpose of Reservoir Modeling

During the discovery stage of a field, information that we have about the field on one hand is scanty at best and is disjointed on the other hand. This makes the integration and assemblage of these pieces of information even a more arduous task. Accordingly, a reservoir model that is developed during the initial periods of a newly discovered field is expected to provide only a coarse description of the reservoir. As field development progresses and production from the field continues, more information will become available so that the model can be continually updated and refined. As the constructed models go through such transformations and become more tuned to the reservoir, the user starts to establish a good level of confidence on the results generated by the model. At that point in time, the reservoir model is said to be mimicking the actual behavior of the field and all of its processes effectively. Now, we have a tool which can be used to answer questions such as: What is the distribution of reservoir properties? Where are the reservoir boundaries located? What types of boundaries and boundary conditions are acting upon the reservoir? According to which pattern wells should be drilled? What is the most efficient well spacing? At what rates wells should be produced? When and in which sections of the reservoir we should implement infill drilling? When will the primary recovery period come to an end? What kind of an improved oil recovery technique should be implemented? These are some of the questions that can be answered with the help of a reservoir simulator. Obviously, our level of confidence on the results generated is strictly controlled on the accuracy and quality of the input data.

1.3 Classical Reservoir Engineering Protocols versus Numerical Modeling

"While classical reservoir engineering applications attempt to bring an exact solution to an approximate problem, reservoir modeling generates an approximate solution to an exact problem." This statement provides a good description of the fundamental difference between classical reservoir engineering analysis and reservoir simulation approach. It will be correct to state that classical reservoir engineering protocols such as decline curve analysis, well test analysis, and material balance approach are simply subsets of the reservoir simulation formalism. In other words, it is possible to start with a compositional formulation of a reservoir in three dimensions, and after implementing a series of assumptions it will be possible to collapse the complicated reservoir simulation formulation to a simple material balance equation formulation (Ertekin et al., 2001). In the course of implementing a series of assumptions we will be simplifying and idealizing the problem or in other words approximating it. If it is preferable to maintain the description of the problem intact, then we have to sacrifice from the accuracy of the solution because an exact solution using analytical tools will not be possible and hence we will end up with an approximate solution to an accurately described problem. It is the decision of the reservoir engineer to use conventional reservoir engineering analysis tools or numerical simulation. However, this decision must be based on a number of factors including the availability and extent of the data required, availability of personnel, and time and budgetary considerations.

1.4 Basic Components of a Reservoir Model

The very first phase of the development of a reservoir model is the identification of the fundamental equations and auxiliary equations that are needed to construct the right mathematical formulation together with the constraints and assumptions that make the solution tractable. Obviously, the ultimate goal is to ensure that the proposed formulation captures the actual physics and the processes that are in place.

1.4.1 Fundamental Equations of the Mathematical Model

In the development of mathematical formulations describing the fluid flow dynamics in porous media, following set of five fundamental equations and principles are used to represent the overall framework of the model.

The continuity equation. The continuity equation is the application of the principle of conservation of mass. The conservation of mass is applied to a hypothetical enclosure fixed in space (porous medium); we call the space enclosed the *control volume* and its boundary the *control surface*. The function of the control volume is to represent the system's macroscopic behavior by taking into consideration the net average of the microscopic effects. Fig. 1.1 shows the concept of control volume and control surface. It should be noted that shape of the control volume depends on the coordinate system selected to describe the flow geometry. For example, in rectangular coordinates the control volume has the shape of a rectangular prism, whereas the control volume depicted in Fig. 1.1 is drawn for a curvilinear grid system.

The momentum equation. This equation represents a force balance on the system and introduces the Newton's second law of motion to fluid system. It simply states that rate of momentum accumulation in a volume element is equal to rate of momentum in minus rate of momentum out, plus sum of forces acting on the system. The momentum equation reduces to a force balance where (mass) × (acceleration) is equal to summation of pressure, and viscous and gravitational forces acting on the system. When some system-specific approximations such as laminar flow, negligible inertial effects are in place, the momentum equation is introduced into the formulation through Darcy's law.

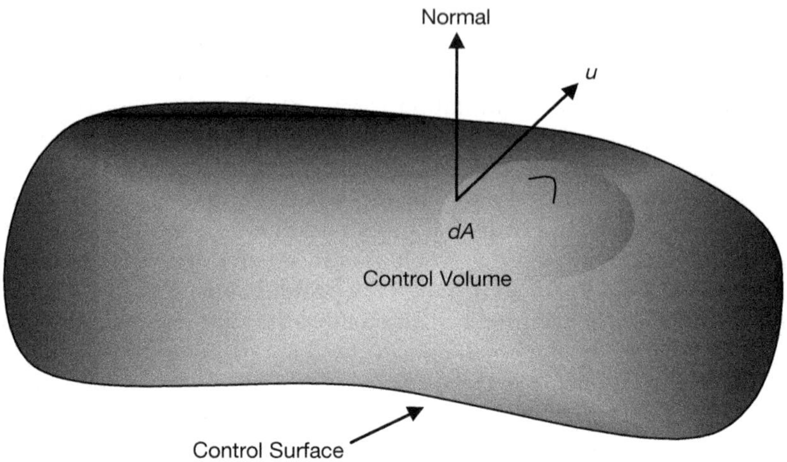

Figure 1.1 Control volume and control surface.

The constitutive equation. In modeling, the constitutive equation describes the rheological behavior of a fluid and it is captured in reservoir engineering through Darcy's law which describes the rate of fluid movement within the system for a Newtonian fluid.

The equation of state. The equation of state describes the thermodynamical state of the fluid. For example, the variation of density with pressure and temperature, and mole fraction distribution of different components among oleic, aqueous, and vapor phases are all calculated with the help of an equation of state. For pure liquids, the use of universal definition of compressibility can also be considered as an equation of state under isothermal conditions. For pure gases, real gas law provides a powerful equation for the calculation of the density of a gas at different pressure and temperature conditions.

The energy balance. The energy equation is similar to the continuity equation and introduces the law of conservation of energy. The use of energy balance becomes very important in nonisothermal systems such as thermal recovery processes. In representing the processes that do not involve thermal recovery operations the energy effects, in general, are neglected.

The mathematical model is not completed unless the associated *boundary* and *initial* conditions are specified. Sometimes it is even more crucial to establish strategies how to attack the problem numerically depending upon the nature of the boundary conditions. In reservoir simulation, boundary conditions are specified at the internal boundaries (well points) and at the external boundaries (physical limits) of the reservoir. Boundary conditions are algebraic conditions on the values of the functions (dependent variables such as pressures and saturations).

Boundary conditions can be as simple as requiring that the dependent variables have certain numerical values. For example, specifying a constant pressure production scheme at the well point is in agreement with this definition. Similarly, when a no-flow boundary condition is specified at the external boundaries, it is simply meant that gradient of the pressure across that boundary is set to be equal to zero. As it will be shown later, a constant flow rate specification at the well point is also equivalent to specification of the gradient of the pressure at the sandface. Boundary conditions, as the phrase implies, brings boundness to the solution. In other words, the solution that is sought would lie between the values that are specified at the boundaries. Initial conditions provide a starting value for the dependent variables.

One final item that needs to be watched closely is that the number of equations and the number of unknowns at any point in space and time must agree. When this check is satisfied, then one will be able to say that the mathematical formulation is capable of describing a well-posed problem.

1.4.2 The Numerical Model

When the mathematical model (formulation) is completed, we will be seeking solutions to the formulated process(es). In most of our applications, we will be seeking solutions for the pressure and saturation functions as a function of time and space. The correct utilization of the equations and principles described results in the mathematical formulation of the problem. Typically, the mathematical model is composed of partial differential equations and some algebraic equations (boundary and initial conditions can also be included in this group). The partial differential equations describe the processes that are taking place and help us in tracking the changes in the values of the dependent variables as the processes continue. Mass and energy balances fall into this category.

A second group of equations that appear in the mathematical formulation are algebraic equations and they are known as auxiliary equations of the process formulated. The role of these equations is to bring a *closure* to the formulation. The capillary pressure equations, saturation relationship, and equilibrium equations all fall into this category. While these equations explicitly describe some of the existing relationships among some of the principal unknowns, at the same time they ensure that in the inventory of the equations and unknowns there is one to one correspondence. Since equations of the mathematical model are composed of differential equations and algebraic equations that need to be solved simultaneously, it will be necessary to convert the differential equations into the algebraic forms. This is usually achieved with the help of numerical tools, and in reservoir simulation the most widely used numerical tools are *finite-differences* and *finite-volumes* techniques. As it will be shown later, these techniques transform the continuous form of the partial differential equations into algebraic forms which are expressed as some preselected discrete points in the space and time domain.

1.4.3 The Computer Model

At the end of the development of the numerical model, one invariably ends up with a large number of algebraic equations (linear or nonlinear) that are expressed in terms of a large number of unknowns. In some of the field application studies the number of equations and unknowns that need to be solved simultaneously may easily exceed several millions. This is when one will need the computational power. To present the problem to the computing device a computer program needs to be developed so that the computing device can take the necessary orders to execute certain algorithmic protocols. The computer model is a collection of these algorithmic protocols. Depending upon the sophistication level of the computer code, these computer programs can be quite challenging to develop from scratch. In general, computer programs are available commercially and they are provided with a user's manual to the practicing simulation engineer and/or scientist.

1.5 Why Model the Reservoir?

1.5.1 Newly Discovered Fields

The need to model the reservoir arises from the fact that in a hydrocarbon recovery project, which in most cases is capital intensive, the risk associated with the selected development plan must be assessed and minimized. Thus, a petroleum engineer desires to obtain an accurate performance prediction of the reservoir under various operating conditions. This can be achieved through the proper use of reservoir simulation concepts. Factors contributing to the risk, such as the complexity of the reservoir, variations of fluid and rock-fluid properties, and the complexity of hydrocarbon recovery mechanisms, are taken into consideration in reservoir simulation through the input data for the simulator.

Reservoir simulation can be used in almost all stages of field operation, ranging from newly discovered fields to the improved recovery projects. In a newly discovered field, the obtainable information is scanty. Moreover, it is perhaps even disjointed because pieces of information are obtained from different parts of the field. The first task is to integrate the available pieces of information as accurately as possible in order to construct an overall picture of the reservoir. In planning the field development, a reservoir simulator can be utilized to establish the most optimum field development plan, such as number of wells to be drilled, well spacing, and well locations. It can also be used to establish optimum operation strategies, recovery mechanisms, and take into account future needs for improved recovery.

1.5.2 Mature Fields

After the field has been fully developed and produced for a period of time, the need for further field operations might arise in order to increase or maintain the recovery of the reservoir. Such field operations can be the in-fill-drilling program in which the appropriate locations and number of in-fill wells are investigated for the improved recovery applications. The questions of when and how such operations should be implemented can be answered by reservoir simulation study.

1.5.3 Project Screening

Petroleum projects are capital-intensive projects in nature. Thus, the implementation of the projects must be executed after thorough and accurate studies. Reservoir modeling is a practical tool which the engineer can use to design and conduct tests that can sufficiently answer some of the questions that engineers are confronted with. The results obtained from performing modeling exercises on different projects provide supporting information pertaining to whether or not such project should be implemented. In other words, reservoir modeling can effectively serve as a powerful tool for project screening.

1.5.4 Modeling as a Dynamic Process

Reservoir modeling is a dynamic process in which the reservoir model is continuously refined as field development and operation progresses and more information becomes available. This process yields a more accurate reservoir description and hence increases the predictive capability of the model. An example of such process is a history matching process in numerical simulation, in which the production history from the well is used in tuning some relevant properties that contain uncertainty, such as relative permeability. Through this process, the property is adjusted until the results from the simulator match the production data from the well. The obtained model is then believed to have a good level of confidence and hence can be used to forecast the performance of the reservoir under different operation scenarios.

1.6 Summary

Reservoir modeling is a powerful tool in the development and management of hydrocarbon reservoirs. It is an effective vehicle through which we can examine the past and future performance of a reservoir. Utilizing our understanding of the physical principles governing the fluid flow dynamics in porous media and the available data, we put our best effort forward to express the problem in terms of mathematical equations and solve them for the purpose of exploring the future of reservoir for various numbers of scenarios under consideration. In achieving this, it should be noted that both mathematics and physics will go hand-to-hand.

One should not forget that in our attempts toward representing the reservoir we are introducing significant approximations. Therefore, it is of paramount importance to keep in mind the various assumptions and approximations that are introduced into the mathematical representation, the limitations of the analytical or numerical tools, and their overall effects on the results generated.

A modeling study, while challenging and demanding, provides ample amount of opportunity to learn more of the reservoir systems that we are dealing with. It is in the hands of the reservoir engineer and scientist to use this powerful tool effectively to reap the maximum benefits from it.

Reference

Ertekin, T., J. Abou-Kassem, and G. King, *Basic Applied Reservoir Simulation*, SPE Textbook Series Vol. 7, Society of Petroleum Engineers, Richardson, TX, 2001.

CHAPTER 2

Mathematics of Reservoir Engineering

The phrase "Mathematical Representation" has been used to connote a number of strictly different things. It can be used, for instance, to describe simple empirical correlations of experimental data, while by the same token, it has been used to describe systems of highly nonlinear partial differential equations governing the behavior of a rather complex system. Here, we define a mathematical model as a set of differential and/or algebraic equations with representative initial and boundary data describing the behavior of a physical system exposed to internal and/or external forces which change the state of equilibrium. It will be necessary to focus on the word "describing" used in the previous statement as it is used rather loosely here since usually our mathematical description is limited by our understanding of the physics of the problem as well as the various simplifying assumptions made.

Although the physics of a problem can take place in widely differing domains and scales such as heat transfer, mass transfer, and electromagnetism, there is one basic cause of all effects, accordingly the mathematical descriptions of the resultant effects are analogous. Developing a formalism for a given problem is essentially translating the controlling physics into some set of mathematical expressions. The basic process of formulating a mathematical model consists of the following three major stages:

1. *The physics of the problem:* This stage is the most difficult one because it requires a solid and sound foundation and hence otherwise the described physics will not capture the actual phenomena and efforts will result in wrong mathematical representation.
2. *Inclusion of the basic laws of physics:* Most often than not, we assume that the system under consideration can be described as a specific process—a basic law of physics need to be invoked. For example, a continuity equation written for a mass transport problem will not describe the flow in porous media until the velocity terms in the continuity equation are supported by Darcy's law.
3. *Simplifying to a treatable problem:* Most often than not, simplifying assumptions are made to enable us to tackle the resulting mathematical model. With the definition of the appropriate initial and boundary conditions, we finalize the complete representation of the problem.

Once the formulation of the problem is completed, then we start to focus on the solution of the problem. Solutions based on hard-computing are categorized as analytical and numerical solutions. Analytical solutions are continuous in nature and they provide exact solutions to an approximate problem (approximate nature of the problem arises due to a set of simplifying assumptions such as homogenous property distribution and removal of nonlinearities). Numerical solutions, however, are not continuous and they are obtained at preselected discrete locations and times. Numerical solutions provide approximate answers to a problem, which is represented almost in its exact form.

2.1 The Generalized Transport Equation (GTE)

2.1.1 The Lagrangian View

Let us consider the transport of some physical quantity, for example, mass, momentum, and energy in a material volume, moving in space with a bulk velocity. Let α be the amount of this intensive quantity, expressed per unit volume within a system, and A its total extensive amount ($A = \iiint_{V(t)} \alpha dV$). We seek to describe the net rate of change of A inside a system of arbitrary volume $V(t)$ found within a flow field. The volume $V(t)$ is a dynamic time-dependent region defined by a boundary surface area $S(t)$—which is a surface also deforming with time. The principle of conservation of this quantity states:

$$\begin{gathered}[\text{Rate of storage within } V(t)] = \\ [\text{Rate of outflow/inflow into } V(t)] + \\ [\text{Rate of quantity generation within } V(t)]\end{gathered} \tag{2.1}$$

Let us look at the individual entries of Eq. (2.1):

$$[\textit{Rate of storage in } V(t)] = \frac{DA}{Dt} \tag{2.2}$$

where D/Dt is known as the substantial or convective derivative and can be defined, in Cartesian coordinates, as:

$$\frac{DA}{Dt} = \frac{\partial A}{\partial t} + v_x \frac{\partial A}{\partial x} + v_y \frac{\partial A}{\partial y} + v_z \frac{\partial A}{\partial z} = \frac{\partial A}{\partial t} + \vec{V} \cdot \nabla A \tag{2.3}$$

The substantial derivative as defined in Eq. (2.3) is a special kind of time derivative which is known as the "derivative following the motion." In Eq. (2.3), v_x, v_y and v_z are the components of the local velocity term $\vec{V}$. This Lagrangian derivative describes the total rate of change of the extensive property $A = \iiint_{V(t)} \alpha dV$of a fluid as it moves through space. $\frac{\partial A}{\partial t}$ is the local time derivative of this quantity, and $\vec{V} \cdot \nabla A$ is the convective derivative or change due to velocity-driven transport.

Now, we can move to the second term of Eq. (2.1). Let $\vec{J}$ be the quantity flux vector, that is, the rate of transport across the boundary surface $S(t)$. We write:

$$[\text{Rate of outflow/inflow into } CV] = -\iint_{S(t)} \vec{J} \cdot \vec{n}\, dS \tag{2.4}$$

In Eq. (2.4), we are considering an elemental surface area dS as the bounding surface of the deforming system. The magnitude of the flux of extensive quantity across a surface area dS with an outward normal $\vec{n}$ is $\vec{J} \cdot \vec{n} dS$, which equals zero if $\vec{J}$ is in the plane of dS, or $-|\vec{J}|\, dS$ if $\vec{J}$ is normal to dS.

We now move to the third term of Eq. (2.1), letting Γ be the generation rate of α per unit volume inside the system, we can write:

$$[\text{Rate of quantity generation within } V(t)] = \iint_{V(t)} \Gamma dV \tag{2.5}$$

Now, we can substitute Eqs. (2.2), (2.4), and (2.5) into Eq. (2.1) to obtain:

$$\frac{DA}{Dt} = -\iint_{S(t)} \vec{J} \cdot \vec{n} dS + \iint_{V(t)} \Gamma dV \tag{2.6}$$

or

$$\frac{D}{Dt} \iiint_{V(t)} \alpha dV + \iint_{S(t)} \vec{J} \cdot \vec{n} dS - \iint_{V(t)} \Gamma dV = 0 \tag{2.7}$$

By reviewing Eq. (2.7), one can offer the following remarks:

- Equation (2.7) offers a conservative equation for A in Lagrangian coordinates. In the Lagrangian point of view, the formulation tracks what happens to a group of fluid particles within the flow domain, starting from their initial location at $t = 0$ as time evolves. The formulation analyzes what happens to the properties of this deforming system, which typically contains a controlled mass of particles, by following it. From the point of view of Lagrangian coordinates, we analyze a fixed mass of fluid moving in space, with variations in the fluid system volume, i.e., $V(t)$, enclosed within the deformable surface $S(t)$.
- Most of the time, we prefer to work in the Eulerian coordinates. In the Eulerian point of view, the formulation defines a space of finite dimensions (the "control volume," or *CV*) and analyzes what happens with its properties as time evolves. In the Eulerian point of view, analysis is carried out on the flow in and out of an element of a fixed size (*CV*), defined by fixed boundaries (the "control surface," or *CS*), and mass inside the *CV* varies. It will be necessary to transform Eq. (2.7) into the Eulerian coordinate system if the point of view of Euler is adopted.
- To make this transformation, one needs to change the substantial or Lagrangian view of the volume integral to a volume integral of Eulerian derivatives. To do this, it will be necessary to invoke the Reynolds transport theorem (*RTT*). The *RTT* is a transformation that will allow connecting control-mass (Lagrangian) formulations to control-volume (Eulerian) formulations.
- We need to transform the surface integral or the second term Eq. (2.7) to a volume integral in order to do this. We need to invoke the divergence theorem.
- With the invocation of the *RTT* and divergence theorem, one can collect all the terms together under a single volume integral, integrated over the arbitrary *CV*. Since the control volume is arbitrary here, the integrand of the equation will be made equal to zero.
- Setting this integrand to zero produces the general transport equation for the quantity α.

2.1.2 Divergence Theorem and Differential Operators

The divergence theorem is an often-used statement in vector calculus that relates the divergence behavior of a vector field inside a volume to the flux of said vector field through the surface that defines the volume. It allows evaluating volumetric integrals in terms of surface integrals and vice versa. The theorem states:

$$\iiint_V \nabla \cdot \vec{B}\, dV = \iint_S \vec{B} \cdot \vec{n}\, dS \tag{2.8}$$

where $\vec{B}$ is a vector field and $\nabla\cdot$ is the divergence operator. In this integral, dV is a volume element in the domain (see Fig. 2.1).

We should recall that for a given scalar function $f(x,y,z)$ the gradient *grad f* is the vector factor function defined by:

$$grad\, f = \frac{\partial f}{\partial x}\hat{i} + \frac{\partial f}{\partial y}\hat{j} + \frac{\partial f}{\partial z}\hat{k} \tag{2.9}$$

In writing Eq. (2.9), we assume that f is differentiable. One can define the following differential operator in Cartesian coordinates:

$$\nabla = \frac{\partial}{\partial x}\hat{\boldsymbol{i}} + \frac{\partial}{\partial y}\hat{j} + \frac{\partial}{\partial z}\hat{\boldsymbol{k}} \tag{2.10}$$

Then, $grad\, f = \nabla f$. The gradient ∇f is a directional derivative. As shown in Eq. (2.9), it is indeed a vector. It carries important information: it is a vector in the direction in which the function $f(x,y,z)$ is increasing most rapidly with distance. Its magnitude equals the rate of that increase and represents the slope of a graph of the function in that direction. The information about the rate of change of $f(x,y,z)$ with respect to distance is found in ∇f and such information play an important role in the quantification of fluxes in fluid flow when f represents a potential.

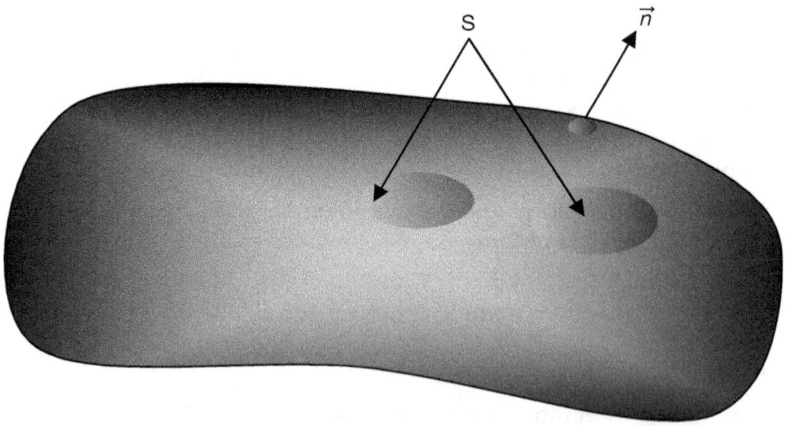

Figure 2.1 Illustration of the divergence theorem.

Let $\vec{V}(x,y,z)$ be a differentiable vector function and v_x, v_y and v_z be the components of $\vec{V}$. Then the function

$$div\vec{V} = \frac{\partial v_x}{\partial x} + \frac{\partial v_y}{\partial y} + \frac{\partial v_z}{\partial z} \tag{2.11}$$

is called the divergence of $\vec{V}$ in Cartesian coordinates or the divergence of the vector defined by $\vec{V}$. A common notation for the divergence of $\vec{V}$ is $\nabla\vec{V}$:

$$div\vec{V} = \nabla \cdot \vec{V} = \left(\frac{\partial}{\partial x}\hat{i} + \frac{\partial}{\partial y}\hat{j} + \frac{\partial}{\partial z}\hat{k}\right) \cdot \left(v_x\hat{i} + v_y\hat{j} + v_z\hat{k}\right) \tag{2.12}$$

As implied in Eq. (2.11), and the dot product in Eq. (2.12), the divergence $\nabla\vec{V}$ is a scalar quantity. This scalar quantifies the net outflow or flux generation per unit volume at every point of the $\vec{V}$ field. In physical terms, when $\vec{V}$ is the mass flux, its divergence directly represents the net rate of loss of mass or rate of decrease of mass per unit volume at every point of interest where the divergence is calculated. Based on this, the divergence theorem quoted in Eq. (2.8):

$$\iiint_V \nabla \cdot \vec{B} dV = \iint_S \vec{B} \cdot \vec{n} dS$$

basically states that the summation of all net outflows of vector $\vec{B}$ measured at every point inside an arbitrary volume $\left(\iiint_V \nabla \cdot \vec{B} dV\right)$ must equal the net flux of $\vec{B}$ evaluated over the entire surface enclosing such arbitrary volume $\left(\iint_S \vec{B} \cdot \vec{n} dS\right)$.

If $f(x,y,z)$ is a twice-differentiable function , then:

$$div(grad\, f) = \nabla \cdot \left(\frac{\partial f}{\partial x}\hat{i} + \frac{\partial f}{\partial y}\hat{j} + \frac{\partial f}{\partial z}\hat{k}\right)$$

$$div(grad\, f) = \frac{\partial^2 f}{\partial x^2} + \frac{\partial^2 f}{\partial y^2} + \frac{\partial^2 f}{\partial z^2} \tag{2.13}$$

which leads to

$$div(grad\, f) = \nabla \cdot (\nabla f) = \nabla^2 f \tag{2.14}$$

where $\nabla^2 f$ is the Laplacian of $f(x,y,z)$ which is the calculation of the divergence or net outflow per unit volume of the ∇f vector field.

Now, we define the curl of a vector field. Let $\vec{V}(x,y,z) = v_x\hat{\imath} + v_y\hat{\jmath} + v_z\hat{k}$ be a differentiable vector function. Then, the function:

$$curl\vec{V} = \nabla \times \vec{V} = \begin{vmatrix} \hat{i} & \hat{j} & \hat{k} \\ \frac{\partial}{\partial x} & \frac{\partial}{\partial y} & \frac{\partial}{\partial z} \\ v_x & v_y & v_z \end{vmatrix}$$

$$\nabla \times \vec{V} = \left(\frac{\partial V_z}{\partial y} - \frac{\partial V_y}{\partial z}\right)\hat{i} + \left(\frac{\partial V_x}{\partial z} - \frac{\partial V_z}{\partial x}\right)\hat{j} + \left(\frac{\partial V_y}{\partial x} - \frac{\partial V_x}{\partial y}\right)\hat{k} \tag{2.15}$$

is called the curl of the vector function, which is a vector describing the rotation of the vector $\vec{V}$ in the three-dimensional space. If $\vec{V}$ is velocity and $\nabla \times \vec{V} = 0$, the flow is said to be irrotational.

Divergence ($\nabla\cdot$), gradient (∇), Laplacian (∇^2), and curl ($\nabla\times$) are important operators involving differentiation on either scalar or vector fields that routinely appear in differential equations of fluid flow.

2.1.3 Reynolds Transport Theorem, Euler Formulations, and Generalized Transport Equation

The *RTT* uses Leibniz's rule for the calculation of derivatives of integrals applied to multiple dimensions. When Leibniz's rule and the chain rule are applied to the definition of the substantial derivative DA/Dt, with $A = \iiint_{V(t)} \alpha dV$ and $\alpha = \alpha(t,\vec{x})$, over the dynamic time-dependent $V(t)$ region defined by the boundary $S(t)$, one obtains:

$$\frac{D}{Dt}\iiint_{V(t)} \alpha(t,\vec{x})dV = \iiint_{V(t)} \frac{\partial \alpha}{\partial t} dV + \iint_{S(t)} \left(\alpha\vec{V}\right)\cdot \vec{n} dS \tag{2.16}$$

which provides a convenient link between Lagrangian and Eulerian formulations. The left-hand side provides the total rate change of the extensive property A as computed by the Lagrangian formulation, and the right-hand side provides Eulerian derivatives counterparts. The first term in the right-hand side is the local time rate of change of A as would be experienced by an observer inside the volume $V(t)$, and the second term is the rate of transport of A through the moving boundary $S(t)$. Here we assume that the velocity of the surface equals the velocity of the flow; if the observer is moving at a different velocity relative to the fluid velocity, Eq. (2.16) would not hold by using $\vec{V}$ as the fluid velocity.

Application of the divergence theorem to Eq. (2.16) yields:

$$\frac{D}{Dt}\iiint_{V(t)} \alpha dV = \iiint_{V(t)} \left[\frac{\partial \alpha}{\partial t} + \nabla\cdot\left(\alpha\vec{V}\right)\right] dV \tag{2.17}$$

Equations (2.16) and (2.17) are statements of the *RTT*. The *RTT* as expressed by Eq. (2.17) combined with the conservation statement in Eq. (2.6) leads to:

$$\iiint_{V(t)} \left[\frac{\partial \alpha}{\partial t} + \nabla\cdot\left(\alpha\vec{V}\right)\right] dV = -\iiint_{V(t)} \left(\nabla\cdot\vec{J}\right) dV + \iiint_{V(t)} \Gamma dV \tag{2.18}$$

where the divergence theorem was applied in Eq. (2.6) to introduce the $\nabla\cdot\vec{J}$ flux term. When grouping terms one ends up with:

$$\iiint_{V(t)} \left[\frac{\partial \alpha}{\partial t} + \nabla\cdot\left(\alpha\vec{V}\right) + \nabla\cdot\vec{J} - \Gamma\right] dV = 0 \tag{2.19}$$

Since in Eq. (2.19) the integration is taken over an arbitrary volume, such a general statement can only be true if its integrand vanishes, that is:

$$\frac{\partial \alpha}{\partial t} + \nabla\cdot\left(\alpha\vec{V}\right) + \nabla\cdot\vec{J} - \Gamma = 0 \tag{2.20}$$

Equation (2.20) is the *GTE*. It is described as a fully generalized form in view of the following:

- It is independent of the coordinate system since no particular geometry has been specified yet
- In the development of the equation, no constitutive relationship has been offered; hence, this same equation can be used to describe transfer of heat, mass, and momentum in any physical domain

We now examine and describe each of the entries of Eq. (2.20):

- $\frac{\partial \alpha}{\partial t} \equiv$ Rate of accumulation of α, also known as the accumulation/depletion or storage term.
- $\nabla \cdot \left(\alpha \vec{V}\right) \equiv$ Convective or advective term due to velocity field. Here we use convective or advective terms interchangeably. This term captures velocity-driven transport.
- $\nabla \cdot \vec{J} \equiv$ Diffusive or dispersive transport, where $\vec{J}$ is second-order tensor. This is a second-order term which can describe phenomena such as molecular diffusion and mechanical dispersion in mass transport, viscous dissipation in momentum transport, or heat conduction and irreversible work in energy transport. It becomes important when there is transport due to other mechanisms other than velocity-driven advection.
- $\Gamma \equiv$ Generation term, mass per unit system volume.

We can carry out the expansion of the divergence to express this equation in terms of other familiar terms. The product rule applied to the divergence of a scalar field α and a vector field $\vec{V}$ yields:

$$\nabla \cdot \left(\alpha \vec{V}\right) = \alpha\left(\nabla \cdot \vec{V}\right) + \vec{V} \cdot \nabla \alpha \tag{2.21}$$

Accordingly, Eq. (2.18) can be written as:

$$\frac{\partial \alpha}{\partial t} + \vec{V} \cdot \nabla \alpha + \alpha \nabla \cdot \vec{V} + \nabla \cdot \vec{J} = \Gamma \tag{2.22}$$

Equations (2.20) and (2.22) are *GTE*s written in the Eulerian coordinate system (i.e., for a fixed control volume) that can be used to describe transport of heat, mass, and momentum for multiphase, multicomponent systems as long as a definition of α is provided. These *GTE*s provide a conservation statement for any intrinsic system variable α in terms of associated convective $\left(\nabla \cdot (\alpha \vec{V})\right)$ and diffusive $(\nabla \cdot \vec{J})$ fluxes and net sources (Γ).

2.2 General Transport Equation Applied to Porous Media

In Chap. 5, we will derive the reservoir transport equations starting from a differential balance applied to a multicomponent fluid system. To illustrate the interchangeability of the differential and integral approaches, here we will illustrate the applicability of the general transport equation in Eq. (2.20) to the same type of reservoir problems and for the case of single-phase flow.

As stated earlier, the general transport equation can be reduced to familiar transport equations by providing a proper definition for α. For the case of mass conservation in a porous media, A is the total fluid mass (M_f) in our fixed (Eulerian) control volume. Since V_b represents the bulk volume of this control volume, one writes for the intensive property α:

$$\alpha = \frac{M_f}{V_b} \tag{2.23}$$

Since $\phi = \dfrac{V_p}{V_b}$ (where V_p is the pore volume):

$$\alpha = \frac{M_f}{(V_P \,/\, \phi)} = \frac{\phi M_f}{V_P} \tag{2.24}$$

If one assumes single-phase full pore occupancy, $V_p = V_f$. Thus, with $\rho = \dfrac{M_f}{V_f}$ representing the single-phase fluid density:

$$\alpha = \phi\rho \tag{2.25}$$

Substitution of Eq. (2.25) in Eq. (2.20) yields:

$$\frac{\partial(\phi\rho)}{\partial t} + \nabla\cdot\left(\rho\phi\vec{V}\right) + \nabla\cdot\vec{J} = \Gamma \tag{2.26}$$

where $\vec{V}$ in the advective term in Eq. (2.26) represents the fluid's interstitial velocity or actual fluid velocity within the pores. In flow in porous media, the advective fluid velocity is calculated in terms of Darcy's law, which for single-phase flow is written as:

$$\vec{v} = -\frac{\overline{\overline{k}}}{\mu}\nabla\Phi \tag{2.27}$$

Darcian velocities in Eq. (2.27) are volume-averaged—that is, they assume that flow occurs across the entire cross section of the porous media. Interstitial velocity is the Darcy's velocity divided by the effective porosity of the medium:

$$\vec{V} = \frac{\vec{v}}{\phi} \tag{2.28}$$

In Eq. (2.27), $\overline{\overline{k}}$ is the permeability tensor (to be discussed in more detail in Chap. 3), and Φ is the flow potential that triggers advective flow (also discussed in more detail in Chap. 3). This flow potential Φ is a scalar quantity and hence is defined by its numerical value at any point in space. The presence of the gradient of potential in space is what causes Darcian flow. The flow direction is in the decreasing direction of flow. Just as in special cases of mass, heat, and momentum transfer, a linear relationship can be established between flow and potential gradient, as indicated in Eq. (2.27). This linear relationship is essentially the basis of Fourier's law of heat conduction, Fick's law of mass diffusion, and also Darcy's law for bulk advective flow porous media.

If pore fluid flow is assumed to be advection-dominated (i.e., only Darcian flow occurs) and hence molecular diffusion and mechanical dispersion are neglected $\left(\nabla \cdot \vec{J} = 0\right)$, Eq. (2.26) gives:

$$\frac{\partial(\phi\rho)}{\partial t} - \nabla \cdot \left(\rho \frac{\overline{\overline{k}}}{\mu} \nabla \Phi\right) = \Gamma \tag{2.29}$$

where Γ is the mass generation term per unit volume, directly related to mass injection or production due to the presence of wells $\left(\Gamma = \frac{\rho q_{xyz}}{V_b}\right)$. In Eq. (2.29), the flow potential term Φ will collapse to pressure term p if gravitational effects are ignored. If $\nabla \Phi = \nabla p$

$$\nabla \cdot \left(\rho \frac{\overline{\overline{k}}}{\mu} \nabla p\right) + \frac{\rho q_{xyz}}{V_b} = \frac{\partial(\phi\rho)}{\partial t} \tag{2.30}$$

For a three-dimensional Cartesian system of constant porosity (Chap. 3 discusses the concept of pore compressibility), Eq. (2.30) becomes:

$$\frac{\partial}{\partial x}\left(\frac{\rho}{\mu} k_x \frac{\partial p}{\partial x}\right) + \frac{\partial}{\partial y}\left(\frac{\rho}{\mu} k_y \frac{\partial p}{\partial y}\right) + \frac{\partial}{\partial z}\left(\frac{\rho}{\mu} k_z \frac{\partial p}{\partial z}\right) + \frac{\rho q_{xyz}}{V_b} = \phi \frac{\partial \rho}{\partial t} \tag{2.31}$$

Chapter 3 also discusses the permeability tensor $\overline{\overline{k}}$. and the conditions that allow its transformation into its principal components k_x, k_y, and k_z. With the implementation of appropriate equations of state, the system of equations will be written in terms of pressure as the dependent variable. Equations (2.30) and (2.31) can be solved in the presence of specified boundary conditions at the internal (well locations) and external (physical limits of the reservoir) boundaries. The temporal derivative on the right-hand side also requires the specification of initial conditions. The analytical treatment of this equation will be shown in Chap. 6, and its numerical treatment in Chap. 7.

2.3 Review of Equations of Flow in Porous Media

2.3.1 Single-Phase Flow Equations

Chapter 5 will further investigate Eq. (2.30) as applied to single-phase incompressible, slightly compressible, and compressible flow applications and will show that the resulting equations are just subsets of the more general compositional model derived in Chap. 5. Here we briefly introduce the most common single-phase models based on the *GTE*-derived Eq. (2.31).

Incompressible Flow

For incompressible flow, one considers the following type of fluid behavior:

$$\rho = \text{constant and } \mu = \text{constant}$$

Under such conditions, Eq. (2.31) collapses to:

$$\frac{\partial}{\partial x}\left(k_x \frac{\partial p}{\partial x}\right) + \frac{\partial}{\partial y}\left(k_y \frac{\partial p}{\partial y}\right) + \frac{\partial}{\partial z}\left(k_z \frac{\partial p}{\partial z}\right) + \frac{\mu q_{xyz}}{V_b} = 0 \tag{2.32}$$

Now, consider a special case where reservoir has homogenous (k = constant) and isotropic ($k_x = k_y = k_z = k$) permeability characteristics. Then Eq. (2.32) collapses to:

$$\frac{\partial^2 p}{\partial x^2} + \frac{\partial^2 p}{\partial y^2} + \frac{\partial^2 p}{\partial z^2} + \frac{\mu q_{xyz}}{kV_b} = 0 \tag{2.33}$$

Equation (2.33) is Poisson's equation. Furthermore, if there is no source or sink term ($q_{xyz} = 0$), then:

$$\frac{\partial^2 p}{\partial x^2} + \frac{\partial^2 p}{\partial y^2} + \frac{\partial^2 p}{\partial z^2} = 0 \tag{2.34}$$

Equation (2.34) is Laplace's equation. Note that Eq. (2.33) can be transformed into Eq. (2.34) using a simple transformation as described below.

Let:

$$M(x,y,z) = p(x,y,z) + \frac{1}{2}\frac{\mu q_{xyz}}{kV_b}x^2 \tag{2.35}$$

We see that

$$\frac{\partial M(x,y,z)}{\partial x} = \frac{\partial p}{\partial x} + \frac{\mu q_{xyz}}{kV_b}x \tag{2.36}$$

and,

$$\frac{\partial^2 M(x,y,z)}{\partial x^2} = \frac{\partial^2 p}{\partial x^2} + \frac{\mu q_{xyz}}{kV_b} \tag{2.37}$$

Similarly,

$$\frac{\partial^2 M(x,y,z)}{\partial y^2} = \frac{\partial^2 p}{\partial y^2} \tag{2.38}$$

$$\frac{\partial^2 M(x,y,z)}{\partial z^2} = \frac{\partial^2 p}{\partial z^2} \tag{2.39}$$

After substituting Eqs. (2.37), (2.38), and (2.39) in Eq. (2.33), we get:

$$\frac{\partial^2 M}{\partial x^2} + \frac{\partial^2 M}{\partial y^2} + \frac{\partial^2 M}{\partial z^2} = 0 \tag{2.40}$$

After solving for $M(x,y,z)$ one can easily transform the solution to $p(x,y,z)$ using Eq. (2.35) so that:

$$p(x,y,z) = M(x,y,z) - \frac{1}{2}\frac{\mu q_{xyz}}{kV_b}x^2 \tag{2.41}$$

Now, going back to Eq. (2.32), this time let us assume that reservoir permeability distribution exhibits anisotropic but homogeneous characteristics. Then, the flow equation can be written as:

$$k_x\frac{\partial^2 p}{\partial x^2} + k_y\frac{\partial^2 p}{\partial y^2} + k_z\frac{\partial^2 p}{\partial z^2} + \frac{\mu q_{xyz}}{V_b} = 0 \tag{2.42}$$

This time we can implement the following transformations on the independent variables x, y, z:

$$X = \frac{x}{\sqrt{k_x}};Y = \frac{y}{\sqrt{k_y}};Z = \frac{z}{\sqrt{k_z}} \tag{2.43}$$

From Eq. (2.43), one can obtain the following relationships:

$$\frac{\partial X}{\partial x} = \frac{1}{\sqrt{k_x}};\frac{\partial Y}{\partial y} = \frac{1}{\sqrt{k_y}};\frac{\partial Z}{\partial z} = \frac{1}{\sqrt{k_z}} \tag{2.44}$$

In addition, we note the following chain rule applications:

$$\frac{\partial^2 p}{\partial x^2} = \frac{\partial}{\partial x}\left(\frac{\partial p}{\partial x}\right) = \frac{\partial}{\partial X}\left(\frac{\partial p}{\partial X}\frac{\partial X}{\partial x}\right)\frac{\partial X}{\partial x} \tag{2.45}$$

which yields,

$$\frac{\partial^2 p}{\partial x^2} = \frac{\partial}{\partial X}\left(\frac{1}{\sqrt{k_x}}\frac{\partial X}{\partial x}\right)\frac{1}{\sqrt{k_x}} \tag{2.46}$$

or,

$$\frac{\partial^2 p}{\partial x^2} = \frac{1}{k_x}\frac{\partial^2 p}{\partial X^2} \tag{2.47}$$

and,

$$\frac{\partial^2 p}{\partial y^2} = \frac{1}{k_y}\frac{\partial^2 p}{\partial Y^2} \tag{2.48}$$

$$\frac{\partial^2 p}{\partial z^2} = \frac{1}{k_z}\frac{\partial^2 p}{\partial Z^2} \tag{2.49}$$

Substitution of Eqs. (2.47), (2.48), and (2.49) into Eq. (2.46) results in:

$$\frac{\partial^2 p}{\partial X^2}+\frac{\partial^2 p}{\partial Y^2}+\frac{\partial^2 p}{\partial Z^2}+\frac{\mu q_{xyz}}{V_b}=0 \tag{2.50}$$

It should be mentioned that when the dependent variable is solved, its distribution will be obtained in (*X*, *Y*, *Z*) space and with a simple back transformation, that is, $x = X\sqrt{k_x}$, $y = Y\sqrt{k_y}$ and $z = Z\sqrt{k_z}$ solution can be brought back to the (*x*, *y*, *z*) space.

Slightly Compressible Flow

In slightly compressible fluid flow, fluid properties such as ρ and μ are weak functions of pressure; therefore, they introduce some weak nonlinearities to the transport equation. On the left-hand side of the equation, change of pressure in the flow direction is rather small. In most conventional reservoirs, away from the immediate vicinity of the well, pressure gradients are not expected to be larger than 0.1 psi/ft. Expanding Eq. (2.31):

$$\begin{aligned}&\left(\rho\frac{\partial}{\partial x}\left(\frac{k_x}{\mu}\frac{\partial p}{\partial x}\right)+\frac{k_x}{\mu}\frac{\partial p}{\partial x}\frac{\partial \rho}{\partial x}\right)+\left(\rho\frac{\partial}{\partial y}\left(\frac{k_y}{\mu}\frac{\partial p}{\partial y}\right)+\frac{k_y}{\mu}\frac{\partial p}{\partial y}\frac{\partial \rho}{\partial y}\right)\\&+\left(\rho\frac{\partial}{\partial z}\left(\frac{k_z}{\mu}\frac{\partial p}{\partial z}\right)+\frac{k_z}{\mu}\frac{\partial p}{\partial z}\frac{\partial \rho}{\partial z}\right)+\frac{\rho q_{xyz}}{V_b}=\phi\frac{\partial \rho}{\partial p}\frac{\partial p}{\partial t}\end{aligned} \tag{2.51}$$

Since $\frac{\partial \rho}{\partial p}=\rho c$ (from the definition of compressibility, see Chap. 3), Eq. can be simplified as, keeping in mind, $\frac{\partial \rho}{\partial x}=\frac{\partial \rho}{\partial p}\frac{\partial p}{\partial x}$

$$\begin{aligned}&\frac{\partial}{\partial x}\left(\frac{k_x}{\mu}\frac{\partial p}{\partial x}\right)+c\frac{k_x}{\mu}\left(\frac{\partial p}{\partial x}\right)^2+\frac{\partial}{\partial y}\left(\frac{k_y}{\mu}\frac{\partial p}{\partial y}\right)+c\frac{k_y}{\mu}\left(\frac{\partial p}{\partial y}\right)^2\\&+\frac{\partial}{\partial z}\left(\frac{k_z}{\mu}\frac{\partial p}{\partial z}\right)+c\frac{k_z}{\mu}\left(\frac{\partial p}{\partial z}\right)^2+\frac{\rho q_{xyz}}{V_b}=\phi c\frac{\partial p}{\partial t}\end{aligned} \tag{2.52}$$

Since c is a significantly small quantity and pressure gradients in space are also small, it will be reasonable to assume that on the left-hand side of Eq. (2.52):

$$c\left[\frac{k_x}{\mu}\left(\frac{\partial p}{\partial x}\right)^2+\frac{k_y}{\mu}\left(\frac{\partial p}{\partial y}\right)^2+\frac{k_z}{\mu}\left(\frac{\partial p}{\partial z}\right)^2\right]\ll\frac{\partial}{\partial x}\left(\frac{k_x}{\mu}\frac{\partial p}{\partial x}\right)+\frac{\partial}{\partial y}\left(\frac{k_y}{\mu}\frac{\partial p}{\partial y}\right)+\frac{\partial}{\partial z}\left(\frac{k_z}{\mu}\frac{\partial p}{\partial z}\right) \tag{2.53}$$

Then, Eq. (2.52) can be written as:

$$\frac{\partial}{\partial x}\left(\frac{k_x}{\mu}\frac{\partial p}{\partial x}\right)+\frac{\partial}{\partial y}\left(\frac{k_y}{\mu}\frac{\partial p}{\partial y}\right)+\frac{\partial}{\partial z}\left(\frac{k_z}{\mu}\frac{\partial p}{\partial z}\right)+\frac{q_{xyz}}{V_b}=\phi c\frac{\partial p}{\partial t} \tag{2.54}$$

In slightly compressible fluids, if the viscosity function is treated as a constant value. Then, one ends up with the final form of the slightly compressible flow equation as:

$$\frac{\partial}{\partial x}\left(k_x\frac{\partial p}{\partial x}\right)+\frac{\partial}{\partial y}\left(k_y\frac{\partial p}{\partial y}\right)+\frac{\partial}{\partial z}\left(k_z\frac{\partial p}{\partial z}\right)+\frac{q_{xyz}\mu}{V_b}=\phi\mu c\frac{\partial p}{\partial t} \tag{2.55}$$

For an isotropic and homogeneous permeability distribution in the system ($k_x = k_y = k_z = k$), Eq. (2.55) reduces to:

$$\frac{\partial^2 p}{\partial x^2} + \frac{\partial^2 p}{\partial y^2} + \frac{\partial^2 p}{\partial z^2} + \frac{q_{xyz}\mu}{V_b} = \frac{\phi\mu c}{k}\frac{\partial p}{\partial t} \tag{2.56}$$

In Eq. (2.56), if there is no source/sink term then one ends up with:

$$\frac{\partial^2 p}{\partial x^2} + \frac{\partial^2 p}{\partial y^2} + \frac{\partial^2 p}{\partial z^2} = \frac{\phi\mu c}{k}\frac{\partial p}{\partial t} \tag{2.57}$$

which is the diffusivity equation for liquids written in Cartesian form.

Compressible Flow (Flow of a Real Gas)

In the case of gas flow in porous media, we use the real gas law as the equation of state. Real gas law indicates that:

$$pV = ZnRT \tag{2.58}$$

or,

$$\rho = \frac{MW}{RT}\frac{p}{Z} \tag{2.59}$$

Since we have a definition of density, we substitute it in the general form of the transport equation in porous media in Eq. (2.31); after simplifications, we end up with:

$$\frac{\partial}{\partial x}\left(\frac{p}{\mu Z}k_x\frac{\partial p}{\partial x}\right) + \frac{\partial}{\partial y}\left(\frac{p}{\mu Z}k_y\frac{\partial p}{\partial y}\right) + \frac{\partial}{\partial z}\left(\frac{p}{\mu Z}k_z\frac{\partial p}{\partial z}\right) + \frac{q_{xyz}}{V_b}\frac{p}{Z} = \phi\frac{\partial}{\partial t}\left(\frac{p}{Z}\right) \tag{2.60}$$

Equation (2.60) is generally considered as a strongly nonlinear partial differential equation because of the appearance of pressure-dependent terms such as p, μ, and Z as coefficients.

The three single-phase flow equations [Eqs. (2.32), (2.57), and (2.60)] are all written at reservoir conditions (i.e., the source term q_{xyz} as written, is expressed at reservoir conditions). However, source/sink term measurements are made at the surface conditions. Accordingly, the three fundamental equations of this section can be rewritten in terms of surface conditions. In making this change, one invokes the concept of formation volume factor (B). Chapter 4 shows that in its simplest formulation, for a single-component, single-phase, this quantity is expressed as:

$$B, \frac{\textit{Volume at reservoir conditions}}{\textit{Volume at surface conditions}} = \frac{\textit{Density at surface conditions}}{\textit{Density at reservoir conditions}}$$

or simply,

$$B = \frac{\rho_{sc}}{\rho} \tag{2.61}$$

In case of incompressible fluids $B = 1$, but to preserve a common form Eq. (2.32) can be written as (dropping subscripts xyz from the q term for the sake of simplicity):

$$\frac{\partial}{\partial x}\left(\frac{k_x}{\mu B}\frac{\partial p}{\partial x}\right)+\frac{\partial}{\partial y}\left(\frac{k_y}{\mu B}\frac{\partial p}{\partial y}\right)+\frac{\partial}{\partial z}\left(\frac{k_z}{\mu B}\frac{\partial p}{\partial z}\right)+\frac{q_{sc}}{V_b}=0 \tag{2.62}$$

Starting with Eq. (2.31) applied to slightly compressible flow, this definition of B leads to:

$$\frac{\partial}{\partial x}\left(\frac{k_x}{\mu B}\frac{\partial p}{\partial x}\right)+\frac{\partial}{\partial y}\left(\frac{k_y}{\mu B}\frac{\partial p}{\partial y}\right)+\frac{\partial}{\partial z}\left(\frac{k_z}{\mu B}\frac{\partial p}{\partial z}\right)+\frac{q_{sc}}{V_b}=\frac{\phi c}{B}\frac{\partial p}{\partial t} \tag{2.63}$$

Finally, for compressible flow, formation volume factor for a real gas can be defined as:

$$B_g=\frac{\dfrac{(p_{sc})(MW)}{Z_{sc}RT_{sc}}}{\left[\dfrac{(p)(MW)}{ZRT}\right]} \tag{2.64}$$

or,

$$B_g=\frac{p_{sc}T}{T_{sc}}\frac{z}{p} \tag{2.65}$$

By substituting Eq. (2.65) into Eq. (2.60) the compressible flow equation at surface conditions can be written as:

$$\frac{\partial}{\partial x}\left(\frac{k_x}{\mu_g B_g}\frac{\partial p}{\partial x}\right)+\frac{\partial}{\partial y}\left(\frac{k_y}{\mu_g B_g}\frac{\partial p}{\partial y}\right)+\frac{\partial}{\partial z}\left(\frac{k_z}{\mu_g B_g}\frac{\partial p}{\partial z}\right)+\frac{q_{sc}}{V_b}=\phi\frac{\partial}{\partial t}\left(\frac{1}{B_g}\right) \tag{2.66}$$

More on the Source/Sink Term

In reservoir simulation equations, the source/sink term can represent a point source at the well. Accordingly, in a continuous partial differential equation, the source/sink term can be expressed as shown below:

$$q_{sc}=q_{sc}\,\delta(x-\zeta)\delta(y-\eta) \tag{2.67}$$

Equation (2.67) is nonzero only at well-point $x=\zeta$ and $y=\eta$. The Dirac delta function (which is also known as unit-impulse function) is defined as:

$$\delta(u)=1,\ \text{if } u=0$$

$$\delta(u)=0,\ \text{if } u\neq 0 \tag{2.68}$$

Accordingly, if there are two wells with coordinates $(x_1 = \zeta_1; y_1 = \eta_1)$ and $(x_1 = \zeta_2; y_1 = \eta_2)$ then the flow equation for incompressible flow, Eq. (2.62), will be written as:

$$\frac{\partial}{\partial x}\left(\frac{k_x}{\mu B}\frac{\partial p}{\partial x}\right) + \frac{\partial}{\partial y}\left(\frac{k_y}{\mu B}\frac{\partial p}{\partial y}\right) + \frac{\partial}{\partial z}\left(\frac{k_z}{\mu B}\frac{\partial p}{\partial z}\right)$$
$$+\frac{q_{sc1}}{V_b}\delta(x-\zeta_1)\delta(y-\eta_1) + \frac{q_{sc2}}{V_b}\delta(x-\zeta_2)\delta(y-\eta_2) = 0 \tag{2.69}$$

In these equations, the term q_{sc} is positive for injection wells and negative for production wells.

2.3.2 Extension of Flow Equations to Multiphase Flow

In hydrocarbon reservoirs, it is very seldom to see that single-phase flow is taking place. In reality, it is much more typical to see two-phase or three-phase flow conditions. Any of the following situations can represent a reservoir flow condition:

- Two-phase flow (oil and gas)
- Two-phase flow (oil and water)
- Two-phase flow (gas and water)
- Three-phase flow (oil, gas, and water)

Two-phase flow conditions (as well as single-phase flow) are subsets of the three-phase flow. Therefore, we can extend the mathematical formulations assuming that three-phase flow conditions prevail.

Here we summarize the associated development detailed in Chap. 5 for the black oil fluid model. In single-phase flow problems, we construct a mathematical formulation and end up with a single governing partial differential equation. In a three-phase flow formulation, there will be at least three governing partial differential equations in black oil formulation (see Chap. 4). The mathematical formulation is written for all fluid components flowing in all reservoir phases. Accordingly, it will be necessary to develop a partial differential equation for each of the flowing fluid components. For surface oil and water flow equations, this is quite straight forward. For the case of surface gas, it becomes necessary to accommodate the transport of gas in three different modes: in solution with oil, in solution with water, and as in the free gas phase. On the left-hand side of the flow equations when Darcy's law is substituted it will have the relative permeability terms and on the right-hand side of the equation within the time derivative the saturation terms will appear. Accordingly, oil, water, and gas pseudo-component flow equations will be obtained as follows:

1. Oil Equation written in surface units:

$$\frac{\partial}{\partial x}\left(\frac{k_x k_{ro}}{\mu_o B_o}\frac{\partial \Phi_o}{\partial x}\right) + \frac{\partial}{\partial y}\left(\frac{k_y k_{ro}}{\mu_o B_o}\frac{\partial \Phi_o}{\partial y}\right) + \frac{\partial}{\partial z}\left(\frac{k_z k_{ro}}{\mu_o B_o}\frac{\partial \Phi_o}{\partial z}\right) + \frac{1}{V_b}q_{osc} = \frac{\partial}{\partial t}\left(\frac{\phi S_o}{B_o}\right) \tag{2.70}$$

2. Water Equation written in surface units:

$$\frac{\partial}{\partial x}\left(\frac{k_x k_{rw}}{\mu_w B_w}\frac{\partial \Phi_w}{\partial x}\right)+\frac{\partial}{\partial y}\left(\frac{k_y k_{rw}}{\mu_w B_w}\frac{\partial \Phi_w}{\partial y}\right)+\frac{\partial}{\partial z}\left(\frac{k_z k_{rw}}{\mu_w B_w}\frac{\partial \Phi_w}{\partial z}\right)+\frac{1}{V_b}q_{wsc}=\frac{\partial}{\partial t}\left(\frac{\phi S_w}{B_w}\right) \tag{2.71}$$

3. Gas equation written in surface units:

$$\begin{aligned}
&\frac{\partial}{\partial x}\left(\frac{k_x k_{rg}}{\mu_g B_g}\frac{\partial \Phi_g}{\partial x}+R_{so}\frac{k_x k_{ro}}{\mu_o B_o}\frac{\partial \Phi_o}{\partial x}+R_{sw}\frac{k_x k_{rw}}{\mu_w B_w}\frac{\partial \Phi_w}{\partial x}\right)\\
&+\frac{\partial}{\partial y}\left(\frac{k_y k_{rg}}{\mu_g B_g}\frac{\partial \Phi_g}{\partial y}+R_{so}\frac{k_y k_{ro}}{\mu_o B_o}\frac{\partial \Phi_o}{\partial y}+R_{sw}\frac{k_y k_{rw}}{\mu_w B_w}\frac{\partial \Phi_w}{\partial y}\right)\\
&+\frac{\partial}{\partial z}\left(\frac{k_z k_{rg}}{\mu_g B_g}\frac{\partial \Phi_g}{\partial z}+R_{so}\frac{k_z k_{ro}}{\mu_o B_o}\frac{\partial \Phi_o}{\partial z}+R_{sw}\frac{k_y k_{rw}}{\mu_w B_w}\frac{\partial \Phi_w}{\partial z}\right)\\
&+\frac{1}{V_b}\left(q_{gsc}+R_{so}\,q_{osc}+R_{sw}\,q_{wsc}\right)=\frac{\partial}{\partial t}\left(\frac{\phi S_g}{B_g}+R_{so}\frac{\phi S_o}{B_o}+R_{sw}\frac{\phi S_w}{B_w}\right)
\end{aligned} \tag{2.72}$$

In Eqs. (2.70) through (2.72), six principal unknowns including p_o, p_w, p_g, S_o, S_w and S_g can be recognized. Then, in order to end up with a well-posed problem, it will be necessary to have three more equations. These equations are described below.

4. Saturation Relationship:

$$S_o + S_w + S_g = 1.00 \tag{2.73}$$

5. Oil/Water Capillary Pressure Relationship:

$$p_{cow}(S_w) = p_o - p_w \tag{2.74}$$

6. Gas/Oil Capillary Pressure Relationship:

$$p_{cgo}(S_g) = p_g - p_o \tag{2.75}$$

In Eqs. (2.70) through (2.72) p_o, p_w, and p_g terms are included in the definition of the flow potentials for each phase (G is positive downward depth from the datum level):

$$\Phi_o = p_o - \frac{1}{144}\frac{g}{g_c}\rho_o G \tag{2.76}$$

$$\Phi_w = p_w - \frac{1}{144}\frac{g}{g_c}\rho_w G \tag{2.77}$$

$$\Phi_g = p_g - \frac{1}{144}\frac{g}{g_c}\rho_g G \tag{2.78}$$

If the multiphase problem is a two-phase problem, for example, if oleic and gaseous phases are flowing together, the water equation will be removed, oil equation will stay as it is, and gas equation will be further simplified by setting $R_{sw} = 0$ and $S_w = 0$. This will result in two partial differential equations (one for the oil phase and one for the gas phase) and Eq. (2.74) will not exist as there is no oil/water interface. For such a flow problem one will end with four equations and four unknowns. Similarly, equations that are mentioned below can be collapsed to single-phase flow equation (e.g., slightly compressible flow or single-phase flow of oil). Following simplifications need to be added:

$$\begin{Bmatrix} S_o = 1 \\ S_w = 0 \\ S_g = 0 \end{Bmatrix} \tag{2.79}$$

$$\begin{Bmatrix} k_{ro} = 1 \\ k_{rw} = 0 \\ k_{rg} = 0 \end{Bmatrix} \tag{2.80}$$

Also, it should be noted that since

$$\rho_o = \rho_{o_{sc}} e^{c(p-p_o)} \tag{2.81}$$

Then, for a single-component, single-phase oil (see Chap. 4)

$$B_o = \frac{\rho_{o_{sc}}}{\rho_o} \tag{2.82}$$

Or,

$$B_o = \frac{1}{1 + c(p - p_o)} \tag{2.83}$$

which assumes that $B_o = 1$ when $p = p_o$ (see Chap. 4). Substituting the constraints as expressed by Eqs. (2.79) and (2.80) and using the definition of B_o, as given in Eq. (2.83), one will end up with the following single-phase flow equation:

$$\frac{\partial}{\partial x}\left(\frac{k_x k_{ro}}{\mu_o B_o}\frac{\partial \Phi_o}{\partial x}\right) + \frac{\partial}{\partial y}\left(\frac{k_y k_{ro}}{\mu_o B_o}\frac{\partial \Phi_o}{\partial y}\right) + \frac{\partial}{\partial z}\left(\frac{k_z k_{ro}}{\mu_o B_o}\frac{\partial \Phi_o}{\partial z}\right) + \frac{1}{V_b} q_{osc} = \phi c \frac{\partial p_o}{\partial t} \tag{2.84}$$

2.3.3 Extension of Flow Equations to Compositional Flow

Here we summarize the more detailed discussions about compositional modeling equations that will be encountered in Chap. 5. Figure 2.2 shows the elemental volume and the transport of fluid component "I" ($I = 1, 2, \ldots, N_c$) in the "x" direction (similar terms can be written along the "y" and "z" directions as well). As shown in Fig. 2.2, in compositional representation, individual components in the mixture are tracked. The

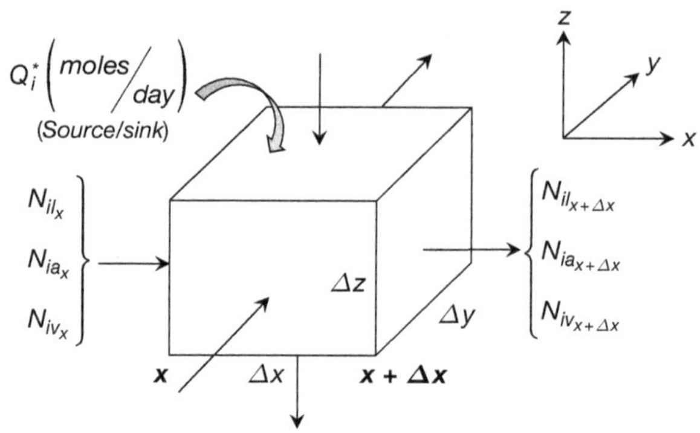

FIGURE 2.2 Mass flow across the faces of an elemental volume.

components can be transported within the oleic (l), aqueous (a), and vapor (v) phases. Allowing component "i" to be transported in the oleic phase (l), in the aqueous phase (a), and in the vapor phase (v), it will be possible to write a molar balance for the component "i" over time "t", so that:

$$[\textit{Net throughout of component "i"}] = [\textit{Net change in component "i"}] \tag{2.85}$$

with the following definitions:

$Q_i^* =$ external injection or production of component "i" (moles/day)
$C_{il} =$ molar concentration of component "i" in the l-phase (moles/ft^3)
$C_{ia} =$ molar concentration of component "i" in the a-phase (moles/ft^3)
$C_{iv} =$ molar concentration of component "i" in the v-phase (moles/ft^3)

and introducing molar fluxes such as $n_{ilx}\left(\dfrac{\text{moles}}{\text{ft}^3-\text{day}}\right)$ together with the phase velocities v_{lx}, v_{ax} and v_{vx}, one can write the following definitions:

$$n_{ilx} = C_{il}v_{lx} = (x_{il}C_l)v_{lx} = x_{il}\frac{\rho_l}{(MW)_l v_{lx}} \tag{2.86}$$

and,

$$N_{ilx} = X_{il}\left(\frac{\rho_l}{(MW)_l}\right)v_{lx}\Delta x\Delta z \tag{2.87}$$

In Eqs. (2.86) and (2.87), x_{il} represents the mole fraction of component "i" in the l-phase. After introducing the Darcy's law written for each phase and assuming that each component in a phase travels at the same phase velocity (assuming no molecular diffusion of components within each phase), one can obtain the following form of the compositional flow equation in Cartesian coordinates:

$$\frac{\partial}{\partial x}\left(\frac{k_{rl}}{(MW)_l \mu_l} k_x \frac{\partial \Phi_l}{\partial x} + \frac{k_{ra}}{(MW)_a \mu_a} k_x \frac{\partial \Phi_a}{\partial x} + \frac{k_{rv}}{(MW)_v \mu_v} k_x \frac{\partial \Phi_v}{\partial x}\right)$$
$$+ \frac{\partial}{\partial y}\left(\frac{k_{rl}}{(MW)_l \mu_l} k_y \frac{\partial \Phi_l}{\partial y} + \frac{k_{ra}}{(MW)_a \mu_a} k_y \frac{\partial \Phi_a}{\partial y} + \frac{k_{rv}}{(MW)_v \mu_v} k_y \frac{\partial \Phi_v}{\partial y}\right)$$
$$+ \frac{\partial}{\partial z}\left(\frac{k_{rl}}{(MW)_l \mu_l} k_z \frac{\partial \Phi_l}{\partial z} + \frac{k_{ra}}{(MW)_a \mu_a} k_z \frac{\partial \Phi_a}{\partial z} + \frac{k_{rv}}{(MW)_v \mu_v} k_z \frac{\partial \Phi_v}{\partial z}\right)$$
$$+ \frac{Q_i^*}{V_b} = \frac{\partial}{\partial t}\left[\phi\left(S_l \frac{x_{il}\rho_l}{(MW)_l} + S_a \frac{x_{ia}\rho_a}{(MW)_a} + S_v \frac{x_{iv}\rho_v}{(MW)_v}\right)\right] \tag{2.88}$$

Equation (2.88) can be written for each individual component. The number of independent variables in the system as defined by Eq. (2.88) for a three-phase flow problem can be calculated as follows:

x_{il}, x_{ia}, x_{iv} $(i = 1, 2, 3, \ldots, N_c)$	$= 3N_c$
p_l, p_a, p_v	$= 3$
S_l, S_a, S_v	$= 3$
Total number of dependent variables	$= 3N_c + 6$

To solve such a system uniquely, it will be necessary to have $3N_c + 6$ linearly independent equations. These $3N_c + 6$ equations are:

Molar balance [Eq. (2.88)] $(i = 1, 2, 3, \ldots, N_c)$	$= N_c$
Saturation relationship, $S_l + S_a + S_v = 1$	$= 1$
Capillary pressure relationship between "a" and "l" phases, $(p_{cla} = p_l - p_a)$	$= 1$
Capillary pressure relationship between "v" and "l" phases, $(p_{cvl} = p_v - p_l)$	$= 1$
Summation of mole fractions, $\sum_{i=1}^{N} x_{il} = 1; \sum_{i=1}^{N} x_{ia} = 1; \sum_{i=1}^{N} x_{iv} = 1$	$= 3$
Equilibrium ratio constant $(i = 1, 2, 3, \ldots N_c)$ $K_{iva} = \frac{x_{iv}}{x_{ia}}; K_{ila} = \frac{x_{il}}{x_{ia}}$	$= 2N_c$
Total number of equations:	$= 3N_c + 6$

As shown above, the number of equations and the number of dependent variables (unknowns) are equal to each other so that the system has mathematical closure. Equation represents the compositional form of the flow equations in porous media so that it can be collapsed to the previously studied single-phase and multiphase flow equations, as will be shown in detail in Chap. 5.

2.3.4 Coordinate Transformations

Let us consider the generalized curvilinear coordinates and rectilinear coordinates shown in Fig. 2.3. The point "P" in the space in Fig. 2.3 is associated with rectangular coordinates (x,y,z) or curvilinear coordinates (u_1, u_2, u_3). Transformation equations from one set of coordinates to the other set can be described by:

$$\begin{cases} x = x(u_1,u_2,u_3) \\ y = y(u_1,u_2,u_3) \\ z = z(u_1,u_2,u_3) \end{cases} \tag{2.89}$$

If u_2 and u_3 are held constant as u_1 varies, the position vector of point P,

$$\vec{r} = x\vec{i} + y\vec{k} + z\vec{k} \tag{2.90}$$

describes a curve called the u_1 coordinate; in a similar way, one can define the u_2 and u_3 coordinate curves through the point P. Also, note that the vectors $\dfrac{\partial \vec{r}}{\partial u_1}, \dfrac{\partial \vec{r}}{\partial u_2}, \dfrac{\partial \vec{r}}{\partial u_3}$

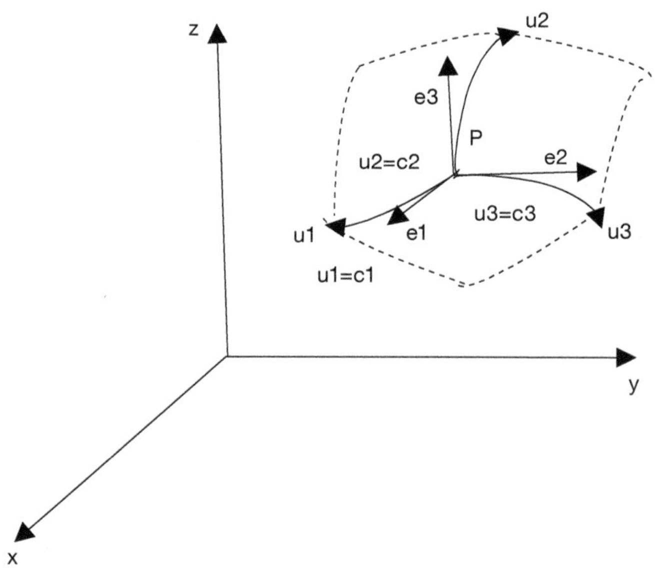

FIGURE 2.3 Relationship between rectangular and curvilinear coordinates.

represent tangent vectors to the u_1, u_2, and u_3 coordinate curves. If $\vec{e_1}$, $\vec{e_2}$ and $\vec{e_3}$ are the unit tangent vectors to curves u_1, u_2, and u_3, respectively, then one can state that

$$\frac{\partial \vec{r}}{\partial u_1} = z_1 \vec{e_1}$$

$$\frac{\partial \vec{r}}{\partial u_2} = z_2 \vec{e_2}$$

$$\frac{\partial \vec{r}}{\partial u_3} = z_3 \vec{e_3} \tag{2.91}$$

In Eq. (2.91), z_1, z_2, and z_3 are called scaling factors, with $z_1 = \left|\frac{\partial \vec{r}}{\partial u_1}\right|$, $z_2 = \left|\frac{\partial \vec{r}}{\partial u_2}\right|$ and $z_3 = \left|\frac{\partial \vec{r}}{\partial u_3}\right|$. Also, if $\vec{e_1}$, $\vec{e_2}$ and $\vec{e_3}$ are mutually perpendicular, the curvilinear coordinate system is an orthogonal coordinate system.

Example 2.1 In the following development, we show the transformation of Laplace's equation from rectilinear (x, y) coordinates to radial-cylindrical (r, θ) coordinates.

Starting with Laplace's equation in rectangular coordinates,

$$\nabla^2_{x,y}\Phi = \frac{\partial^2 \Phi}{\partial x^2} + \frac{\partial^2 \Phi}{\partial y^2} \tag{2.92}$$

Noting that:

$$d\Phi(x,y) = \left(\frac{\partial \Phi}{\partial r}\right)dr + \left(\frac{\partial \Phi}{\partial \theta}\right)d\theta = d\Phi \tag{2.93}$$

with,

$$d\frac{\partial \Phi}{\partial x} = \frac{\partial \Phi}{\partial r}\frac{\partial r}{\partial x} + \frac{\partial \Phi}{\partial \theta}\frac{\partial \theta}{\partial x} \tag{2.94}$$

and,

$$\frac{\partial^2 \Phi}{\partial x^2} = \frac{\partial}{\partial x}\left(\frac{\partial \Phi}{\partial x}\right) = \frac{\partial}{\partial x}\left(\frac{\partial \Phi}{\partial r}\frac{\partial r}{\partial x} + \frac{\partial \Phi}{\partial \theta}\frac{\partial \theta}{\partial x}\right) \tag{2.95}$$

or,

$$\frac{\partial^2 \Phi}{\partial x^2} = \frac{\partial^2 \Phi}{\partial r \partial x}\frac{\partial r}{\partial x} + \frac{\partial \Phi}{\partial r}\frac{\partial^2 r}{\partial x^2} + \frac{\partial^2 \Phi}{\partial \theta \partial x}\frac{\partial \theta}{\partial x} + \frac{\partial \Phi}{\partial \theta}\frac{\partial^2 \theta}{\partial x^2} \tag{2.96}$$

Now,

$$\frac{\partial^2 \Phi}{\partial r \partial x} = \frac{\partial}{\partial x}\left(\frac{\partial \Phi}{\partial r}\right) \tag{2.97}$$

Since:

$$d\left(\frac{\partial \Phi}{\partial r}\right) = \left[\frac{\partial}{\partial r}\left(\frac{\partial \Phi}{\partial r}\right)\right]dr + \left[\frac{\partial}{\partial \theta}\left(\frac{\partial \Phi}{\partial r}\right)\right]d\theta \tag{2.98}$$

Then,

$$\frac{\partial}{\partial x}\left(\frac{\partial \Phi}{\partial r}\right) = \left[\frac{\partial}{\partial r}\left(\frac{\partial \Phi}{\partial r}\right)\right]\frac{\partial r}{\partial x} + \left[\frac{\partial}{\partial \theta}\left(\frac{\partial \Phi}{\partial r}\right)\right]\frac{\partial \theta}{\partial x} \tag{2.99}$$

or,

$$\frac{\partial^2 \Phi}{\partial x \partial r} = \frac{\partial^2 \Phi}{\partial r^2}\frac{\partial r}{\partial x} + \frac{\partial^2 \Phi}{\partial \theta \partial r}\frac{\partial \theta}{\partial x} \tag{2.100}$$

Similarly,

$$d\left(\frac{\partial \Phi}{\partial \theta}\right) = \left[\frac{\partial}{\partial r}\left(\frac{\partial \Phi}{\partial \theta}\right)\right]dr + \left[\frac{\partial}{\partial \theta}\left(\frac{\partial \Phi}{\partial \theta}\right)\right]d\theta \tag{2.101}$$

Then,

$$\frac{\partial}{\partial x}\left(\frac{\partial \Phi}{\partial \theta}\right) = \frac{\partial^2 \Phi}{\partial r \partial \theta}\frac{\partial r}{\partial x} + \frac{\partial^2 \Phi}{\partial \theta^2}\frac{\partial \theta}{\partial x} \tag{2.102}$$

Recalling the relationship between radial–cylindrical and rectangular coordinates:

$$r = \left(x^2 + y^2\right)^{\frac{1}{2}} \text{ and } \theta = \tan^{-1}\left(\frac{y}{x}\right) \tag{2.103}$$

Then, one can obtain the following definitions:

$$\frac{\partial r}{\partial x} = \frac{\partial}{\partial x}\left[\left(x^2 + y^2\right)^{\frac{1}{2}}\right] \text{ or } \frac{\partial r}{\partial x} = \frac{1}{2}\left[\left(x^2 + y^2\right)^{-\frac{1}{2}}\right] \tag{2.104}$$

which can be reduced to:

$$\frac{\partial r}{\partial x} = \frac{x}{\left(x^2 + y^2\right)^{\frac{1}{2}}} \text{ or } \frac{\partial r}{\partial x} = \frac{x}{r} \tag{2.105}$$

Now, one can express $\frac{\partial^2 r}{\partial x^2}$ as follows:

$$\frac{\partial^2 r}{\partial x^2} = \frac{\partial}{\partial x}\left(\frac{x}{r}\right) = x\frac{\partial}{\partial x}\left(\frac{1}{r}\right) + \frac{1}{r} \tag{2.106}$$

or,

$$\frac{\partial^2 r}{\partial x^2} = \frac{1}{r} + x\left[-\frac{1}{r^2}\frac{\partial r}{\partial x}\right] = \frac{1}{r} + x\left[-\frac{1}{r^2}\left(\frac{x}{r}\right)\left(\frac{x}{r}\right)\right] \tag{2.107}$$

or,

Substituting for $r^2 = x^2 + y^2$ on the right-hand side of the above equation gives:

$$\frac{\partial^2 r}{\partial x^2} = \frac{y^2}{r^3} \tag{2.108}$$

Now, it will be necessary to recall that:

$$d(\tan^{-1} w) = \frac{1}{1 + w^2}dw \tag{2.109}$$

or,

$$\frac{d}{dx}(\tan^{-1} w) = \frac{1}{1 + w^2}\frac{dw}{dx} \tag{2.110}$$

Accordingly,

$$\frac{\partial \theta}{\partial x} = \frac{d}{dx}\left[\tan^{-1}\left(\frac{y}{x}\right)\right] = \frac{1}{1 + \left(\frac{y}{x}\right)^2}\frac{d}{dx}\left(\frac{y}{x}\right) \tag{2.111}$$

Then,

$$\frac{\partial \theta}{\partial x} = \frac{y}{\frac{x^2 + y^2}{x^2}}\frac{\partial}{\partial x}\left(\frac{1}{x}\right) \tag{2.112}$$

After carrying out the necessary simplifications,

$$\frac{\partial \theta}{\partial x} = \frac{x^2 y}{r^2}\left(-\frac{1}{x^2}\right) \tag{2.113}$$

or, finally:

$$\frac{\partial\theta}{\partial x} = -\frac{y}{r^2} \tag{2.114}$$

Now, one can work on the second-order derivative:

$$\frac{\partial^2\theta}{\partial x^2} = \frac{\partial}{\partial x}\left(-\frac{y}{r^2}\right) = -y\left(-2r^{-3}\frac{\partial r}{\partial x}\right) = \frac{2y}{r^3}\left(\frac{x}{r}\right) \tag{2.115}$$

or,

$$\frac{\partial^2\theta}{\partial x^2} = \frac{2xy}{r^4} \tag{2.116}$$

Now, one can make substitutions for $\frac{\partial^2\Phi}{\partial x^2}$ and $\frac{\partial^2\Phi}{\partial y^2}$ entries of the Laplace's equation in the rectilinear coordinates. First, we work on the $\frac{\partial^2\Phi}{\partial x^2}$ term [using Eqs. (2.98) through (2.116)]:

$$\frac{\partial^2\Phi}{\partial x^2} = \frac{x^2}{r^2}\frac{\partial^2\Phi}{\partial r^2} - \frac{xy}{r^3}\frac{\partial^2\Phi}{\partial r\partial\theta} - \frac{xy}{r^3}\frac{\partial^2\Phi}{\partial r\partial\theta} + \frac{y^2}{r^4}\frac{\partial^2\Phi}{\partial\theta^2} + \frac{2xy}{r^4}\frac{\partial\Phi}{\partial\theta} + \frac{y^2}{r^3}\frac{\partial\Phi}{\partial r}$$

which can be reduced to:

$$\frac{\partial^2\Phi}{\partial x^2} = \frac{x^2}{r^2}\frac{\partial^2\Phi}{\partial r^2} - \frac{2xy}{r^3}\frac{\partial^2\Phi}{\partial r\partial\theta} + \frac{y^2}{r^4}\frac{\partial^2\Phi}{\partial\theta^2} + \frac{y^2}{r^3}\frac{\partial\Phi}{\partial r} + \frac{2xy}{r^4}\frac{\partial\Phi}{\partial\theta} \tag{2.117}$$

In an analogous manner $\frac{\partial^2\Phi}{\partial x^2}$ term can be expressed as:

$$\frac{\partial^2\Phi}{\partial y^2} = \frac{y}{r^2}\frac{\partial^2\Phi}{\partial r^2} + \frac{2xy}{r^3}\frac{\partial^2\Phi}{\partial r\partial\theta} + \frac{x^2}{r^4}\frac{\partial^2\Phi}{\partial\theta^2} + \frac{y^2}{r^3}\frac{\partial\Phi}{\partial r} - \frac{2xy}{r^4}\frac{\partial\Phi}{\partial\theta} \tag{2.118}$$

Substituting Eqs. (2.117) and (2.118) into $\frac{\partial^2\Phi}{\partial x^2} + \frac{\partial^2\Phi}{\partial y^2} = 0$ yields:

$$\left(\frac{x^2}{r^2} + \frac{y^2}{r^2}\right)\frac{\partial^2\Phi}{\partial r^2} - \frac{2xy}{r^3}\frac{\partial^2\Phi}{\partial r\partial\theta} + \frac{2xy}{r^3}\frac{\partial^2\Phi}{\partial r\partial\theta} + \left(\frac{x^2}{r^4} + \frac{y^2}{r^4}\right)\frac{\partial^2\Phi}{\partial\theta^2}$$
$$+ \left(\frac{x^2}{r^3} + \frac{y^2}{r^3}\right)\frac{\partial\Phi}{\partial\theta} + \left(\frac{2xy}{r^4} - \frac{2xy}{r^4}\right)\frac{\partial\Phi}{\partial\theta} = 0 \tag{2.119}$$

which simplifies to:

$$\frac{\partial^2\Phi}{\partial r^2} + \frac{1}{r^2}\frac{\partial^2\Phi}{\partial\theta^2} + \frac{1}{r}\frac{\partial\Phi}{\partial r} = 0 \tag{2.120}$$

or one ends up with:

$$\frac{1}{r}\left[r\frac{\partial^2\Phi}{\partial r^2}+\frac{\partial\Phi}{\partial r}\right]+\frac{1}{r^2}\frac{\partial^2\Phi}{\partial\theta^2}=0 \tag{2.121}$$

Equation (2.120) is an expanded form of:

$$\frac{1}{r}\frac{\partial}{\partial r}\left(r\frac{\partial\Phi}{\partial r}\right)+\frac{1}{r^2}\frac{\partial^2\Phi}{\partial\theta^2}=0 \tag{2.122}$$

which is also another form of the Laplace's equation in radial-cylindrical coordinates (r and θ directions only).

CHAPTER 3

Reservoir Engineering Fundamentals

Transport in porous media is probably the one area of reservoir engineering which needs an integrated and thorough analysis. All conventional fluid flow equations are based on two important concepts, which are Darcy's flow equation and mass balances (or material balances). Darcy's law is the fundamental building block of the transport equations used in reservoir analysis, and it is always used coupled with mass conservation statements. In this chapter, we will have a brief review of reservoir engineering concepts which are utilized in describing fluid flow in porous media.

3.1 The Porous Medium as a Continuum

The microscopic scrutiny of a porous medium reveals that its local properties may vary widely depending on the volume over which the scrutiny is performed. Instead of a microscopic description, the usual way of approaching a description of a porous media and the fluids within it is to use the *continuum approach*. This approach is statistical in nature and assigns properties such as porosity and permeability at each point in space. Fluid properties and porous medium properties are treated as varying "continually" in space (hence, continuum approach). Each point of the continuum is assigned with property values that characterize the volumetric average behavior of the medium around that point.

Figure 3.1 reflects on the ideas mentioned above. For the concept of continuum, the choice for an appropriate representative volume over which average is performed to obtain properties such as porosity is crucial. Microscopic effects dominate if the representative elementary volume is chosen too small. Macroscopic effects dominate if the representative elementary volume is chosen too large. The size of the representative volume has to be *"just right."* All of our development and analyses in this book are made within the domain of continuum.

3.2 The Fundamental Equations

As briefly introduced in Chap. 1, the solution of a fluid flow problem in applied sciences makes use of up to five *fundamental equations* to obtain the description of the system behavior. These equations are:

- *The Continuity Equation:* It describes how mass is accumulated and transmitted within the system as a result of fluid flow.

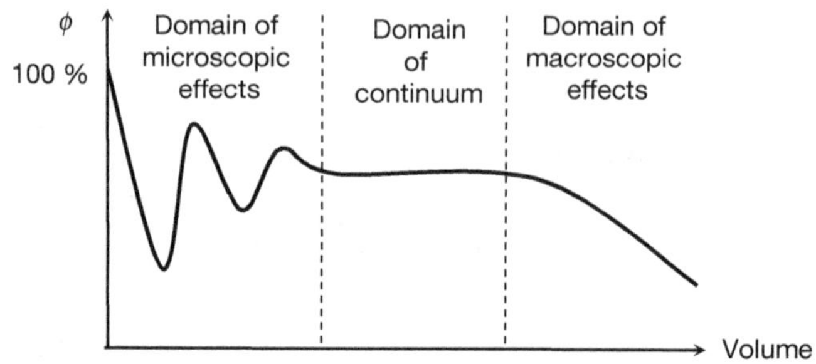

Figure 3.1 The domain of continuum.

- *The Momentum Equation:* It describes how momentum is accumulated and transmitted within the system through the action of forces. It is based on a balance of viscous, pressure, and gravitational forces and follows from the application of the second law of Newton.
- *The Constitutive Equation:* It describes the response of the fluid due to an external effect. In other words, it describes the deformation of the fluid as a result of normal and tangential forces applied to it. The most popular constitutive model is the Newtonian fluid model, in which the viscosity of the fluid is assumed constant. In reservoir modeling, Darcy's law is typically the constitutive equation of choice describing fluid viscous movement within the system for a Newtonian fluid.
- *The Equation of State:* It describes the thermodynamic state of the fluid and how fluid density changes with pressure and temperature.
- *The Energy Equation:* It describes how energy is accumulated and transmitted within the system as a result of energetic interactions. The energy equation is especially important for the study of thermal recovery operations in the reservoir.

In reservoir engineering, the three single most important equations are: the *continuity equation*, the *equation of state*, and *Darcy's law*. We do not use the energy equation because we assume isothermal conditions within the reservoir and hence only changes with pressure are important. We take care of both the *momentum equation* and *constitutive equation* through the introduction of Darcy's law. Darcy's law can be derived from a statistical averaging of the Stokes' equation which is a momentum equation that applies for flow of Newtonian fluids where inertial effects are considered negligible. Therefore, the application of Darcy's law to a fluid flow problem presumes a flow regime in which *viscous forces* are predominant (i.e., laminar flow) and the fluid is Newtonian (constitutive equation or model).

3.3 Fluid and Rock Compressibility Concepts

For single-phase liquids, the compressibility term describes how fluid density changes with pressure can be accomplished through the use of universal definition compressibility, which can lead to an equation of state for slightly compressible fluids under isothermal conditions. For single-phase gases, real gas law provides the means for the calculation of the density and fluid compressibility at different pressure and temperature conditions. For multicomponent, multiphase system, more involved thermodynamic treatment is required, which is covered in Chap. 4. Here we review the concepts of fluid, rock, and total compressibility.

3.3.1 Liquids and Fluid Compressibility

In many single-phase reservoir engineering applications, the concept of fluid compressibility is of paramount importance to describe fluid behavior with pressure. In the case of single-phase liquids and gases, the most common engineering approximations for fluid behavior involve the use of the concept of fluid compressibility. The universal description of compressibility is:

$$c = \pm\frac{1}{\varepsilon}\left(\frac{d\varepsilon}{dp}\right)_T \tag{3.1}$$

i.e., it measures the relative or fractional change of ε (a volumetric quantity) as a function of a unit change in pressure at isothermal conditions. It should be kept in mind that the value of isothermal fluid compressibility is always a positive number. The sign in front of the right-hand side of Eq. (3.1) is determined by the sign of $d\varepsilon/dp$. If $d\varepsilon/dp$ is negative, then, in order to end up with a positive compressibility value, the definition has to carry a negative sign.

For example, Eq. (3.1) is often written in terms of volume ($\varepsilon = V$):

$$c = -\frac{1}{V}\left(\frac{dV}{dp}\right)_T \tag{3.2}$$

and this definition is written with a negative sign. This is because as pressure increases, mechanical stability of fluids and rock materials implies that volume would decrease ($dV < 0$), yielding a negative dV/dp.

The universal description of compressibility could also be written in terms of density ($\varepsilon = \rho$):

$$c = \frac{1}{\rho}\left(\frac{d\rho}{dp}\right)_T \tag{3.3}$$

In this case, a positive sign is used. As pressure increases ($d\rho > 0$) the density of the rock or fluid will increase; hence, $d\rho/dp$ will be positive. The convertibility of Eqs. (3.2) and (3.3) can be demonstrated using the fundamental definition of density:

$$\rho = \frac{m}{V} \quad \rightarrow \quad V = \frac{m}{\rho} \tag{3.4}$$

The expression of fluid compressibility as a function of density can be directly derived by substituting Eq. (3.4) into Eq. (3.2):

$$c = -\frac{1}{m/\rho}\left(\frac{\partial(m/\rho)}{\partial p}\right)_T \tag{3.5}$$

or, simplifying:

$$c = -\rho\left(\frac{\partial}{\partial p}\left(\frac{1}{\rho}\right)\right)_T = -\rho\left(-\frac{1}{\rho^2}\right)\left(\frac{\partial \rho}{\partial p}\right)_T \tag{3.6}$$

It is clear that Eq. (3.6) becomes:

$$c = \frac{1}{\rho}\left(\frac{\partial \rho}{\partial p}\right)_T \tag{3.7}$$

The Equations (3.2) and (3.7) hold for any fluid, both liquids and gases. However, for liquids (i.e., oils and water), we typically assume that isothermal compressibility is constant within the pressure range of interest. This allows us to solve Eq. (3.7) using the separation of variables technique and integrating the both sides,

$$c\int_{p_{sc}}^{p} \partial p = \int_{\rho_{sc}}^{\rho} \frac{1}{\rho}\partial \rho \tag{3.8}$$

which yields:

$$c(p - p_{sc}) = \ln(\rho/\rho_{sc}) \tag{3.9}$$

Now, exponentiating both sides of Eq. (3.9), one obtains:

$$\rho = \rho_{sc}\exp[c(p - p_{sc})] \tag{3.10}$$

Another approximation is introduced here. We recall the *Taylor Series* expansion of the exponential function:

$$e^x = 1 + x + \frac{x^2}{2!} + + \frac{x^3}{3!} + \ldots \tag{3.11}$$

The values of isothermal liquid compressibility are small and they are usually within the range $c = 10^{-6} - 10^{-7}$ psi^{-1}. Therefore, it is reasonable to truncate the Taylor Series expansion of the exponential in Eq. (3.10) after the second term to obtain:

$$\rho = \rho_{sc}[1 + c(p - p_{sc})] \tag{3.12}$$

The Equation is another form of an equation of state only applicable to slightly compressible liquids. It describes the changes in liquid density as a function of pressure.

In many reservoir applications, the concept of formation volume factor (B) is often used to model fluid behavior. Chapter 4 shows that the simplest formulation of the formation volume factor concept is the one that applies to single-component, single-phase fluids. For those cases, formation volume factors can be written as the ratio of fluid density at standard conditions to fluid density at reservoir conditions (see Chap. 4):

$$B = \frac{\rho_{sc}}{\rho} \tag{3.13}$$

Single-phase fluid compressibility could be then alternatively expressed as:

$$c = \frac{1}{\rho}\left(\frac{\partial \rho}{\partial p}\right) = B\frac{\partial}{\partial p}\left(\frac{1}{B}\right) \tag{3.14}$$

If one embraces the assumptions leading to it, Eq. (3.12) can then be alternatively written as:

$$B = \frac{1}{1 + c(p - p_0)} \tag{3.15}$$

which assumes $B = 1$ at $p = p_0$. If the value of B is not 1 at $p = p_0$, the proper B^0 reference should be used at $p = p_0$ and Eq. (3.15) would be rewritten as:

$$B = \frac{B^0}{1 + c(p - p_0)} \tag{3.16}$$

3.3.2 Gases and Real Gas Law

In the case of gases, isothermal fluid compressibility cannot be assumed constant because it significantly varies with pressure. In other words, Eq. (3.8) is not valid for a real gas. For a real gas, the relationship between density and pressure can be derived from the real-gas equation of state:

$$pV = ZnRT \tag{3.17}$$

Recalling the definition of molecular weight, $MW = m/n$, we can write:

$$pV = Z\,(m/MW)\,RT \tag{3.18}$$

Introducing the definition of density [Eq. (3.4)], we obtain:

$$\rho = \frac{(p)(MW)}{ZRT} \tag{3.19}$$

where Z = gas compressibility factor, R = universal gas constant = 10.73 psia-ft^3/R-lbmol, T = temperature (°R). Equation (3.19) is the equation of state for gases. It describes the changes in gas density as a function of pressure.

We may also derive a useful relationship between isothermal gas compressibility, c_g and the Z-factor. Starting with the definition of fluid compressibility with respect to density in (Eq. 3.7), and introducing Eq. (3.19), one obtains:

$$c = \frac{1}{\dfrac{(p)(MW)}{ZRT}}\left(\frac{\partial}{\partial p}\left(\frac{(p)(MW)}{ZRT}\right)\right)_T = \frac{Z}{p}\frac{\partial}{\partial p}\left(\frac{p}{Z}\right)_T \tag{3.20}$$

Taking the derivative of the quotient, one ends up with:

$$c = \frac{Z}{p}\left[\frac{Z - p\left(\frac{\partial Z}{\partial p}\right)_T}{Z^2}\right] = \frac{Z}{P}\left[\frac{1}{Z} - \frac{p}{Z^2}\left(\frac{\partial Z}{\partial p}\right)_T\right] \tag{3.21}$$

Therefore,

$$c = \frac{1}{p} - \frac{1}{Z}\left(\frac{\partial Z}{\partial p}\right)_T \tag{3.22}$$

The Equation is the definition of the isothermal compressibility for a real gas. Notice that for an "ideal" gas:

$$Z = 1 \tag{3.23}$$

and hence,

$$\left(\frac{\partial Z}{\partial p}\right)_T = 0 \tag{3.24}$$

As a result, Eq. (3.22) collapses to:

$$c = \frac{1}{p} \tag{3.25}$$

which is the definition of the compressibility for an ideal gas. Finally, if the concept of formation volume factor (B) is used for a single-component, single-phase (dry) gas:

$$B_g = \frac{\rho_{sc}}{\rho} \tag{3.26}$$

Thus, using Eq. (3.19), formation volume factor for a real (dry) gas can be calculated as:

$$B_g = \frac{\dfrac{(p_{sc})(MW)}{Z_{sc}RT_{sc}}}{\left(\dfrac{(p)(MW)}{ZRT}\right)} \tag{3.27}$$

when one implements B_g (RCF/SCF) units. In practical field units, we rather write:

$$B_g\left(\frac{RB}{SCF}\right) = \frac{p_{sc}TZ}{\alpha_c T_{sc} p} \tag{3.28}$$

where $\alpha_c = 5.615$ ft^3/bbl.

3.3.3 Rock Compressibility and Total Compressibility

In many reservoir engineering applications, it is necessary to consider not only the compressibility of oil, water, and gas, but also the compressibility of rock grains. When dealing with reservoir rock compressibility, three kinds of compressibility can be identified:

1. Rock matrix (rock grain or solid matrix) compressibility:

$$c_s = -\frac{1}{V_s}\frac{dV_s}{dp} \tag{3.29}$$

   where rock matrix volume (V_S) shrinks upon an increase in pore pressure.

2. Pore volume compressibility:

$$c_p = \frac{1}{V_p}\left(\frac{dV_p}{dp}\right)_T \tag{3.30}$$

   The Equation represents the fractional change in pore volume per unit change in reservoir pressure. Note that (dV_S) equals to negative of (dV_p).

3. Rock bulk compressibility:

$$c_b = -\frac{1}{V_b}\left(\frac{dV_b}{dp}\right)_T \tag{3.31}$$

   The Equation represents the fractional change in the total (bulk) volume of the reservoir rock per unit change in the reservoir pressure. Significant changes in bulk volume can occur due to subsidence. However, most usually, $dV_s = -dV_p$ when bulk volume changes are not significant. In such a case, given that $V_p = \phi V_b$, we end up with $c_p = c_\phi$ since:

$$c_p = \frac{1}{V_p}\left(\frac{dV_p}{dp}\right)_T = \frac{1}{\phi}\left(\frac{d\phi}{dp}\right)_T = c_\phi \tag{3.32}$$

Since reservoir rocks are saturated with oil, gas and water it is usually necessary to incorporate the concept of total compressibility. Total compressibility is the fractional change in pore and fluid volume with a unit change in pressure. Total compressibility is defined as:

$$c_t = c_\phi + S_o c_o + S_w c_w + S_g c_g \tag{3.33}$$

The total fluid compressibility is defined as:

$$c_f = \left(S_o c_o + S_w c_o + S_g c_g\right) \tag{3.34}$$

Therefore,

$$c_t = c_\phi + c_f \tag{3.35}$$

3.4 Porosity and Porosity Distribution

The grains that make up the reservoir rock are typically all irregular in shape. The extent of irregularity or lack of roundness will determine the value of porosity of a reservoir rock. Also, some pore space during the sedimentation and cementation may become isolated and these isolated pores do not contribute to fluid flow. The pore spaces in reservoir rock can be formed in two ways: (i) pore space development during sedimentation, and (ii) pore space development as a subsequent stress regime and/or ground water action. The first type of porosity is termed primary porosity (or original porosity) and the second type is termed as secondary. Most sandstone porosity is original porosity while most limestone and dolomite reservoir rocks have secondary porosity, and hence, they are commonly called *dual-porosity* reservoir systems.

Porosity is measured as a percent of total rock volume. The porosity that appears in flow equations is called an effective porosity (continuous or interconnected porosity). The effective porosity has real significance in rocks containing hydrocarbons because it is only from this type of porosity that fluids can move and be recovered.

In a partial differential equation that describes the flow of a fluid in porous media, the porosity term appears on the *RHS* of the equation under the time derivative. For example, if ϕ is a function of pressure, then, the single-phase slightly compressible flow equation in one-dimension can be written as (see Chap. 5):

$$\frac{\partial}{\partial x}\left(\frac{k_x}{\mu B}\frac{\partial p}{\partial x}\right)=\frac{1}{\alpha_c}\frac{\partial}{\partial t}\left(\frac{\phi}{B}\right) \tag{3.36}$$

where $\alpha_c = 5.615$ in practical field units. Equation (3.36) for slightly compressible fluids was first shown in Chap. 2 and will be revisited in Chap. 5. The right-hand side of Eq. (3.36) shows the appearance of the porosity explicitly as part of the "storage/depletion" term. If porosity is not a function of pressure, then it can be treated as a constant in the time domain (since pressure is a function of time) and can be brought in front of the time derivative. On the other hand, if porosity is a function of pressure (due to compressible formation material grains), then one can expand the right-hand side of Eq. (3.36) as follows:

$$\frac{\partial}{\partial t}\left(\frac{\phi}{B}\right)=\left[\frac{1}{B}\frac{\partial \phi}{\partial t}+\phi\frac{\partial}{\partial t}\left(\frac{1}{B}\right)\right] \tag{3.37}$$

or, applying chain rule,

$$\frac{\partial}{\partial t}\left(\frac{\phi}{B}\right)=\left[\frac{1}{B}\frac{\partial \phi}{\partial p}+\phi\frac{\partial}{\partial p}\left(\frac{1}{B}\right)\right]\frac{\partial p}{\partial t} \tag{3.38}$$

In this discussion, we are assuming that bulk volume, V_b, does not change with pressure (no compaction in the formation). Therefore, compressibility expressions for liquids and pore volume stated in Eqs. (3.14) and (3.32) can be written as:

$$\frac{\partial}{\partial p}\left(\frac{1}{B}\right)=\frac{1}{B}c_f \tag{3.39}$$

$$\frac{\partial \phi}{\partial p} = \phi c_{\phi} \tag{3.40}$$

At this point, one ends up with Eq. (3.38) expressed in the following form:

$$\frac{\partial}{\partial t}\left(\frac{\phi}{B}\right) = \frac{\phi}{B}\left[c_{\phi} + c_f\right]\frac{\partial p}{\partial t} \tag{3.41}$$

By setting $c_t = c_{\phi} + c_f$, i.e., the previously discussed definition of total compressibility, we can come up with the final form of the equation:

$$\frac{\partial}{\partial x}\left(\frac{k_x}{\mu B}\frac{\partial p}{\partial x}\right) = \frac{\phi c_t}{B\alpha_c}\frac{\partial p}{\partial t} \tag{3.42}$$

A similar manipulation can be implemented on the right-hand side of the single-phase compressible flow equation as shown below:

$$\frac{\partial}{\partial x}\left(\frac{k_x}{\mu_g B_g}\frac{\partial p}{\partial x}\right) = \frac{1}{\alpha_c}\frac{\partial}{\partial t}\left(\frac{\phi}{B_g}\right) \tag{3.43}$$

Recall that gas formation volume factor B_g in practical field units is defined as [Eq. (3.28)]:

$$B_g\left(\frac{RB}{SCF}\right) = \frac{p_{sc}Tz}{\alpha_c T_{sc} p} \tag{3.44}$$

Then, the right-hand side of Eq. (3.43) can be rewritten as:

$$\frac{1}{\alpha_c}\frac{\partial}{\partial t}\left(\frac{\phi}{B_g}\right) = \frac{T_{sc}}{p_{sc}T}\frac{\partial}{\partial t}\left(\phi\frac{p}{z}\right) \tag{3.45}$$

Now, one can expand the right-hand side of Eq. (3.45) as follows:

$$\frac{1}{\alpha_c}\frac{\partial}{\partial t}\left(\frac{\phi}{B_g}\right) = \frac{T_{sc}}{p_{sc}T}\left[\frac{p}{z}\frac{\partial \phi}{\partial t} + \phi\frac{\partial}{\partial t}\left(\frac{p}{z}\right)\right] \tag{3.46}$$

Expanding Eq. (3.46) further,

$$\frac{1}{\alpha_c}\frac{\partial}{\partial t}\left(\frac{\phi}{B_g}\right) = \frac{T_{sc}}{p_{sc}T}\left[\frac{p}{z}\frac{\partial \phi}{\partial p} + \phi\frac{\partial}{\partial p}\left(\frac{p}{z}\right)\right]\frac{\partial p}{\partial t} \tag{3.47}$$

Resolving the $\frac{\partial}{\partial p}\left(\frac{p}{z}\right)$ derivative,

$$\frac{1}{\alpha_c}\frac{\partial}{\partial t}\left(\frac{\phi}{B_g}\right) = \frac{T_{sc}}{p_{sc}T}\left[\frac{p}{z}\frac{\partial \phi}{\partial p} + \phi\frac{z - p\left(\frac{\partial z}{\partial p}\right)}{z^2}\right]\frac{\partial p}{\partial t} \tag{3.48}$$

or

$$\frac{1}{\alpha_c}\frac{\partial}{\partial t}\left(\frac{\phi}{B_g}\right)=\frac{T_{sc}}{p_{sc}T}\left[\frac{p}{z}\frac{\partial \phi}{\partial p}+\phi\frac{p}{z}\left(\frac{1}{p}-\frac{1}{z}\frac{\partial z}{\partial p}\right)\right]\frac{\partial p}{\partial t} \tag{3.49}$$

In the above equation, we recognize the definition of gas compressibility, c_g [Eq. (3.22)] and pore compressibility [Eq. (3.32)]:

$$c_g=\frac{1}{p}-\frac{1}{z}\left(\frac{\partial z}{\partial p}\right) \tag{3.50}$$

$$\frac{\partial \phi}{\partial p}=\phi c_\phi \tag{3.51}$$

Then, one will have:

$$\frac{1}{\alpha_c}\frac{\partial}{\partial t}\left(\frac{\phi}{B_g}\right)=\frac{T_{sc}}{p_{sc}T}\left[\frac{p}{z}c_\phi\phi+\frac{p}{z}c_g\phi\right]\frac{\partial p}{\partial t} \tag{3.52}$$

and finally one obtains:

$$\frac{1}{\alpha_c}\frac{\partial}{\partial t}\left(\frac{\phi}{B_g}\right)=\frac{T_{sc}}{p_{sc}T}c_t\phi\frac{p}{z}\frac{\partial p}{\partial t} \tag{3.53}$$

Note that in the above equation, the total compressibility is defined as:

$$c_t=c_g+c_\phi \tag{3.54}$$

This development reveals the final form of the compressible flow equation as:

$$\frac{\partial}{\partial x}\left(\frac{k_x}{\mu_g B_g}\frac{\partial p}{\partial x}\right)=\frac{T_{sc}}{p_{sc}T}\phi c_t\frac{p}{z}\frac{\partial p}{\partial t} \tag{3.55}$$

In the above discussion, we have used the slightly compressible and compressible flow equations, respectively, to illustrate how the porosity term is handled when porosity is a function of time. The same discussion will not take place for incompressible flow problem as there is no storage or depletion process taking place in incompressible flow problems and accordingly porosity term does not appear in the incompressible flow equation.

Porosity as a fundamental property in reservoir flow equations may exhibit homogenous or heterogeneous distributions. A homogenous porosity distribution implies porosity is uniformly distributed over the domain of interest. A heterogeneous porosity distribution implies changes in porosity with spatial variations. As a precaution we would like to remind that porosity is not a directional property and it will be wrong and meaningless to make a statement about the isotropic (or anisotropic) nature of porosity distribution. This is easy to understand by simply thinking of porosity as a dimensionless volume. To describe a characteristic volume such as porosity in a three-dimensional space, it will be necessary to use all of the available three directions. In other words, such a description will leave no other free direction, along which a different porosity value can be measured.

3.5 Permeability and Permeability Distributions

The ease with which fluid can move through the interconnected pore spaces of the rock defines the degree of permeability possessed by the rock. Accordingly, high permeability will allow fluids to move rapidly through rocks. A porous material having 1 Darcy of permeability is defined as one that 1 cm^3 of a fluid of 1 centipoise viscosity would flow each second through the porous media with dimensions of 1 cm in length and with 1 cm^2 cross-sectional area perpendicular to the flow direction. The pressure drop across the length of the porous material is 1 atm. The permeability has "dimensions" of area (L^2) and 1 Darcy $\approx 10^{-8}\ cm^2$. The fact that permeability has dimensions of area signals that permeability of a rock can exhibit isotropic or anisotropic characteristics. In a three-dimensional coordinate system, in order to express an area, one will need to use two dimensions of the coordinate system. Thus, depending on which dimensions are considered in describing the area (permeability), it will be possible to come up with three distinct permeability values, e.g., k_x, k_y, k_z. If $k_x = k_y = k_z$, then it will be concluded that reservoir rock is exhibiting isotropic permeability characteristics. Typically, because of the nature of sedimentation the vertical direction permeability, k_z is much less than lateral direction permeabilities. In characterizing the permeability of a reservoir, all of the permutations of homogenous, heterogeneous, isotropic, and anisotropic classifications are valid possibilities. The simplest reservoir can have homogenous and isotropic permeability and the most complex (and the most realistic) reservoir will exhibit heterogeneous and anisotropic permeability characteristics.

3.5.1 Characteristics of Darcy's Equation

It is instructive to write the Darcy's equation for flow in a linear horizontal system, which is always written with respect to the entire cross-sectional area:

$$q = \frac{-Ak\Delta p}{\mu \Delta L} \tag{3.56}$$

Darcy's law states that the superficial velocity ($v = q/A$, A = entire flow cross-sectional area) of a single-phase fluid system is directly proportional to the fluids mobility, k/μ and the pressure gradient, $\Delta p/L$. The assumptions that are inherent to the Darcy's equation are:

- Homogenous rock
- Nonreactive porous medium
- Single-phase flow
- Newtonian fluid
- Laminar flow conditions
- Steady-state flow regime
- Isothermal flow conditions

The negative sign in front of Eq. (3.56) indicates that flow is from high pressure to low pressure ($\Delta p/L < 0$). The differential form of Darcy's equation can be written as, for flow in the x-direction:

$$v_x = -\frac{k_x}{\mu}\frac{\partial \Phi}{\partial x} \tag{3.57}$$

In Eq. (3.57), fluid flow potential gradient $\frac{\partial \Phi}{\partial x}$ is defined as:

$$\frac{\partial \Phi}{\partial x} = \frac{\partial p}{\partial x} - \frac{1}{144}\frac{g}{g_c}\rho\frac{\partial G}{\partial x} \tag{3.58}$$

In Eq. (3.58), the term $\frac{\partial G}{\partial x}$ represents the depth gradient and the depth G is considered as positive downward from the datum level.

When Darcy's equation is written for a multiphase flow system, the relative permeability concept needs to be included so that Eq. (3.24) is rewritten as:

$$v_{j_x} = -\frac{kk_{r_j}}{\mu_j}\frac{\partial \Phi_j}{\partial x} \tag{3.59}$$

where j is a subindex that identifies the phase of interest. Table 3.1 gives a summary of different units that are used in Darcy's equation for gas flow and liquid flow.

Darcy units (gas or liquid)	Field units (gas)	Field units (liquid)
$q = -\frac{kA(\Delta p)}{\mu \Delta L}$	$q = -6.328\frac{kA(\Delta p)}{\mu \Delta L}$	$q = -1.127\frac{kA(\Delta p)}{\mu \Delta L}$
$v = -\frac{k(\Delta p)}{\mu \Delta L}$	$v = -6.328\frac{k(\Delta p)}{\mu \Delta L}$	$v = -1.127\frac{k(\Delta p)}{\mu \Delta L}$
q is in cc/s k is in Darcy A is in cm^2 Δp is in atm μ is in cp ΔL is in cm v is in cm/s	q is in ft^3/day k is in Darcy A is in ft^2 Δp is in psi μ is in cp ΔL is in ft v is in ft/day	q is in bbl/day k is in Darcy A is in ft^2 Δp is in psi μ is in cp ΔL is in ft v is in bbl/day-ft^2
	If k is in perms, (1 Darcy = 1.127 perms) $q = -5.615\frac{kA(\Delta p)}{\mu \Delta L}$ $v = -5.615\frac{k(\Delta p)}{\mu \Delta L}$	If k is in perms, (1 Darcy = 1.127 perms) $q = -\frac{kA(\Delta p)}{\mu \Delta L}$ $v = -\frac{k(\Delta p)}{\mu \Delta L}$

Table 3.1 Darcy's Equation and its Units

3.6 The Flow Potential Concept

The flow potential, defined by M. King Hubert and referred to as Hubert's potential, is basically a combination of pressure and gravitational fields. To derive an expression for flow potential of a fluid at point, P, we define it as the amount of mechanical energy to transform a unit mass from one reference level to an arbitrary level, G, as shown in Fig. 3.2. From a mathematical point, it should be stated that flow "potential" is in reality a "pseudo potential" (as shown later by Equation 3.7.6).

Now, consider one unit of mass of fluid translated from state point P_1 to state point P_2, as shown in Fig. 3.2. Now, in order to write the total work done during this translation, one needs to consider all the following types of work:

- Withdrawing the unit mass of fluid from point P_1:
  Work of collection: $w_1 = -\,p_0 u_0$ (3.60)
  (u_0 is the specific volume at $G = 0$)
- Lifting it to point P_2:
  Potential energy $w_2 = gG$ (3.61)
- Compressing it during the transition:
  Work of compression: $w_3 = -\int_1^2 pdv$ (3.62)
  (limits on the integral refer to state points)
- Rejecting it at point P_2:
  Work of rejection: $w_4 = pu$ (3.63)
- Accelerating it:
  Kinetic energy: $w_5 = \dfrac{v^2}{2g}$ (3.64)

The flow potential is then expressed as the total work done on the unit mass, which is:

$$\Phi = \sum_{i=1}^{5} W_i \tag{3.65}$$

or

$$\Phi^* = -p_0 u_0 + gG - \int_1^2 pdu + pu + \frac{v^2}{2g} \tag{3.66}$$

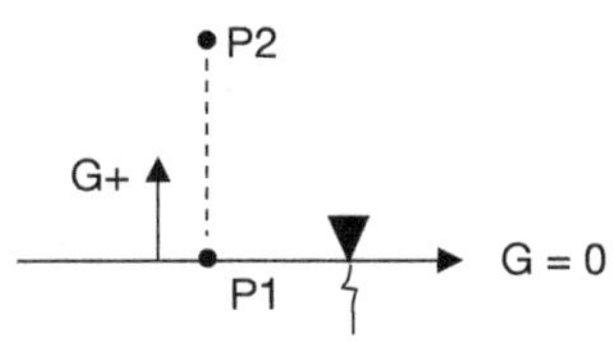

FIGURE 3.2 Changes from state point P_1 to state point P_2.

The Equation is a form of Bernoulli's equation for a unit mass. Integration operation in Eq. (3.66) is carried out using "integration by parts" so that:

$$\int_1^2 pdu = pu \Big|_1^2 - \int_1^2 udp = pu - p_0u_0 - \int_1^2 udp \tag{3.67}$$

Substituting the above equation into Eq. (3.66) and neglecting the kinetic energy gained (this is a good assumption since the velocity term is negligibly small in porous media), one can write:

$$\Phi^* = \int_1^2 udp + gG \tag{3.68}$$

Since specific volume, u, is the inverse of density ($u = 1/\rho$), then the flow potential per unit mass of fluid at some point "p" relative to an arbitrary datum is:

$$\Phi^* = \int_{p_0}^{p} \frac{1}{\rho(p)} dp + gG \tag{3.69}$$

The potential gradient can be written as:

$$\frac{\partial \Phi^*}{\partial x} = \frac{\partial}{\partial x}\left[\int_{p_0}^{p} \frac{1}{\rho(p)} dp + gG\right] \tag{3.70}$$

Operating on the right-hand side of Eq. (3.70) one can write using the chain rule:

$$\frac{\partial \Phi^*}{\partial x} = \frac{\partial}{\partial x}\left[\int_{p_0}^{p} \frac{1}{\rho(p)} dp\right]\frac{\partial p}{\partial x} + g\frac{\partial G}{\partial x} \tag{3.71}$$

or invoking the Leibniz's rule on terms inside the rectangular bracket:

$$\frac{\partial \Phi^*}{\partial x} = \left[\int_{p_0}^{p} \frac{\partial}{\partial p}\frac{1}{\rho(p)} dp\right]\frac{\partial p}{\partial x} + g\frac{\partial G}{\partial x} \tag{3.72}$$

or,

$$\frac{\partial \Phi^*}{\partial x} = \frac{1}{\rho}\frac{\partial p}{\partial x} + g\frac{\partial G}{\partial x} \tag{3.73}$$

The Equation is referred to as Hubbert's potential which is valid for both compressible and incompressible fluids. In developing Eq. (3.73), the depth direction for G was taken positive upward. However, if G is taken positive downward from the datum level, then the sign of G needs to be changed:

$$\frac{\partial \Phi^*}{\partial x} = \frac{1}{\rho}\frac{\partial p}{\partial x} - g\frac{\partial G}{\partial x} \tag{3.74}$$

Multiplying both sides of Eq. (3.74) by ρ one obtains:

$$\rho\frac{\partial \Phi^*}{\partial x} = \frac{\partial p}{\partial x} - g\rho\frac{\partial G}{\partial x} \tag{3.75}$$

At this point, one can redefine the gradient of the flow potential to be consistent with the universal dimensions of flow potential as:

$$\frac{\partial \Phi}{\partial x} = \rho \frac{\partial \Phi^*}{\partial x} \tag{3.76}$$

which results in:

$$\frac{\partial \Phi}{\partial x} = \frac{\partial p}{\partial x} - \frac{1}{144} \frac{g}{g_c} \rho \frac{\partial G}{\partial x} \tag{3.77}$$

Here Eq. (3.77) is written in practical field units. A dimensional analysis as shown below will show that both sides of the equation are dimensionally consistent (the g/g_c term can be taken as equal to unity for practical purposes). This dimensional analysis involves the use of three fundamental dimensions, including mass (M), length (L), and time (T).

$$\left[\frac{M/LT^2}{L}\right] = \left[\frac{M/LT^2}{L}\right] - \left[\frac{L}{T^2}\right]\left[\frac{M}{L^3}\right]\left[\frac{L}{L}\right]$$

$$\left[\frac{M}{L^2T^2}\right] = \left[\frac{M}{L^2T^2}\right] - \left[\frac{M}{L^2T^2}\right]$$

The above analysis confirms that Eq. (3.77) is dimensionally consistent.

3.7 Permeability as a Symmetric Tensor of a Second Rank

The constitutive equation for describing flow in porous media is Darcy's law. Figure 3.3 shows a three-dimensional permeability state at a point within the reservoir with normal and shear components in tensor notation indicating that permeability is a second-order tensor.

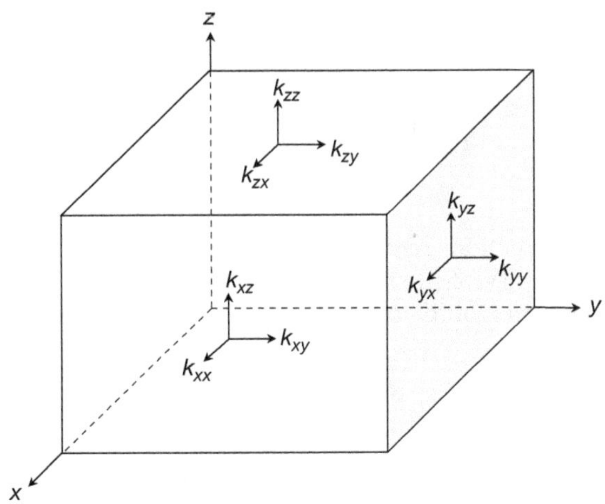

FIGURE 3.3 Permeability as a symmetric tensor.

Permeability can be a complex function of spatial location and direction as it is a measure of the connectivity of pore spaces in a given direction. In general, it is assumed that the coordinate system of study is aligned with the principal axes of the permeability tensor—also known as the principal flow directions. When such an alignment is established the permeability tensor simplifies to a diagonalized permeability matrix. As shown earlier, Darcy's law can be written in terms of the flow potential gradient as follows:

$$\vec{v} = -\frac{\bar{\bar{k}}}{\mu}\nabla\Phi \tag{3.78}$$

In parallel to Eq. (3.78), Darcy's law can be written using the full permeability tensor ($\bar{\bar{k}}$) notation in a matrix form. For a rectangular coordinate system, we write:

$$\begin{bmatrix} v_x \\ v_y \\ v_z \end{bmatrix} = -\frac{1}{\mu}\begin{bmatrix} k_{xx} & k_{xy} & k_{xz} \\ k_{yx} & k_{yy} & k_{xz} \\ k_{zx} & k_{zy} & k_{zz} \end{bmatrix}\begin{bmatrix} \partial\Phi/\partial x \\ \partial\Phi/\partial y \\ \partial\Phi/\partial z \end{bmatrix} \tag{3.79}$$

In Eq. (3.79), viscosity, μ, is not a directional quantity. Equation (3.46) can be expanded as follows:

$$v_x = -\frac{1}{\mu}\left(k_{xx}\frac{\partial\Phi}{\partial x} + k_{xy}\frac{\partial\Phi}{\partial y} + k_{xz}\frac{\partial\Phi}{\partial z}\right) \tag{3.80}$$

$$v_y = -\frac{1}{\mu}\left(k_{yx}\frac{\partial\Phi}{\partial x} + k_{yy}\frac{\partial\Phi}{\partial y} + k_{yz}\frac{\partial\Phi}{\partial z}\right) \tag{3.81}$$

$$v_z = -\frac{1}{\mu}\left(k_{zx}\frac{\partial\Phi}{\partial x} + k_{zy}\frac{\partial\Phi}{\partial y} + k_{zz}\frac{\partial\Phi}{\partial z}\right) \tag{3.82}$$

To simplify the problem, the goal here is to find a coordinate system (x', y', z') in which the permeability tensor reduces to a diagonal form of the type:

$$[k'] = \begin{bmatrix} k_{x'x'} & 0 & 0 \\ 0 & k_{y'y'} & 0 \\ 0 & 0 & k_{z'z'} \end{bmatrix} \tag{3.83}$$

Here the x', y', z' directions refer to the principal flow directions and they are related to the depositional history of the formation. In this case, the flow equations along the principal flow directions would collapse to:

$$v_{x'} = -\frac{A}{\mu}\left(k_{x'x'}\frac{\partial\Phi}{\partial x'}\right) \tag{3.84}$$

$$v_{y'} = -\frac{A}{\mu}\left(k_{y'y'}\frac{\partial\Phi}{\partial y'}\right) \tag{3.85}$$

$$v_{z'} = -\frac{A}{\mu}\left(k_{z'z'}\frac{\partial \Phi}{\partial z'}\right) \tag{3.86}$$

or,

$$\begin{bmatrix} v_{x'} \\ v_{y'} \\ v_{z'} \end{bmatrix} = -\frac{1}{\mu}\begin{bmatrix} k_{x'x'} & 0 & 0 \\ 0 & k_{y'y'} & 0 \\ 0 & 0 & k_{z'z'} \end{bmatrix}\begin{bmatrix} \partial\Phi/\partial x' \\ \partial\Phi/\partial y' \\ \partial\Phi/\partial z' \end{bmatrix} \tag{3.87}$$

This diagonalized form of Darcy's law is the most extensively used in reservoir modeling, given that it greatly simplifies the resulting governing equations.

This time, we can write the original Eq. (3.46) in terms of the following matrix equation:

$$[v] = [k][M] \tag{3.88}$$

where

$$[M] = -\frac{1}{\mu}\begin{bmatrix} \partial\Phi/\partial x \\ \partial\Phi/\partial y \\ \partial\Phi/\partial z \end{bmatrix} \tag{3.89}$$

$$[k] = \begin{bmatrix} k_{xx} & k_{xy} & k_{xz} \\ k_{yx} & k_{yy} & k_{xz} \\ k_{zx} & k_{zy} & k_{zz} \end{bmatrix} \tag{3.90}$$

and the diagonalized Eq. (3.87) is:

$$[v'] = [k'][M'] \tag{3.91}$$

At this stage, we are after a similarity transformation to diagonalize the nine-component matrix $[k]$ in the three-component (diagonalized) form $[k']$. We start by multiplying both sides of Eq. (3.88) with a rectangular $n \times n$ square, non-singular matrix of $[G]$ so that:

$$[G][v] = [G][k][M] \tag{3.92}$$

Since matrix $[G]$ is considered as non-singular, it follows:

$$[G]^{-1}[G] = [G][G]^{-1} = [I] \tag{3.93}$$

In Eq. (3.93), $[I]$ is the $n \times n$ square identity matrix. Introducing Eqs. (3.93) into (3.92) gives:

$$[G][v] = [G][k][G]^{-1}[G][M] \tag{3.94}$$

Identifying the following transformations in the left- and right-hand sides:

$$[G][v] = [v'] \tag{3.95}$$

$$[M'] = [G][M] \tag{3.96}$$

The equation implies that the following similarity transformation applies to the permeability tensor:

$$[k'] = [G][k][G]^{-1} \tag{3.97}$$

and one would end up with

$$[v'] = [k'][M'] \tag{3.98}$$

This equation is of the same form as Eq. (3.91), which was set up for a diagonal matrix. The transformation that allowed this diagonalization is given by Eq. (3.97). Mathematically, it is then possible to find the coordinate system x', y', z' in which the permeability tensor has the form $[k']$, as long as the appropriate transformation $[G]$ defined in Eq. (3.97) is found. This procedure thus involves finding the eigenvalues (λ) and the linearly independent eigenvectors of the original $[k]$ matrix. The eigenvalues of the system are found by applying the eigenvalue equation:

$$\det\left([k] - \lambda[I]\right) = 0 \tag{3.99}$$

The resulting n-linearly independent eigenvectors will then define the new x', y', z' coordinate system or principal flow directions of the problem.

3.8 Wettability and Interfacial Tension

Wetting is the ability of a liquid to establish contact with a solid surface as a result of intermolecular forces when the two are brought together. Wettability represents an interaction between rock and fluids in the reservoir. It is a function of the type of fluid and the type of surface.

The degree of wettability is determined by a force balance between adhesive forces (forces between dissimilar molecules typically found in different phases) and cohesive forces (forces between similar molecules found within the same phase). Wetting affinity of a reservoir rock (water-wet or oil-wet) is determined by measuring the contact angle, as shown in Fig. 3.4. As the contact angle decreases, the wetting characteristics of the liquid/solid interaction increase.

As it can be seen from Fig. 3.4, the contact angle of a liquid on a surface is the result of the mechanical equilibrium (force balance) of a drop resting on a homogenous surface. In other words,

$$\gamma_{sv} - \gamma_{sl} = \cos\theta\gamma_{lv} \tag{3.100}$$

In Eq. (3.100), γ_{sv} is the interfacial tension (*IFT*) or surface tension between the solid and vapor, γ_{sl} is the interfacial tension between the solid and liquid, and γ_{lv} is the interfacial tension between the liquid and vapor phase. Finally, θ represents the wetting or contact angle. Complete wettability would be evidenced by a zero contact angle, and complete non-wetting would be manifested by a contact angle of 180°.

Interfacial tension is a property of the interface between two immiscible phases. The *IFT* concept can relate to the liquid/liquid and liquid/solid phase boundaries (interfaces); while for liquid/gas interfaces, it is most commonly referred to as surface tension, for example, when one of the phases is air. The surface tension of water-hydrocarbon systems may vary between 20 dynes/cm to 70 dynes/cm.

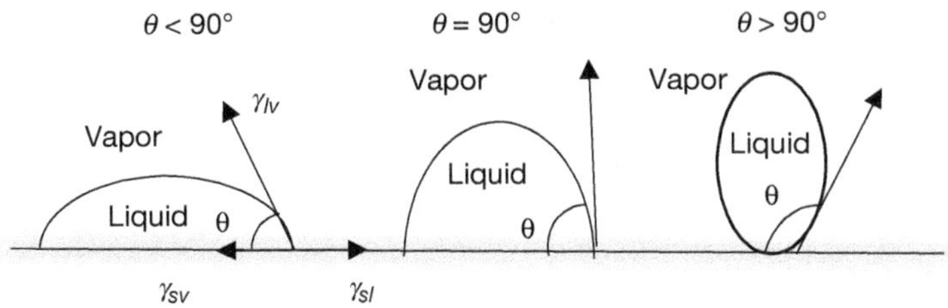

Figure 3.4 Contact angles formed by a liquid drop on a smooth homogenous solid surface.

3.9 The Relative Permeability Concept

Recalling the multiphase transport equations briefly presented in Chap. 2, which will be discussed in more detail in Chap. 5, one can easily recognize that rock and fluid properties such as wettability (contact angle), interfacial tension, surface tension, pore throat diameter, and pore-specific surface area do not appear explicitly in reservoir flow equations written macroscopically. However, the roles of such microscopic properties on multiphase flow in porous media are conveyed to macroscopic transport equations especially by two important functions which are known as relative permeability relationships and capillary pressure relationships. We will review the important aspects of these two relationships.

Since all of the hydrocarbon reservoirs contain water (which may or may not be mobile), it will be necessary to study the permeability characteristics of the porous media to each phase appearing in the pores of the reservoir rock. The ratio of effective permeability to a phase at a particular saturation to the permeability at 100% saturation (absolute permeability) is called as relative permeability to that phase at that saturation. Accordingly,

$$k_{ro} = \frac{k_o}{k} \tag{3.101}$$

$$k_{rg} = \frac{k_g}{k} \tag{3.102}$$

$$k_{rw} = \frac{k_w}{k} \tag{3.103}$$

The relative permeability data for two-phase relative permeability characteristics exhibit a typical shape, as shown in Fig. 3.5.

At 100% saturation of either oil or water phase the relative permeability to that phase is 1.0. However, in reality since water saturation cannot go below the irreducible water saturation S_{wirr}, it is not possible to have the k_{ro} values assigned for water saturations less than the irreducible water saturation. Similarly, when oil saturation reaches the residual oil saturation (or $S_w = 1 - S_{or}$), water saturation cannot be increased any further. Therefore, k_{rw} values at water saturation $S_w > 1 - S_{or}$ cannot be measured either. These two endpoint saturations and the permeability curves between the characteristic water

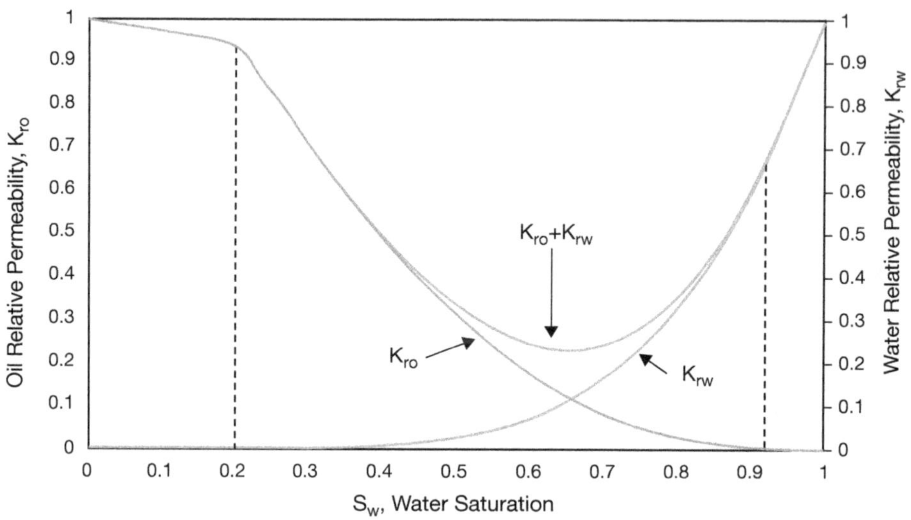

FIGURE 3.5 Relative permeability as a function of water saturation.

saturations bring significant information about the petrophysical characteristics of the reservoir rock as discussed below.

The behavior (shape) of the relative permeability curves between $S_w = S_{wirr}$ and $S_w = 1 - S_{or}$ carry important information about some of the important petrophysical characteristics of the porous media. For example, if the relative permeability curves are all depressed, it is a clear sign of small pore spaces across sectional areas (or small permeabilities). If the curves exhibit linear shapes, then it is the indication of low interfacial tension between the phases. It should also be noted that the summation of relative permeabilities at any particular saturation does not total to 1.0. On the other hand, as the interfacial tension between the phases becomes smaller and the relative permeability curves straighten up, then the summation of the relative permeability values become closer to 1.0. This can be easily seen if one considers the limiting case when interfacial tension in the limit goes to zero, which implies that both phases are completely miscible or the relative permeability is simply equal to the saturation of that phase.

The shapes of the relative permeability curves also carry certain signatures on the wetting affinity of the reservoir rock. This could be best illustrated by looking at the saturation value at the cross-over point of the two curves where $k_{ro} = k_{rw.}$ If the water saturation value at the cross-over point is less than 50%, then we make the observation that at $S_o > S_w$ (at the cross-over point) $k_{ro} = k_{rw}$. Then, it will be reasonable to question that how come at an oil saturation greater than water saturation, k_{ro} is not greater than k_{rw}. Accordingly, it will be correct to assume that at a certain portion of the oil phase which is wetting the surfaces of the rock grains becomes immobile; this is why $k_{ro} = k_{rw}$ when $S_o > S_w$. If the cross-over point for the two curves take place at $S_w > S_o$ (as in the case of Fig. 3.5), then we conclude that wetting affinity of the porous medium is water-wet. As a final point on this discussion, one can deduce that as the cross-over point saturation is increasingly larger than S_o, the water-wet nature of the reservoir rock increases. Conversely, at the cross-over point saturation, if S_w becomes much smaller than S_o,

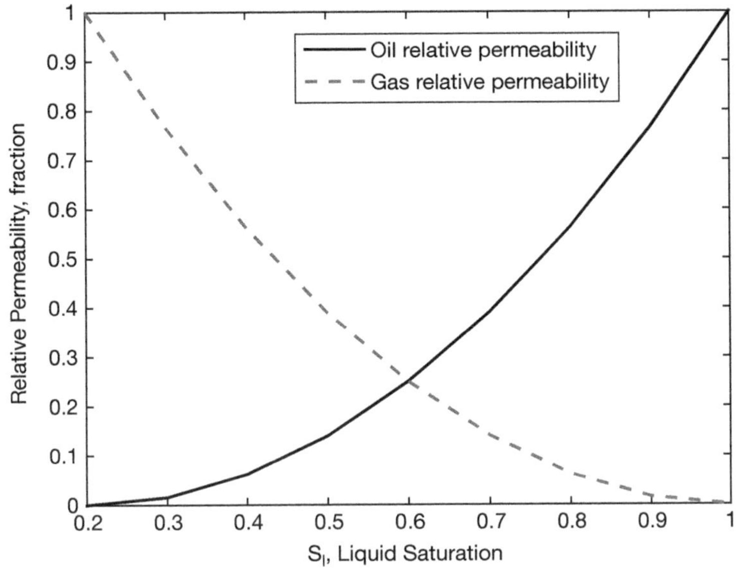

FIGURE 3.6 General relative permeability characteristics.

then we confirm that wetting affinity of the rock is increasingly oil-wet. It should also be noted that the wetting phase relative permeability is concave upwards and the non-wetting phase relative permeability generally takes an S shaped curve. For the non-wetting phase, irreducible saturation is much less than the relative permeability to the non-wetting phase at the irreducible wetting phase saturation.

The aforementioned remarks apply to gas–oil relative permeability characteristics as well. It can be mentioned in Fig. 3.6 that typically k_{ro} and k_{rg} values are plotted as a function of total liquid saturation ($S_l = S_o + S_{wc}$).

Three-phase relative permeability data can be presented in the form of ternary diagrams, as shown in Fig. 3.7. Typically, three-phase relative permeability data for k_{ro} is obtained from two sets of two-phase relative permeability characteristics. These two sets are S_w versus k_{ro} and k_{rw} and S_o versus k_{ro} and k_{rg}. At a particular oil saturation, a three-phase relative permeability model is used to calculate k_{ro} values at different combinations of phase saturations.

Two-Phase Relative Permeability Models

Corey's two-phase (oil-water) relative permeability model uses a normalized saturation function described as (Corey, 1954):

$$S_w^* = \frac{S_w - S_{wirr}}{1 - S_{wirr}} \tag{3.104}$$

Then,

$$k_{rw} = S_w^* \tag{3.105}$$

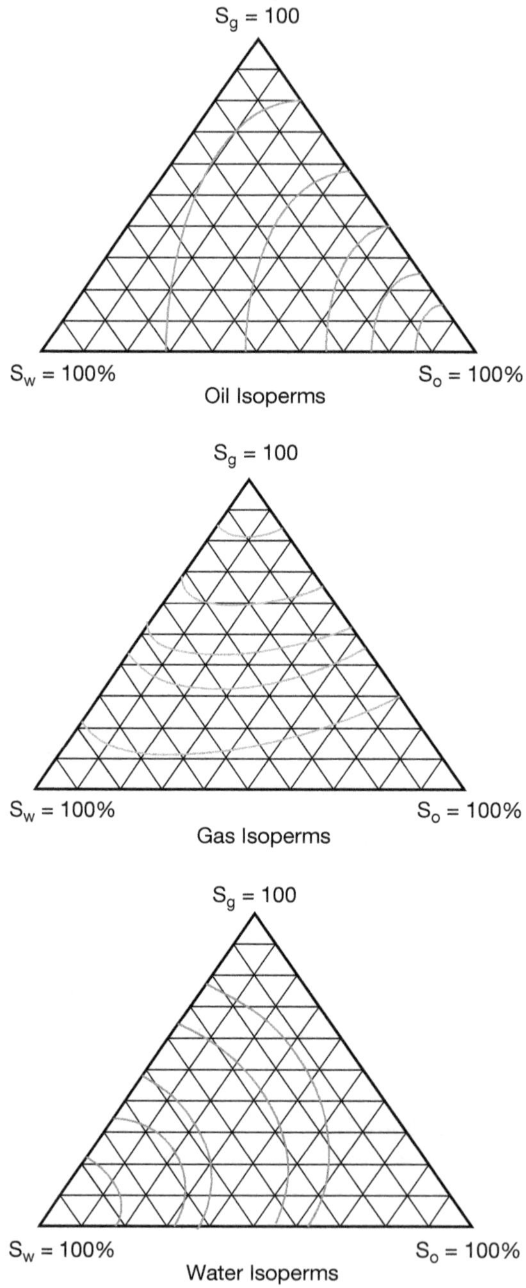

FIGURE 3.7 Representation of three-phase relative permeability data.

and,

$$k_{ro} = \left(1 - S_w^*\right)^2 \times \left[1 - \left(S_w^*\right)^2\right] \tag{3.106}$$

Another widely used two-phase oil/water relative permeability model is Naar and Henderson (1961) model, which suggests the following relationships:

$$k_{rw} = \left(S_w^*\right)^4 \tag{3.107}$$

and:

$$k_{ro} \frac{\left(1 - 2S_w^*\right)^{\frac{3}{2}}}{(2 - \left(1 - 2S_w^*\right)^{\frac{1}{2}}} \tag{3.108}$$

For two-phase gas/oil relative permeability, Brooks and Corey (1964) relative permeability model is also used widely, where:

$$k_{ro} = \left(\frac{S_o - S_{or}}{1 - S_{or}}\right)^{(2+3\lambda)/\lambda} \tag{3.109}$$

and

$$k_{rg} = \left(\frac{1 - S_o}{1 - S_{or}}\right)^2 \left[1 - (\frac{S_o - S_{or}}{1 - S_{or}})^{\frac{2+\lambda}{\lambda}}\right] \tag{3.110}$$

The parameter λ is related to the distribution of pore sizes. For narrow distributions, λ is typically larger than 2 and for wider pore size distributions, λ is less than 2.

Three-Phase Relative Permeability Models

The basic premise under the three-phase relative permeability models is the assumption of the phase distribution in a simple pore channel, as shown in Fig. 3.8.

As shown in Fig 3.8, there exists three interfaces in this simplified representation. The first interface is between the wetting phase (w) and the surfaces of the rock grains. The second interface is between the water phase and the intermediate phase (o), and finally the third interface is between the intermediate phase and the gaseous phase (g). The representation, as described in Fig. 3.8, will lead us to conceptualize that:

$$k_{rw} = f\left(S_w\right) \tag{3.111}$$

and

$$k_{rg} = f\left(S_g\right) \tag{3.112}$$

The functional dependencies shown in Eqs. (3.111) and (3.112) can be explained by intuitive thinking that any additional energy lost at the interface II is for water phase as the water phase does not experience any interaction at the interface III. The length of

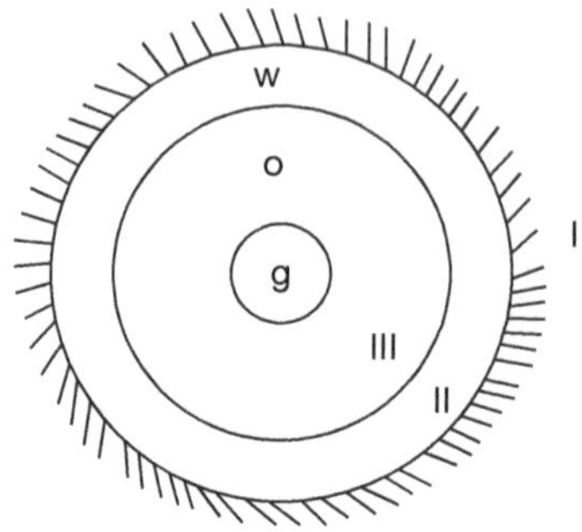

Figure 3.8 Idealized phase distributions in a circular pore channel (a water-wet system is considered).

the interface II is simply described by the water-phase saturation and thus water-phase relative permeability is simply a function of water-phase saturation. Accordingly, in three-phase flow, water-phase relative permeability at a given water saturation can easily be imported from two-phase oil/water relative permeability characteristics. A similar discussion can be structured for the relative permeability to gas. The gas phase only sees the interface III and the length of this interface is simply controlled by the magnitude of the gas saturation. Again, the two-phase oil/gas relative permeability data can be used to assign the k_{rg} values at particular gas saturations in three-phase flow.

Following similar lines of thinking, now one can describe the following functional relationship for k_{ro}:

$$k_{ro} = f\left(S_w, S_g\right) \tag{3.113}$$

It should be noted that Eq. (3.113) does not mean that k_{ro} is simply a function of S_o although one can argue that $S_o = 1 - S_w - S_g$. It should be noted that there are infinite possible combinations of S_w and S_g that will result in the same S_o. However, in each of these constraints S_w and S_g will have different values. Again, as can be easily seen from Fig 3.8, flow of the oil phase is influenced by two interfaces, namely interface II and interface III. Therefore, k_{ro} becomes a function of both S_w and S_g.

The first model of Stone is quite widely used to calculate the k_{ro} at a given S_w and S_g saturation combination where $S_o = 1 - S_w - S_g$ (Stone, 1970):

$$k_{ro}\left(S_w, S_g\right) = \frac{S_o - S_{or}}{1 - S_{wir} - S_{or}} \cdot \frac{k_{row}\left(S_w\right)/k_{rocw}}{1 - \dfrac{S_w - S_{wir}}{1 - S_{wir} - S_{or}}} \cdot \frac{k_{rog}\left(S_g\right)/k_{rocw}}{1 - \dfrac{S_g}{1 - S_{wir} - S_{or}}} \tag{3.114}$$

where k_{rocw} is the oil-phase relative permeability at S_{wc}. Stone (1973) proposed another model which is referred as Stone II model:

$$\frac{k_{ro}\left(S_w, S_g\right)}{k_{rocw}} = \left\{\left[\frac{k_{row}\left(S_w\right)}{k_{rocw}} + k_{rw}\left(S_w\right)\right]\left[\frac{k_{rog}\left(S_g\right)}{k_{rocw}} + k_{rg}\left(S_g\right)\right] - \left[k_{rw}\left(S_w\right) + k_{rg}\left(S_g\right)\right]\right\} \tag{3.115}$$

The entries on the right-hand side of Eqs. (3.114) and (3.115) are available from the two-phase oil/water and the two-phase oil/gas relative permeability data sets. Also, Eq. (3.115) is applicable for $k_{ro}(S_w,S_g) \geq 0$ and if for a given saturation combination $k_{ro}(S_w,S_g)$ becomes negative, then, $k_{ro}(S_w,S_g)$ is set to be equal to zero.

Brooks-Corey "power-law" relationships are also often used to generate oil, water, and gas relative permeabilities and they are listed below (Brooks and Corey, 1964):

$$k_{ro} = k_{ro,max}\left(\frac{S_o - S_{or}}{1 - S_{or} - S_{wirr} - S_{gc}}\right)^{n_o} \tag{3.116}$$

$$k_{rw} = k_{rw,max}\left(\frac{S_w - S_{wirr}}{1 - S_{or} - S_{wirr} - S_{gc}}\right)^{n_w} \tag{3.117}$$

$$k_{rg} = k_{rg,max}\left(\frac{S_g - S_{gc}}{1 - S_{or} - S_{wirr} - S_{gc}}\right)^{n_g} \tag{3.118}$$

The n_o, n_w and n_g entries in Eqs. (3.116), (3.117), and (3.118) may be assigned values between 1 and 6. The maximum relative permeabilities are between 0 and 1. Baker (1988) provided the following three equations which can be used in constructing three-phase relative permeabilities from two-phase relative permeability data using linear interpolation:

$$k_{ro} = \frac{\left[((S_w - S_{wr})k_{ro(W/O)} + \left(S_g - S_{gr}\right)k_{ro(G/O)}\right]}{\left(S_w - S_{wirr}\right) + \left(S_g - S_{gr}\right)} \tag{3.119}$$

$$k_{rw} = \frac{\left[((S_o - S_{or})k_{ro(W/O)} + \left(S_g - S_{gr}\right)k_{ro(G/W)}\right]}{\left(S_o - S_{or}\right) + \left(S_g - S_{gr}\right)} \tag{3.120}$$

and,

$$k_{rg} = \frac{\left[((S_w - S_{wr})k_{rg(G/W)} + (S_o - S_{or})k_{rg(G/O)}\right]}{\left(S_w - S_{wirr}\right) + \left(S_o - S_{or}\right)} \tag{3.121}$$

3.10 Capillary Pressure in Porous Media

The pressure difference existing across an interface separating two immiscible fluids, one of which wets the surface of the rock in preference to other, is known as the capillary pressure. Petroleum engineering convention defines capillary pressure as the pressure in the non-wetting phase (e.g., in a water-wet system) minus the pressure in the wetting phase (e.g., water):

$$p_c = p_{non\text{-}wet} - p_{wet} \tag{3.122}$$

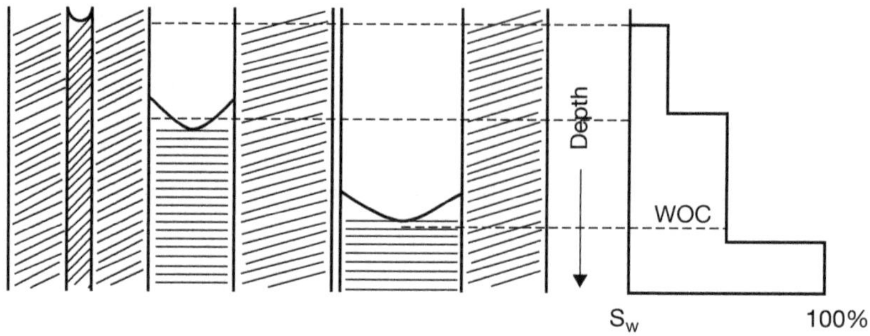

FIGURE 3.9 Saturation distribution in transition zone as a function of pore size distribution.

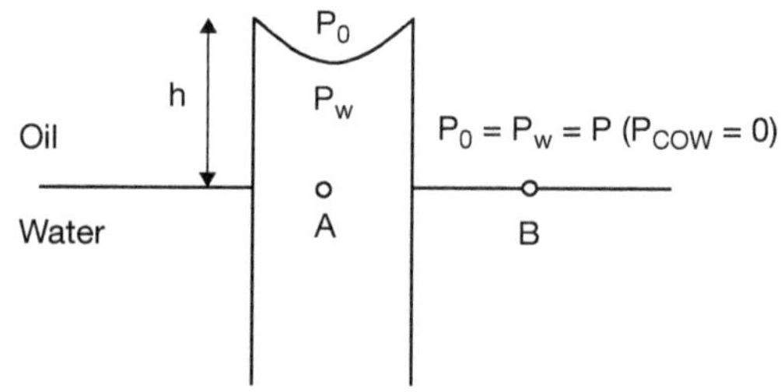

FIGURE 3.10 Capillary rise across the oil–water interface.

The porous media in nature exhibits a rather wide range of pore sizes. It is also well known that if we assume that the porous media in general would behave as a bundle of capillary tubes of various sizes a wetting fluid will rise in smaller capillary tubes more profoundly. Capillary rise provides a conceptual illustration of the interplay between adhesive and cohesive forces in a media with preferential wetting. The magnitude of capillary is a function of not only the capillary size but also the interfacial tension that exists at the surface of the wetting phase (see Fig. 3.9). Figure 3.10 shows an isolated view of a capillary channel residing across the oil–water contact. By using the densities of oil and water and capillary height gained by the water phase one can make the following simple derivation.

$$\text{At point } A,\ p = p_w + \rho_w g h \tag{3.123}$$

$$\text{At point } B,\ p = p_o + \rho_o g h \tag{3.124}$$

Subtraction of Eq. (3.123) from Eq. (3.124) yields:

$$0 = p_o - p_w + gh(\rho_o - \rho_w) \tag{3.125}$$

or

$$p_{cow} = p_o - p_w = gh(\rho_w - \rho_o) \tag{3.126}$$

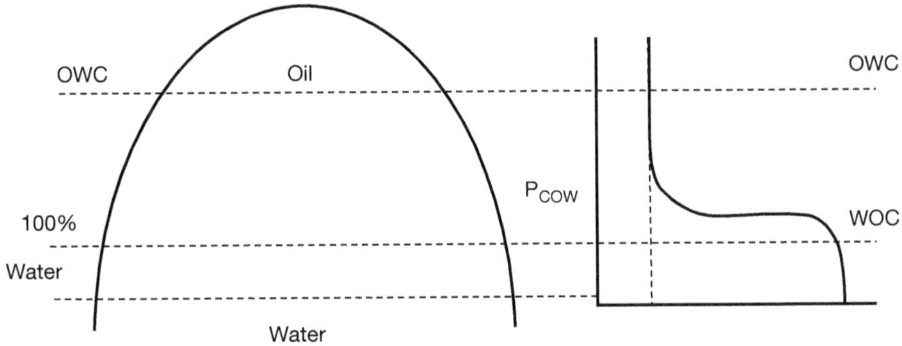

FIGURE 3.11 Height of the transition zone and capillary pressure characteristics.

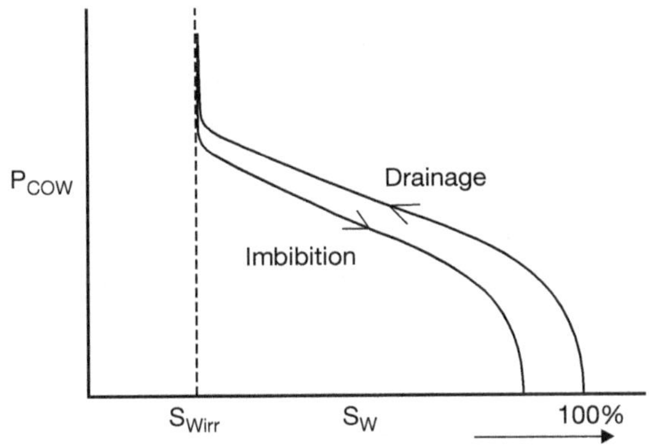

FIGURE 3.12 Hysteresis effect on capillary pressure function.

According to the convention adopted in Eq. (3.122), in Eq. (3.126), oil represents the non-wetting phase and hence water is the wetting phase. Finally, solving for h (height of the transition zone), one obtains:

$$h = \frac{p_{cow}}{\Delta \rho g} \tag{3.127}$$

The implication of Eq. (3.127) is given in terms of fluid distribution, as shown in Fig. 3.11.

When capillary pressure experiments are conducted, it is more often observed a hysteresis effect depending upon whether measurements are done during the drainage (decreasing the wetting plane saturation) or imbibition (increasing the wetting phase saturation) cycle. As shown in Fig. 3.12, imbibition and drainage plots are separated due to hysteresis in contact angle. Capillary pressure curves also carry the signature of the magnitude of permeability of the formulation material, as shown in Fig. 3.13.

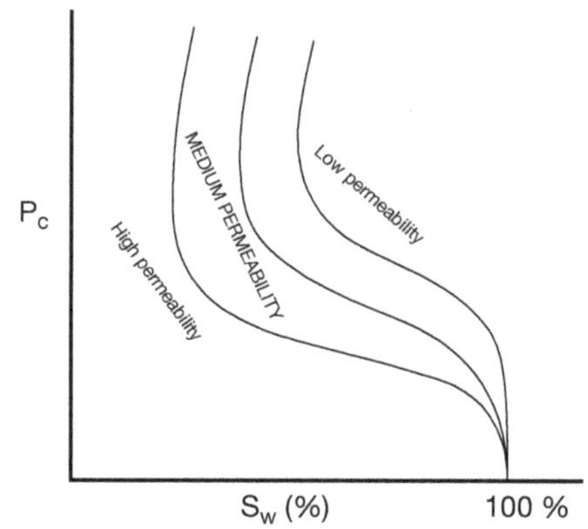

FIGURE 3.13 Variation of capillary pressure curves with formation permeability.

Similar to capillary pressure between oil and water phases a similar discussion can be made on the capillary pressure between a gaseous phase and a liquid phase (oil or water). Such a capillary pressure function is defined as:

$$p_{cgo} = p_g - p_o \tag{3.128}$$

Again, according to the convention stated in conjunction with Eq. (3.122), gas phase represents the non-wetting phase whereas oil is the wetting phase. Eq. (3.128) can be used in calculating the length of the transition zone between oil region and gas cap across the oil–gas contact.

3.11 Skin Factor and Wellbore Flow

The skin factor is a dimensionless quantity that represents a dimensionless pressure drop accounting for deviations in the (idealized) homogeneous reservoir characteristics assumption in the immediate vicinity of the wellbore. More specifically, it accounts for *permeability* heterogeneities in the domain surrounding the wellbore.

During the drilling and completion of a well, the permeability of the formation in the vicinity of the wellbore may be altered. The permeability reduction is referred to as *formation damage* and permeability enhancement is referred to as *stimulation*. Accordingly, a composite system around the wellbore is formed, as shown in Fig. 3.14.

There are some key observations related to Fig. 3.14. If the permeability in the inner zone is less than the original permeability of the formation (damage), the observed pressure drop at the wellbore is more than the one with no damage. If the permeability of the inner zone is enhanced (by acidizing or hydraulic fracturing), then the observed pressure in the wellbore will be less than the predicted by the reservoir models (such as the classical well test models discussed in Chap. 6) when they do not take into account the existing negative skin factor. When we express this pressure difference in terms of dimensionless pressure drop, then we are able to quantify the concept of skin.

As explained above, the skin value results from the difference, in dimensionless terms, between observed (actual) pressure drops in the well and the prediction of idealized wellbore models ($s = 0$), as follows:

$$s = \left(\Delta p_D\right)_{actual} - \left(\Delta p_D\right)_{ideal} \tag{3.129}$$

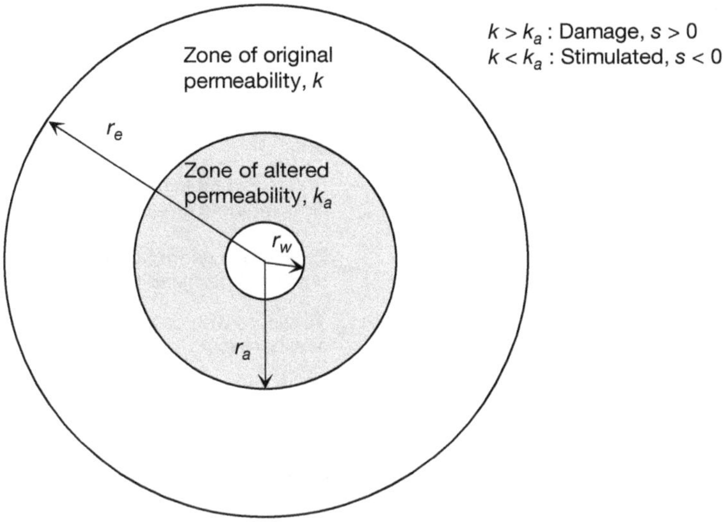

FIGURE 3.14 Altered zone in the vicinity of a wellbore.

In the most ideal situation, the idealized well theory based on homogeneous reservoir characteristics ($s = 0$) would predict exactly the same pressure drop as the one measured (observed) in the field, in which case the skin remains zero. In actual application, observed wellbore pressure drops are expressed as the sum of the ideal prediction and the pressure difference due to skin, as shown below:

$$(\Delta p)_{actual} = (\Delta p)_{ideal} + (\Delta p)_{skin} \tag{3.130}$$

The dimensionless version of Eq. (3.130) would be:

$$(\Delta p_D)_{actual} = (\Delta p_D)_{ideal} + s \tag{3.131}$$

which is a restatement of Eq. (3.129). Using the definition of dimensionless pressure drop, the value of skin and the dimensional pressure drop due to skin are related through the following expression:

$$s = \frac{(\Delta p_{skin})}{p_i q_D} \tag{3.132}$$

We can also consider skin as equivalent to changing the wellbore radius (r_w), as shown by the following expression:

$$r_w\big|_{effective} = r_w\big|_{actual} e^{-s} \tag{3.133}$$

In Eq. (3.133), we see that:

$$\text{If} \quad s > 0 \quad \rightarrow \quad r_w\big|_{effective} < r_w\big|_{actual}$$

$$\text{If} \quad s < 0 \quad \rightarrow \quad r_w\big|_{effective} > r_w\big|_{actual}$$

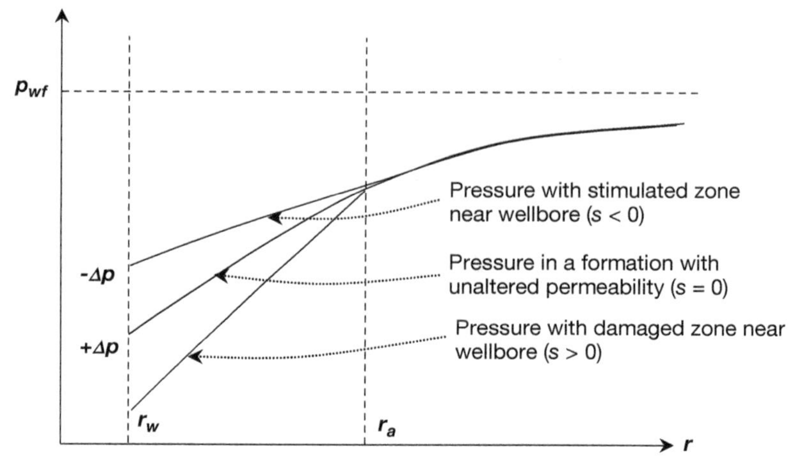

FIGURE 3.15 Schematic representation of positive and negative skins.

In the case of a wellbore with skin, we say that even though the wellbore has an $r_w = r_{w(actual)}$, it is behaving as if its r_w was like $r_{w(effective)}$. The Equation makes the pressure drop in an actual reservoir with skin equal to an ideal reservoir without skin but with a different $r_{w(effective)}$.

Figure 3.15 compares the difference in near-wellbore pressure distribution as a result of the existing skin. In Fig. 3.15, $r = r_a$ represents the zone of altered permeability, as illustrated in Fig. 3.14. Figure 3.15 compares the pressure distribution for the homogeneous formation with those for the formation with a stimulated zone in the vicinity of the wellbore ($s < 0$) and that with a damaged zone near the wellbore ($s > 0$). It is important to see that skin is a localized phenomenon. Therefore, the calculation of the pressure drop at a point of the reservoir away from the wellbore is the same regardless of the skin around that wellbore, as implied in Fig. 3.15 for $r > r_a$.

In analytical reservoir modeling, wellbore flow is analyzed using classical well testing tools discussed in Chap. 6. In those equations, the concept of skin is always accounted for in the vicinity of the well. In numerical reservoir modeling (Chap. 7), wells are hosted by a reservoir gridblock where the sink/source type of wellbore equation must be deployed. A well-known and often-utilized wellbore equation coupling the pressure of the reservoir block hosting the well to the flow generated at the formation face is written in the form suggested by Peaceman (Peaceman, 1978, 1983):

$$q_{i,j,k} = -\frac{2\pi \bar{k}_{i,j,k} h_{i,j,k}\left(p_{i,j,k} - p_{sf_{i,j,k}}\right)}{\mu(p_{i,j,k})B(p_{i,j,k})\left[\ln\left(\frac{r_e}{r_w}\right)_{i,j,k} + s_{i,j,k}\right]} \tag{3.134}$$

When Eq. (3.134) is written for a vertical well (well perpendicular to x-y plane), then following definitions apply:

$$\bar{k}_{i,j,k} = \sqrt{\left(k_{x_{i,j,k}} k_{y_{i,j,k}}\right)} \tag{3.135}$$

$$h_{i,j,k} = (\Delta z)_{i,j,k} \tag{3.136}$$

$$r_{e_{i,j,k}} = 0.28\frac{\left[\left(\frac{k_y}{k_x}\right)^{\frac{1}{2}}(\Delta x)^2 + \left(\frac{k_x}{k_y}\right)^{\frac{1}{2}}(\Delta y)^2\right]^{\frac{1}{2}}}{\left(\frac{k_y}{k_x}\right)^{\frac{1}{4}} + \left(\frac{k_x}{k_y}\right)^{\frac{1}{4}}} \tag{3.137}$$

All the terms in Eqs. (3.134) through (3.137) carry the practical field units. The $s_{i,j,k}$ term that appears in the denominator of Eq. (3.134) is the skin factor. Again, if $s_{i,j,k} > 0$ then a restricted flow condition is imposed on the flow into the wellbore [note that when $s_{i,j,k} > 0$, denominator of Eq. (3.134) becomes larger and the value of $q_{i,j,k}$ becomes smaller]. If $s_{i,j,k} < 0$ then a stimulated flow condition is imposed on the flow into the well. Notice that the skin factor $s_{i,j,k}$ in fact represents a dimensionless pressure, as one can recast Eq. (3.134) as:

$$\left[\ln\left(\frac{r_e}{r_w}\right) + s\right] = \frac{2\pi kh\Delta p}{q\mu B} \tag{3.138}$$

Both sides of Eq. (3.138) are dimensionless. Therefore, the skin factor can be considered simply as a dimensionless pressure group [the *RHS* of Eq. (3.138) is called dimensionless pressure group]. As further discussed in Chap. 6, the skin factor used in wellbore equations [such as in Eq. (3.134)] is typically a composite skin factor under which a number of other factors contribute to the flow restriction and/or flow stimulation. Therefore, one can write the skin as:

$$s_{apparent} = s_{formation} + s_{restricted\ entry} + s_{turbulent\ flow} \tag{3.139}$$

or,

$$\begin{aligned} s_{apparent} &= s_f\,(due\,to\,formation\,damage\,of\,stimulation) \\ &+ s_{perf}\,(due\ to\ restricted\ entry - perforations) \\ &+ s_{IT}\,(due\ to\ inertial - turbulent\ flow\ conditions) \end{aligned} \tag{3.140}$$

References

Baker, L.E. "Three-Phase Relative Permeability Correlations," presented at the *SPE Enhanced Oil Recovery Symposium*, Tulsa, Oklahoma, April 16–21. SPE-17369-MS, 1988.

Brooks, R.H., and A.T. Corey, "Hydraulic Properties of Porous Media." *Hydrology Papers*, No. 3, Colorado State U., Fort Collins, Colorado, 1964.

Corey, A.T., "The Interrelation Between Gas and Oil Relative Permeabilities," *Producers Monthly*, 19:28, 1954.

Naar, J., and J.H. Henderson, "An Imbibition Model: Its Applications to Flow Behavior and the Prediction of Oil Recovery," *SPE J.*, 61, June 1961.

Peaceman, D.W., "Interpretation of Well-Block Pressures in Numerical Reservoir Simulation," *Soc. Pet. Eng.* 1:183–194, June 1978; *Trans. AIME*, 253.

Peaceman, D.W., "Interpretation of Well-Block Pressures in Numerical Reservoir Simulation with Nonsquare Grid Blocks and Anisotropic Permeability," *SPE Journal*, 23:3, June 1983.

Stone, H.L., "Probability Model for Estimating Three-Phase Relative Permeability," *J. Pet. Technol.* 22(2):214–218. SPE-2116-PA, 1970.

Stone, H.L., "Estimation of Three-Phase Relative Permeability and Residual Oil Data." *J. Can. Pet. Technology*. 12(4):53–61, 1973.

CHAPTER 4

Hydrocarbon Fluid Models and Thermodynamics

Thermodynamics is a critical tool required to understand and analyze processes of interest in energy engineering practice. It studies energy transformations and establishes fundamental relationships that must be honored when such transformations take place. Such fundamental thermodynamic laws governing these energy transformations also define and constrain the physical variables fundamental to these processes.

Fluid behavior modeling is particularly important to reservoir engineering. In reservoir engineering, single-component, binary, ternary and multicomponent representations are routinely used to model fluid phase behavior depending on the demands of the modeling application. In this chapter, we examine those reservoir fluid modeling choices and their associated thermodynamic analysis.

4.1 Reservoir Fluid Modeling Choices

Figure 4.1 illustrates the different fluid model representations that may be considered in reservoir engineering. Depending on the complexity required to model the phase behavior of a petroleum mixture, single-component, binary, ternary or multicomponent fluid model representations are available. As one uses less and less components to characterize a fluid system, increased lumping becomes necessary to describe thermodynamic phase behavior in simpler terms.

The simplest choice that can be made to model a petroleum mixture is the use of a single pseudo-component fluid ($N_c = 1$). The use of single-component fluid models completely ignores compositional effects. Such model would not be able to capture any of the complexities involved with phase equilibria (e.g., mixture vaporization and condensation) since evolving hydrocarbon phases inherently have distinct compositions. Implementation of this fluid model is thus restricted to single-phase applications (i.e., oils above bubble point conditions and gases above dew point conditions, and only as long as they remain above saturation conditions and composition remains constant) where single-values of fluid properties (such as phase density, viscosity, and isothermal compressibility) is what is required by transport equations.

In increasing order of thermodynamic complexity, the binary mixture is the simplest fluid model that could be used to model "mixture" *PVT* behavior. This makes the "black-oil fluid" model the simplest thermodynamic mixture possible since it is indeed composed of only two hydrocarbon components ($N_c = 2$). In a black-oil formulation,

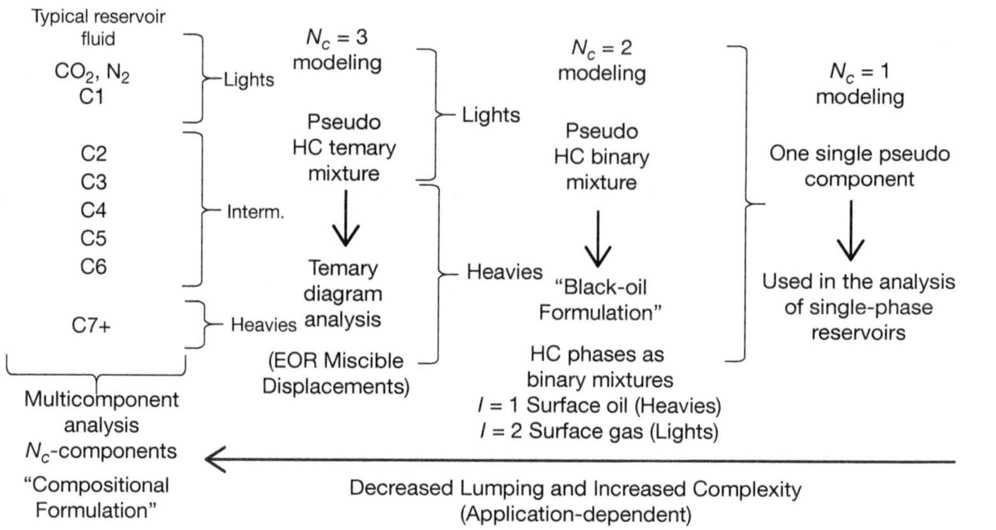

FIGURE 4.1 Reservoir fluid modeling choices and phase behavior complexity.

hydrocarbon phases are indeed "made up" of two hydrocarbon components: a light *pseudo*-component (L), and a heavy *pseudo*-component (H), as shown in Fig. 4.1. These "pure" pseudo-components are to be defined or characterized at surface conditions (i.e., stock tank pressure and temperature conditions), as shown later.

The black-oil fluid model allows for the simplest form of mass transfer among flowing phases: one that occurs within the context of a binary hydrocarbon system. Simplifying this formulation any further would require the study of single-phase flow or the study of a multiphase system where mass transfer between phases is disallowed—as may be the case with the immiscible flow of water and oil. In such a system, phase compositions are mostly fixed for engineering purposes and no component may cross between phases.

Predictably, one notable flaw of black-oil models is that they essentially ignore the potentially significant effect that the exchange of *intermediate* hydrocarbon components among phases can have on reservoir fluid behavior. In black-oil formulations, only surface-gas and surface-oil may transfer between reservoir phases. Also, solution gas inside the oil reservoir must have exactly the same "composition" (i.e., *pure* surface gas pseudo-component) as the gas pseudo-component in the reservoir gas phase. As part of this shortcoming, any gas injected into the reservoir must also match the composition and *PVT* behavior of the produced surface gas. This restriction can severely limit the use of black-oil formulations for the study of other important reservoir processes—such as injection of gases such as CO_2 for enhanced oil recovery (EOR) purposes—which do not adhere to binary mixture mass transfer behavior.

The next natural step in recognizing increased thermodynamic complexity in a reservoir fluid is the use of ternary mixture models ($N_c = 3$). In such fluid models, the existence of a light *pseudo*-component (L), an intermediate *pseudo*-component (I), and a heavy *pseudo*-component (H) is recognized, as illustrated in Fig. 4.1. *Pseudo*-components are made up of lumped fractions of species with similar volatilities. As shown in Fig. 4.1, CO_2, N_2, and C_1 may be considered part of the light (L) pseudo-component, C_2 to C_6

molecules may be lumped together to make up the intermediate (I) pseudo-component, and the heavy fraction C_7^+ would be a natural candidate for becoming the heavy (H) pseudo-component. The ternary lumping of hydrocarbon mixtures allows for the use of equilateral triangle (ternary) diagram representations. These representations can be instrumental in demonstrating additional complexities exhibited by reservoir enhanced recovery processes. Among those, vaporizing and condensing drives that occur during EOR miscible gas-injection are fitting examples of reservoir processes that may be captured by ternary mixture models (and could not by a binary mixture model) because these processes are driven by the exchange of intermediate components between the in situ fluid and the injected gas.

Unsurprisingly, real fluid behavior at a molecular level does not conform to the proposition of molecules within a defined lumped pseudo-component strictly behaving as if they were part of a pure fluid. Molecules would inherently "break away" to some (large or small) degree from their lumped fraction during any phase transition processes. As a result, some loss of thermodynamic information is unavoidable when lumped models are deployed as fluid models. Loss of accuracy in predictions of actual phase compositions and properties, true onsets of phase appearance and disappearance, and true relative amounts of coexisting phases would be (a large or small) part of such lost information.

To minimize this loss of thermodynamic information, fully compositional fluid models ($N_c > 3$) are recommended. This is shown in Fig. 4.1. A reservoir modeler may decide to deploy a higher number of pseudo-component to characterize the behavior of the reservoir mixture; for example, a particular reservoir fluid mixture may be characterized as being composed of C1, C2, C3, C4, C5, C6, C7–18, C19–29, and C30+ (a nine-component system, $N_c = 9$). Increasing the number of fluid components in a fluid formulation is particularly important for the modeling of processes where compositionally driven mass transfer is expected to be significant among reservoir phases. In compositional-based fluid models, fluid behavior is modeled using multicomponent thermodynamics, which necessitates more detailed knowledge of the chemical species conforming the hydrocarbon mixture. Generally speaking, a rigorous composition-based fluid model would always be the best choice when the most accurate description of reservoir fluid behavior is required. However, no naturally occurring petroleum mixture has ever been completely characterized into every single possible molecular species that may be found in it. Reservoir fluids are complex mixtures with numerous chemical components, and even if all species were to be fully defined, the presence of hundreds of components can pose serious challenges in terms of the computational overhead associated with multicomponent thermodynamic calculations. As a result of the trade-off between fluid modeling accuracy and computational expenses, some degree of lumping will be ultimately necessary in fully compositional models. The overarching goal would be to lump the original complex mixture of numerous chemical components into far fewer pseudo-component groups but with minimal loss of thermodynamic information.

Despite their well-known limitations, and the simultaneous availability of fully compositional simulators, black-oil simulators remain very popular in reservoir modeling because they are convenient, require a less detailed fluid characterization, and usually deliver a stable and much faster execution. Therefore, reservoir engineering practioners are very likely to be involved with these models at some point in their career and should be prepared and fully acquainted with inherent assumptions and limitations. We will review the thermodynamic details behind compositional and black-oil fluid models in the next sections.

4.2 Compositional Fluid Modeling: The N_c-Component Fluid

The compositional reservoir fluid model considers the most general scenario: the one where N_c-chemical species (*fluid components*) are found in the reservoir—each of which coexisting in *various phases* within the thermodynamic system. The model considers up to N_c species ($I = 1, 2, \ldots, N_c$) or components distributed among N_p reservoir phases ($J = 1, 2, \ldots, N_p$). As in any thermodynamic system, it is critical to first clearly differentiate what constitutes a "phase," and what constitutes a "component." In here, the subscript "I" denotes fluid "component," and the subscript "J" denotes fluid "phase." In compositional modeling, there is a total of "$N_p * N_c$" fluid compositions (i.e., N_c-compositions per phase). In the case of a two-phase, multicomponent system in Fig. 4.2, $N_p = 2$ and such system would have $2 * N_c$ fluid compositions associated with it.

These fluid compositions can be expressed in a *molar* or *mass* basis, as defined below and depicted in Fig. 4.2. For **molar compositions** or x_{IJ}'s, by definition, one writes:

$$x_{IJ} = \frac{\text{Moles of component "}I\text{" in phase "}J\text{"}}{\text{Total moles in phase "}J\text{"}} = \frac{N_{IJ}}{N_J} \tag{4.1}$$

and for **mass composition** or ω_{IJ}'s, by definition:

$$\omega_{IJ} = \frac{\text{Mass of component "}I\text{" in phase "}J\text{"}}{\text{Total mass in phase "}J\text{"}} = \frac{M_{IJ}}{M_J} \tag{4.2}$$

These *molar* and *mass* compositions are directly related to one another using molecular weight information for the *I-th* component (MW_I) and *J-th* reservoir phase (MW_J), as shown:

$$\omega_{IJ} = x_{IJ} \cdot \frac{MW_I}{MW_J} \tag{4.3}$$

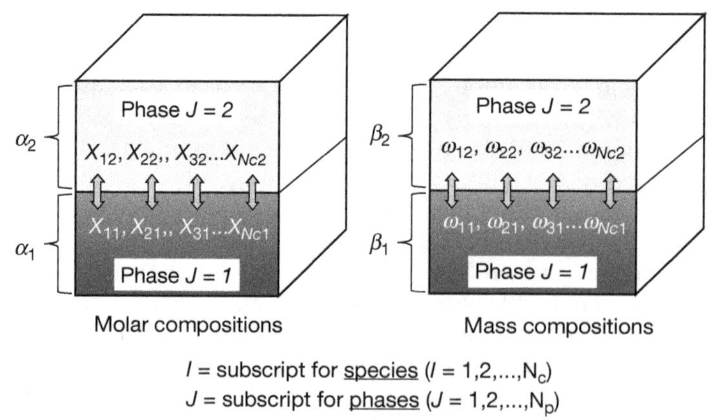

FIGURE 4.2 Reservoir fluid compositional modeling: molar and mass compositions.

where $MW_J = \sum_{I=1}^{N_c} x_{IJ} MW_I$. In Eqs. (4.1) and (4.2), N_{IJ} is the molar amount of each fluid component in phase J and M_{IJ} is the mass amount of each fluid component in phase J, respectively. Total amounts of phase in molar and mass basis (N_J and M_J, respectively) are given by:

$$N_J = \sum_{I=1}^{N_c} N_{IJ} \tag{4.4}$$

$$M_J = \sum_{I=1}^{N_c} M_{IJ} \tag{4.5}$$

with $M_{IJ} = N_{IJ} \cdot MW_I$ and $M_J = N_J \cdot MW_J$, from where Eq. (4.3) can be derived.

By dividing Eqs. (4.4) and (4.5) by total N_J and M_J, respectively, owing to the definitions in Eqs. (4.1) and (4.2), it follows that each reservoir phase must satisfy the following composition constraint:

$$\sum_{I=1}^{N_c} x_{IJ} = 1 \tag{4.6}$$

$$\sum_{I=1}^{N_c} \omega_{IJ} = 1 \tag{4.7}$$

The fractional amounts of total moles or mass found within each of the J phases are denoted as α_J and β_J, respectively. These variables are presented next to each phase in Fig. 4.2, and they quantify the relative amount or extent of each of these phases within the system. These phase amounts are defined as:

$$\alpha_J = \frac{\textit{Moles of phase "J"}}{\textit{Total moles of entire system}} = \frac{N_J}{N_T} \tag{4.8}$$

$$\beta_J = \frac{\textit{Mass of phase "J"}}{\textit{Total mass of entire system}} = \frac{M_J}{M_T} \tag{4.9}$$

where:

$$N_T = \sum_{J=1}^{N_P} N_J = \sum_{I=1}^{N_c} N_I \tag{4.10}$$

$$M_T = \sum_{J=1}^{N_P} M_J = \sum_{I=1}^{N_c} M_I \tag{4.11}$$

In Eqs. (4.10) and (4.11), N_I and M_I correspond to the total (molar and mass, respectively) amounts of I-component within the entire system and across all phases, or:

$$N_I = \sum_{J=1}^{N_p} N_{IJ} \tag{4.12}$$

$$M_I = \sum_{J=1}^{N_p} M_{IJ} \tag{4.13}$$

Based on prevailing values of N_I and M_I, each fluid component is also assigned an "overall" composition within the system, defined as:

$$c_I = \frac{\textit{Moles of component "I" across all phases}}{\textit{Total moles of entire system}} = \frac{N_I}{N_T} \tag{4.14}$$

$$z_I = \frac{\textit{Mass of component "I" across all phases}}{\textit{Total mass of entire system}} = \frac{M_I}{M_T} \tag{4.15}$$

where N_T and M_T are defined above.

The four-equation set in Eqs. (4.10) to (4.11) further imposes constraints on α_J, β_J, c_I, and z_I values. In fact, by dividing through N_T [Eqs. (4.10)] and M_T [Eqs. (4.11)] one has:

$$\sum_{J=1}^{N_P} \alpha_J = 1 \tag{4.16}$$

$$\sum_{I=1}^{N_c} c_I = 1 \tag{4.17}$$

$$\sum_{J=1}^{N_P} \beta_J = 1 \tag{4.18}$$

$$\sum_{I=1}^{N_c} z_I = 1 \tag{4.19}$$

where the definitions in Eqs. (4.8), (4.14), (4.9), and (4.15) have been used, respectively.

Also, after substituting the definitions of x_{IJ} [Eq. (4.1)], α_J [Eq. (4.8)], c_I [Eq. (4.14)] into Eq. (4.12), the following molar *species conservation balances* can be written:

$$c_I = \sum_{J=1}^{N_P} x_{IJ}\alpha_J \tag{4.20}$$

Similarly, after substituting the definitions of ω_{IJ} [Eq. (4.2)], β_J [Eq. (4.9)], and z_I [Eq. (4.15)] into Eq. (4.13), the following mass *species conservation balances* are written for each component in the system:

$$z_I = \sum_{J=1}^{N_P} \omega_{IJ}\beta_J \tag{4.21}$$

When the thermodynamic system in Fig 4.2 is visualized as a pore space, fractional amounts of total moles or mass taken by each of the J phases (α_J and β_J) are directly related to phase saturations (S_J) in the multicomponent system in Fig. 4.2 via phase densities (ρ_J, for mass densities, and $\hat{\rho}_J$, for molar densities). The relationships are:

$$S_J = \frac{V_J}{\sum_{J=1}^{N_P} V_J} = \frac{\frac{\alpha_J}{\hat{\rho}_J}}{\sum_{J=1}^{N_P} \frac{\alpha_J}{\hat{\rho}_J}} = \frac{\frac{\beta_J}{\rho_J}}{\sum_{J=1}^{N_P} \frac{\beta_J}{\rho_J}} \tag{4.22}$$

If one of the fluid components does not participate in the N_c-fluid characterization and thus is not part of the α_J or β_J thermodynamic split (flash calculations), Eq. (4.22) must account for it. This is often done for the water component in reservoir simulation, when water could be treated as a separate phase and component excluded from thermodynamic calculations. In such a case, α_J or β_J calculations would only consider hydrocarbon phases but not water. Equation (4.22) would need to be rewritten as:

$$S_J = (1 - S_w) \cdot \frac{\frac{\alpha_J}{\hat{\rho}_J}}{\sum_{J=1}^{N_P} \frac{\alpha_J}{\hat{\rho}_J}} = (1 - S_w) \cdot \frac{\frac{\beta_J}{\rho_J}}{\sum_{J=1}^{N_P} \frac{\beta_J}{\rho_J}} \tag{4.23}$$

where S_w is the water phase saturation.

4.2.1 Thermodynamic Equilibrium and K_I Ratios

Reservoir compositional modeling assumes that there exists *local* thermodynamic equilibrium at all times and at all points within the reservoir domain. Reservoirs are visualized as a large collection of small subsystems or numerical gridblocks, and each subsystem is assumed as being in *thermodynamic equilibrium*. As a result, at any point in time and space, thermodynamic properties are prescribed within each subsystem by *thermodynamic equilibrium* conditions.

Thermodynamic phases are said to be in *equilibrium* when: 1) There are no pressure differences across any of the phases (mechanical equilibrium), 2) There are no temperature differences across any of the phases (thermal equilibrium); and, 3) There is no net mass transfer across any of the phases (chemical equilibrium). An often-used thermodynamic criteria to impose no-net-mass-transfer conditions in a thermodynamic system is the iso-fugacity criteria. The fugacity is a thermodynamic measure of the escaping tendency of a component in mixture solution. It is directly expressed in terms of the partial pressure concept, and it becomes equal to partial pressure for ideal gas mixtures. There are $N_p * N_c$ fugacities (f_{IJ}) in a compositional system (i.e., N_c-fugacities per phase, one for each component within a phase). Figure 4.3 illustrates this situation for the case of mass transfer among two phases: a reservoir gas phase ($J =$ "g") and a reservoir oil phase ($J = o$). There are $2 * N_c$ fugacity values in such system, and difference in fugacities for a given fluid component in one phase with respect to another phase (f_{Ig} vs. f_{Io}) gives the measure of the potential for transfer of that component between such phases. Equal fugacities of a component between any two phases results in zero net mass transfer of those phases, and phases are said to be in equilibrium once pressure and temperature are also equilibrated. Fugacities of each component in a fluid mixture are often calculated using applicable equations of state (EOSs) for the fluid of interest.

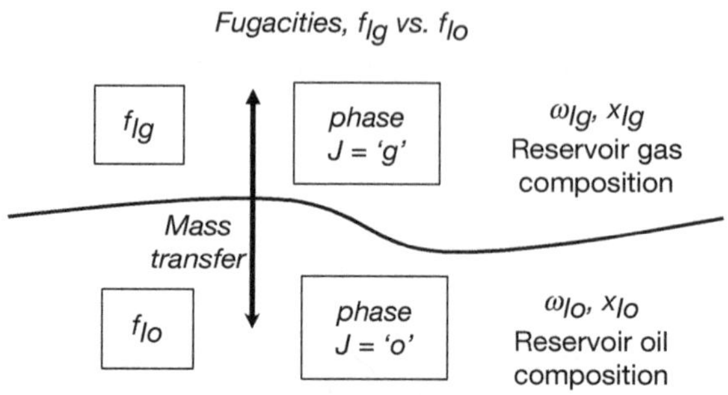

FIGURE 4.3 Fugacities and mass transfer across phases.

Common choices in reservoir engineering are the use of the Soave (1972) or Peng and Robinson (1976) EOSs for such purposes. Comprehensive reviews of the most common equations of state and fugacity calculations deployed in reservoir engineering are found in Danesh (1998), Firoozabadi (1999), Whitson and Brule (2000), Ahmed (2007), and Pedersen et al. (2015).

It should be emphasized that the statement "no net mass transfer" among phases never implies that phase compositions should become equal across interfaces. It rather implies that all phase compositions are to remain unchanged as long as no further changes in pressure, temperature, and overall system composition are experienced by the thermodynamic system. Once $f_{Ig} = f_{Io}$ for all components ($I = 1,2, \ldots N_c$) at a given pressure and temperature, thermodynamic equilibrium is reached and compositions of each phase (w_{IJ}, x_{IJ}) remain fixed or constant and related through the concept of equilibrium ratios (K_I) defined for each component. In mass and molar terms, these definitions are:

$$K_I = \frac{\omega_{Ig}}{\omega_{Io}} \tag{4.24}$$

$$K_I^* = \frac{x_{ig}}{x_{Io}} \tag{4.25}$$

where K_I is the equilibrium ratio of component "I" calculated using mass compositions, and K_I* is the equilibrium ratio of component "I" calculated using molar compositions. In reservoir compositional modeling, equilibrium ratios are routinely calculated using suitable EOSs as part of fugacity calculations (Danesh, 1998; Firoozabadi, 1999; Whitson and Brule, 2000; Ahmed, 2007; Pedersen et al., 2015).

4.2.2 Compositional Modeling and Flash Calculations

Once all equilibrium ratios (K_I's) are known for all species within a multiphase thermodynamic system in equilibrium, compositional material balances presented above can be readily applied to further constrain compositional variables to thermodynamic equilibrium conditions. To generalize the procedure and facilitate bookkeeping, one phase is typically selected as the reference phase in phase equilibrium calculations. All K_I's are then calculated with respect to the designated reference phase. If one (arbitrarily) selects the phase N_p as the reference phase, K_I's for every component in the system may be redefined as:

$$K_{IJ} = \frac{\omega_{IJ}}{\omega_{I,N_p}} \tag{4.26}$$

when written in terms of mass compositions. In Eq. (4.26), K_{IJ}'s may be visualized as the normalization of all ω_{IJ}'s compositions with ω_{I,N_c} used as the normalizing factor. As a result, there would be $N_p - 1$ sets of K_{IJ}'s in a generalized compositional system, i.e., one K_{IJ} set per interface present in the system.

Equation (4.18) can now be rewritten as:

$$\beta_{N_p} = 1 - \sum_{J=1}^{N_p-1} \beta_J \tag{4.27}$$

where the fractional amount of total mass taken by the reference phase (β_{N_p}) has been singled out from the sum. By also singling out the contribution of the reference phase in Eq. (4.21), one writes:

$$z_I = \omega_{I,N_p}\beta_{N_p} + \sum_{J=1}^{N_p-1} \omega_{IJ}\beta_J \tag{4.28}$$

Substituting Eq. (4.27) and the K_{IJ} definition [Eq. (4.26)] into Eq. (4.28):

$$z_I = \omega_{I,N_p}\left(1 - \sum_{J=1}^{N_p-1} \beta_J\right) + \omega_{I,N_p} \sum_{J=1}^{N_p-1} K_{IJ}\beta_J \tag{4.29}$$

from where the composition of the reference phase (ω_{I,N_p}) may be readily calculated as:

$$\omega_{I,N_p} = \frac{z_I}{1 + \sum_{J=1}^{N_p-1} (K_{IJ} - 1)\beta_J} \tag{4.30}$$

Species conservation within a phase requires $\sum_{I=1}^{N_c} \omega_{IJ} = 1$ [Eq. (4.7)], which applies to all phases including the N_p-phase. Thus one writes:

$$\sum_{I=1}^{N_c} \omega_{I,Np} = 1 \tag{4.31}$$

By subtracting Eq. (4.31) from Eq. (4.7) and substituting the K_{IJ} definition [Eq. (4.26)], the following expression is derived:

$$\sum_{I=1}^{N_c} (K_{IJ} - 1)\omega_{I,Np} = 0 \tag{4.32}$$

Equation (4.30) for $\omega_{I,Np}$ can now be introduced into Eq. (4.32) to show:

$$\sum_{I=1}^{N_c} \frac{z_I(K_{IJ} - 1)}{1 + \sum_{J=1}^{N_p-1} (K_{IJ} - 1)\beta_J} = 0 \tag{4.33}$$

which is the multiphase or generalized Rachford–Rice equation presented by Okuno et al. (2010) for multiphase compositional simulation. This equation is to be written $N_p - 1$ times in order to calculate all β_J's ($J = 1, 2, \ldots, N_p - 1$). β_{N_p} is calculated via Eq. (4.27).

Equation (4.33) is a multiphase extension of the widely known Rachford and Rice (1952) flash equation, originally derived for liquid–vapor equilibrium calculations where $N_p = 2$ and only a single set of K_I is needed. The original Rachford–Rice equation for liquid–vapor equilibrium is written as (Rachford and Rice, 1952), which is Eq. (4.33) applied to $N_p = 2$:

$$\sum_{I=1}^{N_c} \frac{z_I(K_I - 1)}{1 + \beta_g(K_I - 1)} = 0 \tag{4.34}$$

Rachford–Rice equations are the workhorse behind all flash calculations performed in compositional reservoir simulation. In flash calculations, multicomponent fluid mixture of known compositions (z_I) or "feed," representing reservoir fluid's overall in situ composition, are separated into a number of fluid phases in equilibrium. As a result, their amounts (β_J's) and distinct compositions (ω_{IJ}'s) can be calculated.

As Eqs. (4.33) and (4.34) indicate, Rachford–Rice flash calculations require that z_I's and all equilibrium ratios K_{IJ}'s should be known, leaving β_J's as the only unknowns left to be solved for. Such iterative solution is usually carried out by deploying nonlinear solvers such as bisection or Newton–Raphson. Once the amounts of each of the flashed or separated phases are known (β_J's), composition of the reference phase (ω_{I,N_p}) can be readily calculated via Eq. (4.30). All remaining phase compositions (ω_{IJ}'s) would follow from the knowledge of ω_{I,N_p} and K_{IJ}'s, using Eq. (4.26). Most importantly, since K_{IJ}'s are seldom known in advance, flash calculations typically entail iterative updates of K_{IJ}'s via fugacity calculations through equations of state (Danesh, 1998; Firoozabadi, 1999; Whitson and Brule, 2000; Ahmed, 2007; Michelsen and Mollerup, 2007; Pedersen et al., 2015).

4.3 Black-Oil Fluid Model and its *PVT* Properties

A popular approach to characterize *PVT* behavior of complex reservoir mixtures in a straightforward manner is to ignore very detailed compositional complexities and simply link fluids' *surface volume* to *reservoir volume* information through the deployment of concepts such as "formation volume factors" and "solution (or solubility) ratios." These concepts were originally proposed for reservoir material balance or mass accounting purposes and date back to the work of Schilthuis in the 1930s. Schilthuis (1936) recognized that petroleum fluids do indeed occupy substantially different volumes at surface and reservoir conditions and such changes must be taken care of doing mass and volume calculations or balances. This approach is the foundation of the classical implementation of a black-oil fluid model.

Petroleum fluid phases change their volume as they travel through the production system due to loss and/or gain of mass coupled with thermal and compressibility effects. These effects can become significant as these fluids move from reservoir conditions (i.e., a high-pressure, high-temperature environment) to surface conditions (i.e., a lower-pressure, lower-temperature environment). Figure 4.4 shows the production of a wellbore fluid stream composed of three comingled *reservoir fluid phases* (reservoir gas, reservoir oil, and reservoir water) and its separation into three *surface fluid phases*

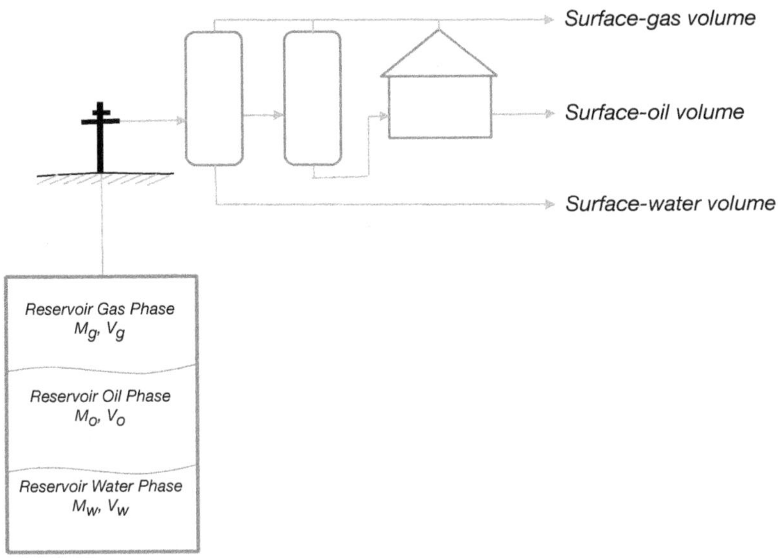

FIGURE 4.4 Overall view of reservoir vs. surface volumes.

(surface gas, surface oil, and surface water). The properties of these reservoir and surface phases are distinctly different both compositionally and volumetrically.

Figure 4.5 provides a closer look of the differences between *reservoir* and *surface* fluid volumes. Production and separation processes release surface components from reservoir phases, allowing them to aggregate and form the overall surface production streams depicted in Fig. 4.5. It is noted that every *surface* fluid component (surface gas, surface oil, surface water) was originally found within a reservoir fluid phase. Therefore, every in-situ reservoir fluid phase in Fig. 4.5 may be visualized as a mixture or combination of surface gas, surface oil, and surface water—they are just mixed at different proportions. In other words, any reservoir fluid phase could be "reconstituted" back by merely remixing surface production volumes in the appropriate amounts.

To capture the volumetric and compositional changes between surface and reservoir phases showcased in Figs. 4.4 and 4.5, "black-oil" simplified fluid formulations are used. The foundation of this fluid formulation is the definition of a set of "formation volume factors" (*FVFs*) (e.g., B_o and B_g) and "solubility ratios" (e.g., R_s and R_v) aimed at the quantification of these volumetric and compositional changes that occur during production. Figure 4.6 presents a reservoir view of these volumetric and compositional changes, and how they can be traced back to the definitions of these "black-oil" fluid *PVT* properties. Definitions in Fig. 4.6 follows the volume variables nomenclature presented in Fig. 4.5.

Following the nomenclature presented in Figs. 4.5 and 4.6, *FVFs* are defined by directly contrasting surface and reservoir volumes via the following volumetric ratios, at any point during production. These *FVFs* can be defined for each of the reservoir phases:

$$B_o\left[\frac{RB}{STB}\right] = \frac{\text{Volume of reservoir oil}}{\text{Surface oil volume from reservoir oil}} = \frac{V_o[RB]}{(V_{Osc})_o[STB]} = \frac{q_o}{(q_{osc})_o} \quad (4.35)$$

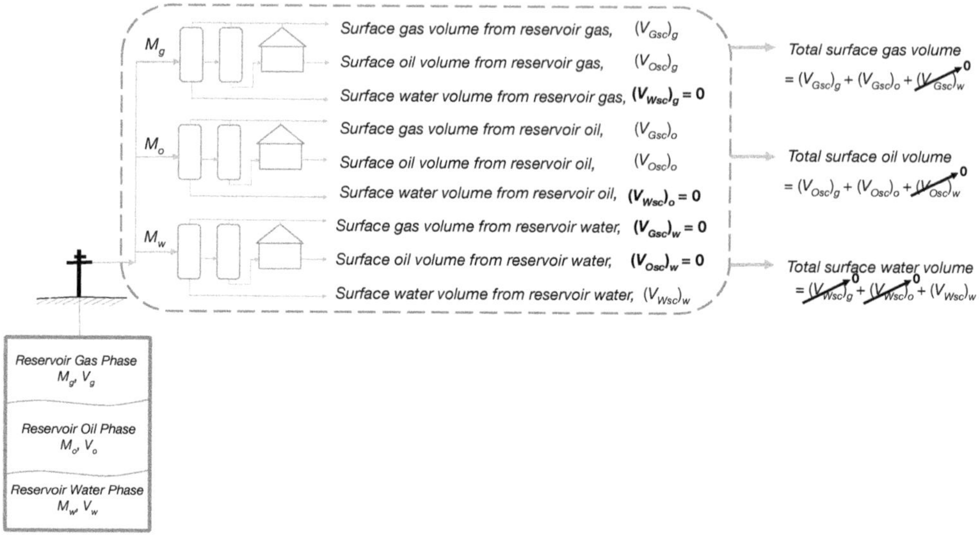

FIGURE 4.5 Surface view of reservoir vs. surface volumes.

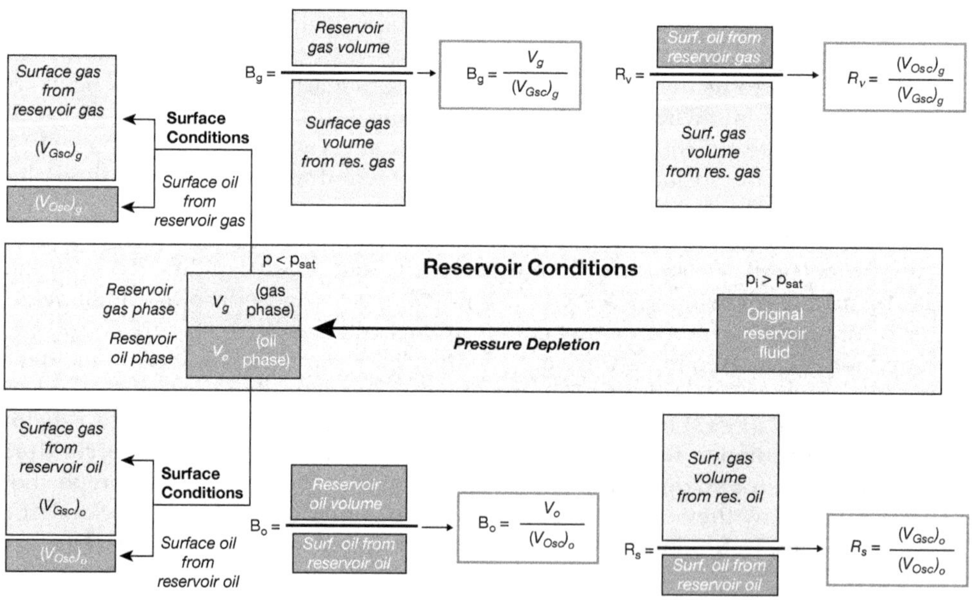

FIGURE 4.6 Reservoir view of reservoir vs. surface volumes for the hydrocarbon phases.

$$B_g\left[\frac{RB}{scf}\right] = \frac{\text{Volume of reservoir gas}}{\text{Surface gas volume from reservoir gas}} = \frac{V_g[RB]}{(V_{Gsc})_g[scf]} = \frac{q_g}{(q_{gsc})_g} \tag{4.36}$$

$$B_w\left[\frac{RB}{STB}\right] = \frac{\text{Volume of reservoir water}}{\text{Surface water volume from reservoir water}} = \frac{V_w[RB]}{(V_{Wsc})_w[STB]} = \frac{q_w}{(q_{wsc})_w} \tag{4.37}$$

Note that these *FVF* definitions also acknowledge that volumetric and volumetric flow rate ratios are equal to each other, as long as both (surface and reservoir) volumetric rates are expressed on the same time basis.

It has been noted that the black-oil fluid model visualizes reservoir fluids as being "composed" of separator fluids. That is, the black-oil fluid model *recombines* surface data to reconstitute reservoir phases. Surface fluids are thus said to be the fluid components or "pseudo-components" in this formulation. These pseudo-components produced at the surface were originally "dissolved" in the reservoir phases. This dissolution and associated reservoir fluid composition is quantified through the definition of "solubility ratios," which provide a measure of the relative amounts of surface constituents contained in the produced reservoir fluids. Two of the most commonly utilized solubility ratios in a black-oil formulation are defined for the reservoir oil phase (ratio of surface gas to surface oil within the reservoir oil) and reservoir gas phase (ratio of surface oil to surface gas within the reservoir gas) as follows:

$$R_s = \frac{\text{Surface gas volume from reservoir oil}}{\text{Surface oil volume from reservoir oil}} = \frac{(V_{Gsc})_o}{(V_{Osc})_o} = \frac{(q_{gsc})_o}{(q_{osc})_o} \tag{4.38}$$

$$R_v = \frac{\text{Surface oil volume from reservoir gas}}{\text{Surface gas volume from reservoir gas}} = \frac{(V_{Osc})_g}{(V_{Gsc})_g} = \frac{(q_{osc})_g}{(q_{gsc})_g} \tag{4.39}$$

where R_s is the solution gas–oil ratio and R_v is the volatilized oil–gas ratio. Again, solubility definitions written above acknowledge that volumetric and volumetric flow rate ratios could be used interchangeably. Altogether, values of B_o, B_g, R_s, and R_v are standard *PVT* properties of hydrocarbon fluids, or "black-oil *PVT* properties." Their values are experimentally calculated via, for example, reservoir fluid studies based on *PVT* laboratory tests, such as Constant Composition Expansion (*CCE*), Differential Liberation Expansion (*DLE*), Constant Volume Depletion (*CVD*), and separator tests. These tests are especially designed based on fluid type and they attempt to mimic the depletion process depicted in Figs. 4.4 and 4.5 for representative reservoir fluid samples collected early in production life of the reservoir (Whitson and Brule, 2000; Walsh and Lake, 2003; McCain, 2017).

Figures 4.7 and 4.8 display typical values of black-oil *PVT* properties for a gas-condensate fluid at a reservoir temperature of 250°*F* ($p_{dew} = 2787.7\ psia$) and a volatile-oil fluid at a reservoir temperature of 200°*F* ($p_{bubble} = 4250\ psia$), respectively. These figures also display the dependencies of these variables with pressure as reported by Zhang and Ayala (2017a). Analogous black-oil B_o, B_g, R_s, and R_v vs. pressure plots can be found in Walsh and Lake (2003), and in multiphase research work based on black-oil *PVT* fluid formulations (e.g., Becker et al., 2016; Zhang and Ayala, 2016; Zhang et al., 2016a; Zhang et al., 2016b; Singh et al., 2017; Zhang and Ayala, 2017a; Zhang and Ayala, 2017b).

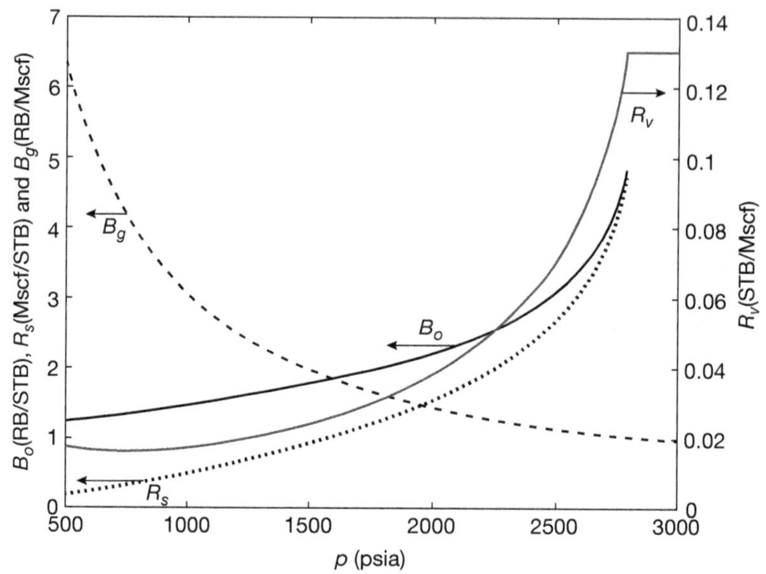

FIGURE 4.7 Typical black oil *PVT* properties of a gas condensate fluid at $T = 250°F$. (Data from Zhang and Ayala, 2017a.)

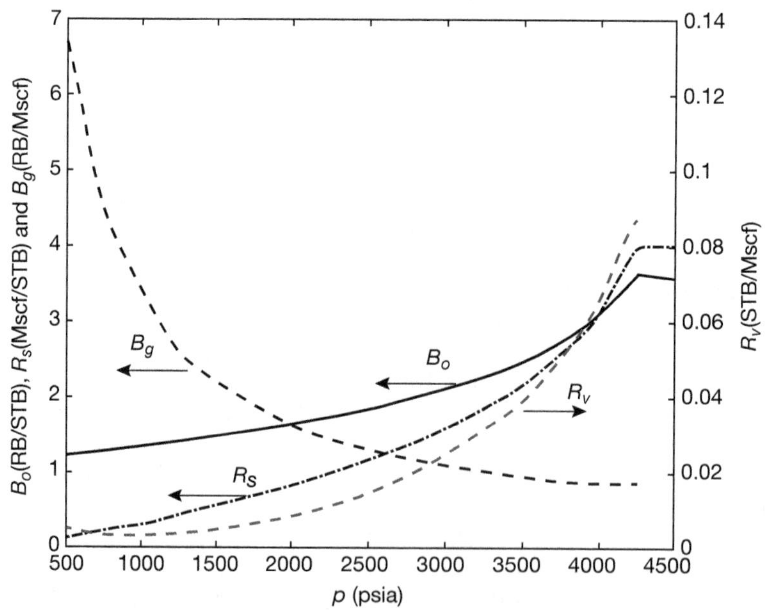

FIGURE 4.8 Typical black oil *PVT* properties of a volatile oil fluid at $T = 200°F$. (Data from Zhang and Ayala, 2017a.)

Early, classical black-oil formulations failed to account for any surface oil within the reservoir gas [i.e., $(V_{Osc})_g = 0$ in Fig. 4.5]. This assumption inevitably mandates that no volatilized oil could be found within the reservoir gas [$R_v = 0$ in Eq. (4.39)], which can prove to be a poor assumption for some reservoir fluids such as volatile oils and retrograde gases. When black-oil formulations disallow the presence of surface oil in the reservoir gas phase, the formulations are known as "classical" or "standard" black-oil models. Such formulations eliminate R_v information from Figs. 4.7 and 4.8, and default it to $R_v = 0$. In general, this is a reasonable assumption for oils found far away from mixture's critical point conditions, but it is an unreasonable assumption for near-critical reservoir fluids. For these more complex near-critical fluids, the thermodynamic behavior of the reservoir gas phase would be poorly represented when forced to behave as a non-condensing dry gas ($R_v = 0$). Thus, classical black-oil formulations should be avoided to model near-critical fluids. "Extended" or "modified" black-oil models should be used in such cases, which are black-oil formulations where no restriction is placed on the amount of surface oil found in the reservoir gas phase ($R_v > 0$).

The common assumption that water and hydrocarbons are not miscible for many reservoir scenarios is also in display in Fig. 4.5 [$(V_{Wsc})_o = (V_{Wsc})_g = 0$ and $(V_{Gsc})_w = (V_{Osc})_w = 0$]. These simplifications imply that no water is found within the reservoir hydrocarbon phases and that no hydrocarbons are present within the reservoir water phase. These restrictive water/hydrocarbon solubility assumptions can be relaxed by introducing the appropriate solubility ratio definition for the reservoir water phase, as done in Eqs. (4.38) and (4.39) for the oil and gas reservoir phases, respectively.

4.3.1 Compositional Calculations in Black-Oil Models

The simple yet useful *black-oil fluid model* is best understood once one clearly differentiates what constitutes a "phase" from what constitutes a fluid "component" in these models—a description already introduced in the preceding section. To reinforce this point, we will now relate B_o, B_g, R_s, and R_v black-oil *PVT* information to actual reservoir phase compositions [ω_{IJ}'s, as defined in Eq. (4.2)]. This allows building a bridge between black-oil fluid models and the compositional fluid models discussed in Sec. 4.2.

By way of illustration, Fig. 4.9 depicts a reservoir with three producing phases (reservoir gas, oil, and water phases), along with their reservoir compositions ω_{IJ}'s, and all resulting surface components released at surface conditions once these reservoir fluids have gone through the surface production system.

As already considered, the black-oil fluid model visualizes fluids at reservoir conditions as being "made up" of separator fluids produced at surface conditions. In other words, a black-oil system allows up to three *pseudo-components* in the formulation when water is considered: $I \in \{G,O,W\}$ ($N_c = 3$), where "I" is the subindex denoting each of the mixture "components." These three fluid pseudo-components correspond to each of the three surface production streams shown at surface conditions in Fig. 4.9, namely:

- *I = "G" (upper case "G") for the surface gas pseudo-component*
- *I = "O" (upper case "O") for the surface oil pseudo-component*
- *I = "W" (upper case "W") for the surface water pseudo-component*

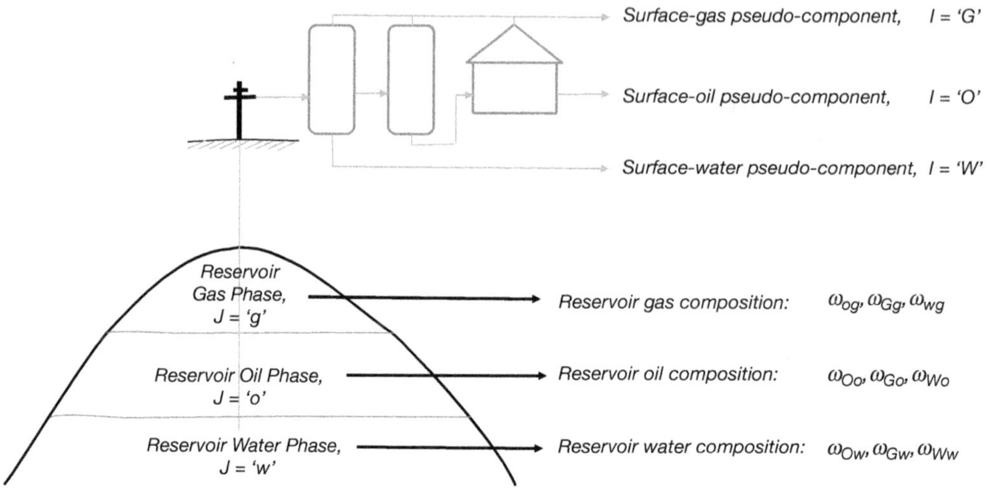

Figure 4.9 Schematic representation of phases and components in a black-oil fluid formulation.

In terms of reservoir fluid phases, the black-oil fluid model *recombines* this surface data to reconstitute them. As shown in Fig. 4.9, up to three reservoir phases can be found in a black-oil formulation: $J \in \{g, o, w\}$ ($N_p = 3$), where "J" is the subindex denoting each of the reservoir "phases,: namely:

- *J = "g" (lower case "g") for the reservoir gas phase*
- *J = "o" (lower case "o") for the reservoir oil phase*
- *J = "w" (lower case "w") for the reservoir water phase*

An important recognition in black-oil fluid modeling is that all reservoir phase compositions (ω_{IJ}'s) can be directly linked and correlated to standard black-oil *PVT* properties (B_o, B_g, R_s, and R_v) for each of the pseudo-components [$I \in \{G, O, W\}$] and reservoir phases [$J \in \{g, o, w\}$] being considered. This is shown next. In these derivations, the acronyms *RB*, *STB*, and *scf* refer to reservoir barrels, stock-tank barrels, and standard cubic feet, respectively.

For **reservoir oil phase** (J = "o") composition (ω_{Oo}, ω_{Go}, ω_{Wo}) calculations:

- **ω_{Oo} is the mass concentration of surface oil in the reservoir oil phase** $\omega_{Oo} = \dfrac{M_{Oo}}{M_o}$. In this definition, M_{Oo} is the mass amount of *surface oil* contained in an arbitrary reservoir oil sample "V_o" taken to the surface and M_o is the total mass of the reservoir oil phase in the produced reservoir oil sample "V_o". In other words:

$$M_o = \alpha_c \rho_o V_o \tag{4.40}$$

where α_c is the volume conversion factor (5.615 *ft*^{3}*/bbl*), ρ_o is the oil phase density at reservoir conditions in *lb/ft*^{3}, and V_o is the produced volume sample of reservoir oil in *RB*. Also,

$$M_{Oo} = \alpha_c \rho_{osc} (V_{osc})_o = \alpha_c \rho_{osc} \frac{V_o}{B_o} \tag{4.41}$$

where ρ_{osc} is the density of the surface oil at surface conditions [*lb/scf*], and $(V_{osc})_o = V_o / B_o$ is the volume that the reservoir oil sample [V_o] takes once at surface conditions, according to the definition of the oil formation volume factor, $B_o = \frac{V_o}{(V_{osc})_o}$ in Eq. (4.35). Therefore:

$$\omega_{Oo} = \frac{M_{Oo}}{M_o} = \frac{\alpha_c \rho_{osc} \frac{V_o}{B_o}}{\alpha_c \rho_o V_o} = \frac{\rho_{osc}}{\rho_o B_o} \tag{4.42}$$

from where the definition of B_o in terms of mass densities also follows $\left(B_o = \frac{\rho_{osc}}{\omega_{Oo} \rho_o} \right)$.

- $\boldsymbol{\omega_{Go}}$ **is the mass concentration of surface gas in the reservoir oil phase**, i.e., $\omega_{Go} = \frac{M_{Go}}{M_o}$. In here, M_{Go} is the mass of surface gas contained in oil reservoir sample V_o once at the surface $= \rho_{gsc} (V_{Gsc})_o = \rho_{gsc} (R_s (V_{Osc})_o)$, using the definition of solution gas oil ratio (R_s, in *scf/STB*) in Eq. (4.38), and ρ_{gsc} is the density of the surface gas at surface conditions [*lb/scf*]. Therefore,

$$\omega_{Go} = \frac{M_{Go}}{M_o} = \frac{\rho_{gsc} R_s \frac{V_o}{B_o}}{\alpha_c \rho_o V_o} = \frac{\rho_{gsc} R_s}{\alpha_c \rho_o B_o} \tag{4.43}$$

- $\boldsymbol{\omega_{Wo}}$ **is the mass concentration of surface water in the reservoir oil phase**, which is customarily assumed to be equal to zero **($\boldsymbol{\omega_{Wo} = 0}$)**.

Since $\omega_{Wo} = 0$, the reservoir oil phase of the black-oil model is a binary mixture of two hydrocarbons, the *surface gas pseudo-component* and the *surface oil pseudo-component*, so that:

$$\omega_{Oo} + \omega_{Go} = 1 \tag{4.44}$$

To enforce overall material balance in the reservoir oil phase in a black-oil model, i.e., $M_o = M_{Oo} + M_{Go}$, requires:

$$\alpha_c \rho_o V_o = \alpha_c \rho_{gsc} \frac{V_o}{B_o} + \rho_{gsc} R_s \frac{V_o}{B_o} \tag{4.45}$$

from where reservoir oil phase density (ρ_o) can be readily calculated:

$$\rho_o = \frac{\alpha_c \rho_{osc} + \rho_{gsc} R_s}{\alpha_c B_o} \tag{4.46}$$

Alternate expressions for ω_{Oo} and ω_{Go} can be obtained by substituting this ρ_o expression:

$$\omega_{Oo} = \frac{\rho_{osc}}{\rho_o B_o} = \frac{\alpha_c \rho_{osc}}{\alpha_c \rho_{osc} + \rho_{gsc} R_s} \tag{4.47}$$

$$\omega_{Go} = \frac{\rho_{gsc} R_s}{\alpha_c \rho_o B_o} = \frac{\rho_{gsc} R_s}{\alpha_c \rho_{osc} + \rho_{gsc} R_s} \tag{4.48}$$

It is noted that $\omega_{Oo}, \omega_{Go} = f(\rho_{osc}, \rho_{gsc}, R_s)$ where ρ_{osc} and ρ_{gsc} are constant values in a black-oil formulation. Thus, $\omega_{Oo}, \omega_{Go} = f(R_s)$—i.e., the concept of "$R_s$" operates as a *proxy for the calculation of oil phase compositions at in situ reservoir conditions,* even as its units [*scf/STB*] would seem to indicate a surface condition specification. These units are the result of the type of pseudo-components being considered by the black-oil formulation.

For **reservoir gas** ($J =$ "g") **composition** ($\omega_{Og}, \omega_{Gg}, \omega_{Wg}$) calculations:

- ω_{Gg} **is the mass concentration of surface gas in the reservoir gas phase,** $\omega_{Gg} = \frac{M_{Gg}}{M_g}$. Here, M_g is the total mass of the reservoir gas phase in a produced reservoir gas sample "V_g" and M_{Gg} is the mass amount of surface gas contained in such reservoir gas sample once at the surface. This is:

$$M_g = \alpha_c \rho_g V_g \tag{4.49}$$

where ρ_g is the gas phase density at reservoir conditions in *lb/ft*^{3}, and V_g is the sample volume of reservoir gas being produced, in *RB*, and:

$$M_{Gg} = \rho_{gsc} (V_{Gsc})_g = \rho_{gsc} \frac{V_g}{B_g} \tag{4.50}$$

where ρ_{gsc} is the density of the surface gas at surface conditions [*lb/scf*], and $(V_{Gsc})_g = V_g / B_g$ (in *scf*) is the surface-gas volume that the gas sample [V_g] generates once at surface conditions. This uses the definition of the gas formation volume factor $B_g = \frac{V_g}{(V_{Gsc})_g}$ in Eq. (4.36), expressed in *RB/scf*. Therefore,

$$\omega_{Gg} = \frac{M_{Gg}}{M_g} = \frac{\rho_{gsc} \frac{V_g}{B_g}}{\alpha_c \rho_g V_g} = \frac{\rho_{gsc}}{\alpha_c \rho_g B_g} \tag{4.51}$$

from where the definition of B_g in terms of mass densities follows: $B_g = \frac{\rho_{gsc}}{\alpha_c \omega_{Gg} \rho_g}$. If the reservoir gas phase were a dry gas, with no volatilized surface oil in it ($R_v = 0$), $w_{Gg} = 1$ and $B_g = \frac{\rho_{gsc}}{\alpha_c \rho_g}$.

- ω_{Og} **is the mass concentration of (volatilized) surface oil in the reservoir gas phase**, i.e., $\omega_{Og} = \frac{M_{Og}}{M_g}$. In this expression, M_{Og} is the mass of surface oil that drops out at the surface out of the reservoir gas sample V_g, which equals to $\alpha_c \rho_{osc}(V_{Osc})_g = \alpha_c \rho_{osc}(R_v(V_{Gsc})_g)$ because of the definition of volatilized oil to gas ratio (R_v, in *STB/scf*) in Eq. (4.39). Thus:

$$\omega_{Og} = \frac{M_{Og}}{M_g} = \frac{\alpha_c \rho_{osc} R_v \frac{V_g}{B_g}}{\alpha_c \rho_g V_g} = \frac{\rho_{osc} R_v}{\rho_g B_g} \tag{4.52}$$

- ω_{Wg} **is the mass concentration of surface water in the reservoir gas phase**, which is also customarily assumed to be equal to zero **($\omega_{Wg} = 0$).**

As a result, the reservoir gas phase is also a binary mixture of two hydrocarbons: the *surface gas pseudo-component* and the *surface oil pseudo-component*, and:

$$\omega_{Og} + \omega_{Gg} = 1 \tag{4.53}$$

which enforces overall material balance in the reservoir gas phase within the black-oil model, i.e., $M_g = M_{Og} + M_{Gg}$. From this overall material balance constraint, reservoir gas phase density (ρ_g) is calculated:

$$\rho_g = \frac{\rho_{gsc} + \alpha_c \rho_{osc} R_v}{\alpha_c B_g} \tag{4.54}$$

from where alternate expressions for ω_{Gg} and ω_{Og} can be obtained by substituting the ρ_g equation into the previous expressions:

$$\omega_{Gg} = \frac{\rho_{gsc}}{\alpha_c \rho_g B_g} = \frac{\rho_{gsc}}{\rho_{gsc} + \alpha_c \rho_{gsc} R_v} \tag{4.55}$$

$$\omega_{Og} = \frac{\rho_{osc} R_v}{\rho_g B_g} = \frac{\alpha_c \rho_{osc} R_v}{\rho_{gsc} + \alpha_c \rho_{osc} R_v} \tag{4.56}$$

It is noted that $\omega_{Gg}, \omega_{Og} = f(\rho_{osc}, \rho_{gsc}, R_v)$. As already stated, ρ_{osc} and ρ_{gsc} are constant values in a black-oil formulation. This $\omega_{Gg}, \omega_{Og} = f(R_v)$ dependency shows that the concept of R_v operates as a *proxy for the estimation of gas phase compositions at in situ reservoir conditions.* For a dry gas with no volatized oil within it ($R_v = 0, \omega_{Og} = 0$), it follows that $\omega_{Gg} = 1$. Early or classical black-oil models disregard R_v values ($R_v = 0$) and basically assume that reservoir gas phase compositions are "known"—or rather conveniently fixed—at values equal to $\omega_{Og} = 0, \omega_{Gg} = 1$ at all times during the producing life of the reservoir.

For **reservoir water** (J = "w") **composition** ($\omega_{Ow}, \omega_{Gw}, \omega_{Ww}$) calculations:

- ω_{Ww} **is the mass concentration of surface water in the reservoir water phase**, which is also customarily assumed to be equal to one **($\omega_{Ww} = 1$)** when hydrocarbon concentration in the water phase is considered negligible ($\omega_{Ow} = \omega_{Gw} = 0$). This assumption may be relaxed by introducing the solubility of gas in water (R_{sw}) into the formulation.

4.3.2 Black-Oil Fluid Thermodynamics and Flash Calculations

As shown earlier, equilibrium ratios (K_i's) are important thermodynamic properties which are necessary to perform flash calculations in multicomponent, multiphase systems. Because in black-oil fluid models ω_{Oo}, $\omega_{Go} = f(R_s)$ and ω_{Gg}, $\omega_{Og} = f(R_v)$, equilibrium ratios (K_i's) are also calculated in a straightforward fashion once R_s and R_v information is available for any black-oil fluid formulation.

Let us consider the waterless system depicted in Fig. 4.10. Using the equilibrium ratio definition and the compositional information from the preceding section, equilibrium ratios for the *surface-gas* and *surface-oil* components become, respectively:

$$K_G = \frac{\omega_{Gg}}{\omega_{Go}} = \frac{1}{R_s} \cdot \frac{\alpha_c \rho_{osc} + \rho_{gsc} R_s}{\rho_{gsc} + \alpha_c \rho_{osc} R_v} \tag{4.57}$$

$$K_O = \frac{\omega_{Og}}{\omega_{Oo}} = R_v \cdot \frac{\alpha_c \rho_{osc} + \rho_{gsc} R_s}{\rho_{gsc} + \alpha_c \rho_{osc} R_v} = R_v R_s K_G \tag{4.58}$$

Thus, equilibrium ratios for each pseudocomponent (K_G and K_O) can be uniquely and explicitly calculated as a function of R_s and R_v, which in turn are unique functions of pressure at prevailing reservoir temperature. As a result, black-oil flash calculations become non-iterative, given that equilibrium ratios do not need to be iterated on in contrast to fully compositional modeling (see Section 4.2.2). Indeed, the application of the Rachford–Rice equation for binary mixtures (Warren and Adewumi, 1993) would yield the following explicit, iterationless expression for the calculation of the mass fraction of vapor for the hydrocarbons in the reservoir (β_g), which corresponds to Eq. (4.34) applied to this $N_c = 2$ system:

$$\beta_g = -\left[\frac{z_O}{K_G - 1} + \frac{z_G}{K_O - 1}\right] \tag{4.59}$$

where z_O, z_G characterize the *overall* mass composition of the in situ fluid, and also:

$$z_O = \beta_g \omega_{Og} + (1 - \beta_g)\omega_{Oo} \tag{4.60}$$

$$z_G = \beta_g \omega_{Gg} + (1 - \beta_g)\omega_{Go} \tag{4.61}$$

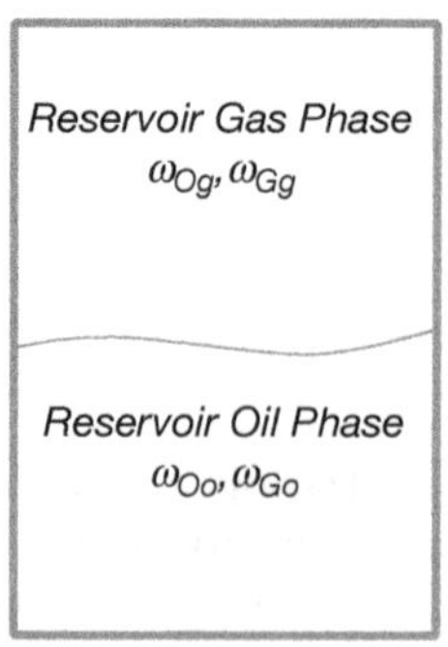

Figure 4.10 Hydrocarbon equilibrium compositions in a black-oil fluid model.

from the application of Eq. (4.21) to this $N_c = 2$, $N_p = 2$ system. It should be kept in mind that the use of flash calculations in any compositional simulation is undertaken with the goal of ultimately estimating of *phase compositions* (ω_{IJ}) at every prevailing pressure (and temperature) condition. Once ω_{IJ} – phase compositions are known, phase properties are calculated using the appropriate equation of state along with all remaining inputs required by the flow equations. In black-oil fluid simulation, however, these flash calculations **do not** need to be performed and equilibrium ratio calculations can be skipped altogether. This is because of the fact that in black-oil thermodynamics, all phase compositions (ω_{Oo}, ω_{Go}, ω_{Wo}, ω_{Og}, ω_{Gg}, ω_{Wg}, ω_{Ow}, ω_{Gw}, ω_{Ww}) and phase densities (ρ_o and ρ_g) can be directly and explicitly calculated as sole functions of pressure without any intermediate flash calculation step. This explains the substantial savings in computer overhead when black-oil thermodynamics is used in lieu of rigorous multicomponent thermodynamics.

Section 4.3.1 demonstrated that once B_o, B_g, R_s, and R_v values are known for a petroleum system, all reservoir oil and gas phase compositions and densities (ω_{Oo}, ω_{Go}, ω_{Gg}, ω_{Og}, ρ_o, ρ_g) can be calculated as sole functions of pressure. Tables 4.1 and 4.2 illustrate this point for the gas condensate fluid presented in Fig. 4.7. Table 4.1 summarizes the B_o, B_g, R_s, and R_v information in Fig. 4.7. Using Eqs. (4.47), (4.48), (4.55), (4.56), (4.46), and (4.54), each column of Table 4.2 is readily generated for the values of ω_{Oo}, ω_{Go}, ω_{Gg}, ω_{Og}, ρ_o, and ρ_g, respectively. In these calculations, ρ_{osc} and ρ_{gsc} values for the gas condensate fluid are 41.978 lb/ft^3 and 0.06591 lb/ft^3, respectively. Tables 4.3 and 4.4 illustrate the same procedure but for the volatile oil fluid presented in Fig. 4.8. Reported ρ_{osc} and ρ_{gsc} for the volatile oil mixture are 49.19 lb/ft^3 and 0.05347 lb/ft^3, respectively.

p (psia)	B_o (RB/STB)	B_g (RB/Mscf)	R_s (Mscf/STB)	R_v (STB/Mscf)
2787.7	4.8750	1.0247	4.73474	0.13053
2600	3.4384	1.0876	3.10384	0.08296
2400	2.8479	1.1821	2.37340	0.06145
2200	2.4819	1.2974	1.89934	0.04794
2000	2.2186	1.4388	1.54753	0.03843
1800	2.0139	1.6148	1.26794	0.03141
1600	1.8468	1.8381	1.03611	0.02613
1400	1.7057	2.1285	0.83821	0.02216
1200	1.5835	2.5191	0.66572	0.01925
1000	1.4756	3.0691	0.51308	0.01730
800	1.3791	3.8953	0.37660	0.01636
600	1.2919	5.2658	0.25394	0.01680
400	1.2131	7.9604	0.14417	0.02010

TABLE 4.1 Summary of black-oil *PVT* properties for gas condensate fluid in Fig. 4.7
Source: Data from Zhang and Ayala (2017a).

p (psia)	ω_{Oo} Eq. (4.47)	ω_{Go} Eq. (4.48)	ω_{Gg} Eq. (4.55)	ω_{Og} Eq. (4.56)	ρ_o (lb/ft³) Eq. (4.46)	ρ_g (lb/ft³) Eq. (4.54)
2787.7	0.43030	0.56970	0.68176	0.31824	20.01	16.80
2600	0.53536	0.46464	0.77119	0.22881	22.80	13.99
2400	0.60108	0.39892	0.81983	0.18017	24.52	12.11
2200	0.65312	0.34688	0.85366	0.14634	25.90	10.60
2000	0.69797	0.30203	0.87917	0.12083	27.11	9.28
1800	0.73825	0.26175	0.89900	0.10100	28.23	8.09
1600	0.77536	0.22464	0.91453	0.08547	29.32	6.98
1400	0.81012	0.18988	0.92658	0.07342	30.38	5.95
1200	0.84306	0.15694	0.93559	0.06441	31.45	4.98
1000	0.87453	0.12547	0.94174	0.05826	32.53	4.06
800	0.90473	0.09527	0.94474	0.05526	33.64	3.19
600	0.93370	0.06630	0.94333	0.05667	34.80	2.36
400	0.96125	0.03875	0.93295	0.06705	36.00	1.58

Table 4.2 Pressure-driven black-oil thermodynamic properties from black-oil *PVT* data for gas condensate fluid case in Table 4.1

p (psia)	B_o (RB/STB)	B_g (RB/Mscf)	R_s (Mscf/STB)	R_v (STB/Mscf)
4250	3.6356	0.8184	4.00000	0.08757
4000	3.0685	0.8757	3.11619	0.06557
3700	2.6683	0.9066	2.50700	0.04780
3400	2.3834	0.9753	2.02417	0.03534
3100	2.1728	1.0737	1.70127	0.02733
2800	1.9879	1.1764	1.39984	0.01993
2500	1.8344	1.3135	1.17183	0.01453
2200	1.7123	1.4561	0.96620	0.01034
1900	1.6027	1.7251	0.76613	0.00753
1600	1.4967	2.0536	0.61094	0.00542
1300	1.4152	2.4955	0.46506	0.00396
1000	1.3195	3.3853	0.30468	0.00330
700	1.2618	4.7520	0.20908	0.00362
400	1.1782	8.6147	0.11141	0.00724

Table 4.3 Summary of black-oil *PVT* properties for volatile oil fluid in Fig. 4.8
Source: Data from Zhang and Ayala (2017a)

p (psia)	ω_{Oo} Eq. (4.47)	ω_{Go} Eq. (4.48)	ω_{Gg} Eq. (4.55)	ω_{Og} Eq. (4.56)	ρ_o (lb/ft³) Eq. (4.46)	ρ_g (lb/ft³) Eq. (4.54)
4250	0.56358	0.43642	0.68853	0.31147	24.01	16.90
4000	0.62373	0.37627	0.74700	0.25300	25.70	14.56
3700	0.67325	0.32675	0.80199	0.19801	27.38	13.10
3400	0.71846	0.28154	0.84562	0.15438	28.73	11.55
3100	0.75225	0.24775	0.87628	0.12372	30.09	10.12
2800	0.78679	0.21321	0.90666	0.09334	31.45	8.93
2500	0.81509	0.18491	0.93018	0.06982	32.90	7.79
2200	0.84243	0.15757	0.94930	0.05070	34.10	6.89
1900	0.87084	0.12916	0.96256	0.03744	35.24	5.73
1600	0.89424	0.10576	0.97276	0.02724	36.75	4.77
1300	0.91741	0.08259	0.97997	0.02003	37.89	3.89
1000	0.94430	0.05570	0.98322	0.01678	39.48	2.86
700	0.96110	0.03890	0.98164	0.01836	40.56	2.04
400	0.97889	0.02111	0.96397	0.03603	42.65	1.15

TABLE 4.4 Pressure-driven black-oil thermodynamic properties from black-oil *PVT* data for volatile oil fluid case in Table 4.3

These tables reinforce the idea that all fluid *PVT* properties (oil formation volume factor, gas formation volume factor, solution gas/oil ratio, volatized oil/gas ratio), phase compositions (ω_{Oo}, ω_{Go}, ω_{Og}, ω_{Gg}) and densities (ρ_o and ρ_g) are sole functions of pressure in a black-oil fluid formulation. The pressure at which these properties are specified corresponds to the pressure of the *reference* phase selected as a primary unknown for the numerical solution. In most numerical simulators, (reservoir) *oil pressure* is routinely used as one of the primary unknowns and thus becomes the reference pressure for *PVT* fluid calculations. The selection of which phase to use for reference pressure is inconsequential when capillary pressures are negligible. However, fluid *PVT* behavior can be substantially altered when capillary pressures are large and careful attention should be paid to the selection of reference phase for such cases (Zhang and Ayala, 2017b).

Because ω_{Oo}, ω_{Go}, ω_{Og} and ω_{Gg} in Tables 4.2 and 4.4 represent *two-phase equilibrium* phase compositions for the pseudo-binary thermodynamic system, they also convey all of the information needed to quantify bubble and dew point variations with pressure for these systems. This is best visualized via **pressure-composition (*p-x* or *p-z*) phase diagrams,** which are powerful aids used in the visualization of *PVT* behavior of binary mixtures. In these diagrams, composition of the volatile component (in this case, I = "G", the surface-gas pseudo-component) is customarily used as the abscissa.

This is shown in Figs. 4.11 and 4.12 for the gas condensate and volatile oil mixtures of Tables 4.2 and 4.4, respectively. These phase envelopes result from directly plotting ω_{Go} *vs. pressure* data (bubble point line) and ω_{Gg} *vs. pressure* data (dew point line) obtained from Tables 4.2 and 4.4. The *bubble point line* corresponds to the two-phase thermodynamic condition at which the first gas bubble is formed in an otherwise

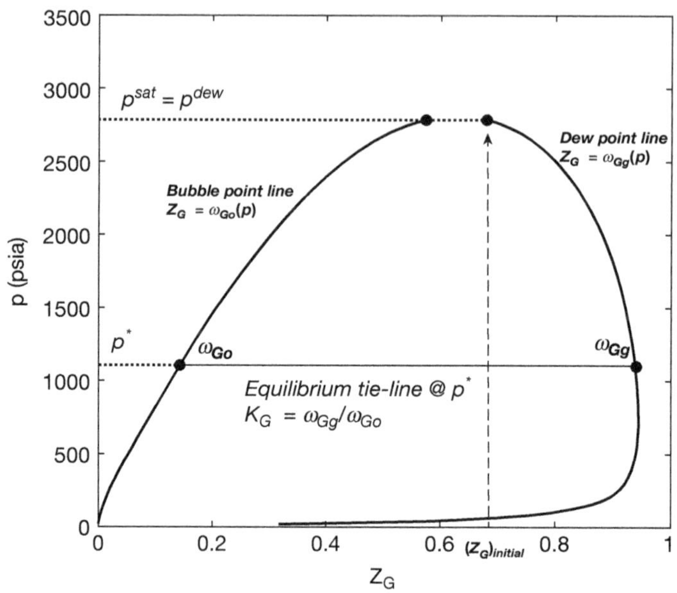

FIGURE 4.11 Pressure vs. composition diagram of pseudo-binary black-oil fluids: gas-condensate case in Table 4.2.

wholly liquid, saturated system. At the bubble point, overall fluid composition (z_G) equals saturated liquid composition ($z_G = \omega_{Go}$) at the prevailing pressure, i.e., the ω_{Go} *vs.* p line in Figs. 4.11 and 4.12 is the *bubble point* line of this fluid. Concurrently, the *dew point line* corresponds to the two-phase thermodynamic condition at which a first liquid droplet is formed in an otherwise wholly gaseous, saturated system. At the dew point, overall fluid composition (z_G) equals saturated vapor composition ($z_G = \omega_{Gg}$) at the prevailing pressure, i.e., the ω_{Gg} *vs.* p line in Figs. 4.11 and 4.12 is the *dew point* line.

In Fig. 4.11, p^{sat} corresponds to p_{dew} of the original reservoir fluid at reservoir temperature since the *PVT* table was experimentally derived for a rich natural gas mixture (retrograde gas). In Fig. 4.12, p^{sat} corresponds to p_{bubble} of the original reservoir fluid at prevailing reservoir temperature since the *PVT* table was experimentally derived for a reservoir oil mixture sample. Any two-phase condition in these reservoirs must occur inside these envelopes. At any pressure level $p^* < p^{sat}$ in Figs. 4.11 and 4.12, equilibrium compositions ω_{Go} and ω_{Gg} and equilibrium ratios $K_G = \omega_{Gg}/\omega_{Go}$ [and $K_O = (1 - \omega_{Gg})/(1 - \omega_{Go})$] are fixed and fully defined. Phase splits (flash calculations) would follow from the application of overall material conservation for the volatile pseudo-component. Written in terms of the "lever rule", β_g calculations can be expressed as:

$$\beta_g = \frac{z_G - \omega_{Go}}{\omega_{Gg} - \omega_{Go}} \tag{4.62}$$

with $\beta_o = 1 - \beta_g$. The lever rule prescribes that the relative amounts of reservoir gas phase (β_g) and oil phase (β_o) are direct functions of overall composition z_G of the in situ fluid. From direct application of the lever rule, it is clear that $\beta_g = 0$ (bubble point

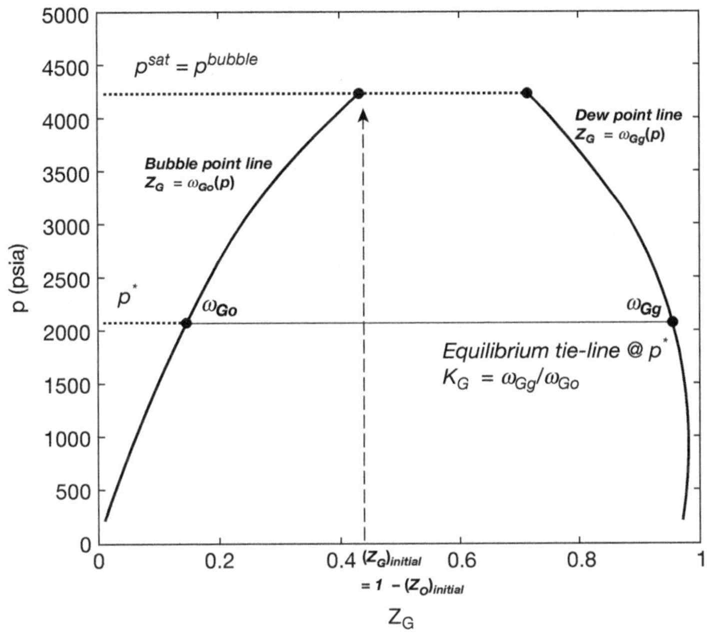

FIGURE 4.12 Pressure vs. composition diagram of pseudo-binary black-oil fluids: volatile oil case in Table 4.4.

condition) when $z_G = \omega_{Go}$ and $\beta_g = 1$ (dew point condition) when $z_G = \omega_{Gg}$. For any z_G-value in between, $0 < \beta_g < 1$ as one moves along the equilibrium tie-line ($\omega_{Go} < z_G < \omega_{Gg}$) at the prevailing pressure. The relationship among equilibrium mass fractions and oil and gas saturations (S_o and S_g), according to Eq. (4.23), is given by:

$$S_o = \frac{V_o}{V_o + V_g + V_w} = (1 - S_w) \cdot \frac{\dfrac{\beta_o}{\rho_o}}{\dfrac{\beta_o}{\rho_o} + \dfrac{\beta_g}{\rho_g}} \tag{4.63}$$

with $S_g = 1 - S_o - S_w$. Reservoir oil and gas densities needed in this relationship are also prescribed by pressure, as shown in Tables 4.2 and 4.4.

The observation that all black-oil *PVT* properties are *uniquely* defined once pressure has been specified (Tables 4.2 and 4.4) is a direct consequence of the application of Gibbs' phase rule to black-oil (binary) thermodynamics. In the absence of capillary effects, when a binary hydrocarbon mixture ($N_c = 2$) splits into two hydrocarbon phases ($N_p = 2$) at a fixed temperature (isothermal depletion), the Gibbs' phase rule predicts:

$$F = (N_C - N_P + 2) - 1 = 1 \tag{4.64}$$

In other words, the Gibbs' phase rule prescribes that, as long as two phases exist, only one degree of freedom (F = 1) is left for this system. Thus, the specification of any other intensive thermodynamic variable (e.g., pressure) completely defines the thermodynamic state and fixes the values of all remaining intensive fluid properties—including saturated or equilibrium phase compositions (ω_{Oo}, ω_{Go}, ω_{Og}, ω_{Gg}), equilibrium densities (ρ_o, ρ_g), and related *PVT* properties (B_o, B_g, R_s, R_v).

This is to say, there *can only be one unique value* of B_o, B_g, R_s, R_v, ω_{Oo}, ω_{Go}, ω_{Og}, ω_{Gg}, ρ_o, ρ_g at each unique pressure level for a true binary mixture. According to this, and regardless of how such values were to be attained experimentally, the same B_o, B_g, R_s, R_v, ... values should be obtained if one were to repeat the experiment while keeping the same mixture sample, reservoir pressure values, and surface separation conditions. As a result, B_o, B_g, R_s, R_v values calculated through a *CVD* (constant-volume depletion) laboratory test, where the liquid hydrocarbon phase is kept immobile inside the *PVT* cell, should be the same even if the experiment is repeated and the liquid phase is mobilized (removed) from the cell. This is, unsurprisingly, an approximation because no petroleum mixture is truly binary. Nothing really prevents observed B_o, B_g, R_s, R_v values from changing when a different *PVT* cell experiment is conducted for the same fluid. However, assuming the same B_o, B_g, R_s, R_v vs. pressure dependency for a given black-oil fluid, regardless of path, is consistent with the thermodynamic framework imposed by the binary mixture assumption behind black-oil fluid modeling.

Also, because surface production in black-oil modeling (see Fig. 4.9) is made up of (pseudo-) "pure" single-phase components, properties of all produced fluids would be subject to the constraints of thermodynamics of pure substances. Thermodynamic properties of single-phase, *pure* substances are unique functions of pressure and temperature as per Gibbs' phase rule. As a result, standard (surface) fluid densities ρ_{osc}, ρ_{gsc}, and ρ_{wsc} may only take ***unique*** values because they are defined at fixed (surface) pressure and temperature conditions (stock-tank conditions; typically, 14.7 *psia* and 60°*F*). This is, again, an approximation because *standard (surface) oil density* (and its specific gravity) may change in time due to compositional changes in the produced stock-tank oil. However, these changes in stock-tank (surface) oil gravity or density are explicitly neglected in black-oil fluid formulations.

4.3.3 Black-Oil *PVT* Data from Compositional and Density Information

It should be clear by now that black-oil *PVT* property definitions could be written in terms of *densities* and *compositions* rather than volumes as more traditionally defined. For hydrocarbon fluids, reservoir volumes (V_o, V_g) and surface volumes (V_{Osc}, V_{Gsc}) can be expressed in terms of reservoir densities (ρ_o, ρ_g), surface densities (ρ_{Osc}, ρ_{Gsc}), and phase compositions (ω_{IJ}'s). It is also noted that $V_{Isc} = V_{Jsc}$ and $\rho_{Isc} = \rho_{Jsc}$ at the surface because $\omega_{IJ} = 1$ for every phase produced at surface conditions in the black-oil model. All hydrocarbon volumes in Fig. 4.5 may be written in terms of mass compositions and densities as follows:

$$V_o[RB] = \frac{M_o}{\alpha_c \rho_o} \tag{4.65}$$

$$V_g[RB] = \frac{M_g}{\alpha_c \rho_g} \tag{4.66}$$

$$(V_{Osc})_o[STB] = \frac{(\omega_{Oo} M_o)}{\alpha_c \rho_{osc}} \tag{4.67}$$

$$(V_{Gsc})_o[scf] = \frac{(\omega_{Go} M_o)}{\rho_{gsc}} \tag{4.68}$$

$$(V_{Osc})_g[STB] = \frac{(\omega_{Og} M_g)}{\alpha_c \rho_{osc}} \tag{4.69}$$

$$(V_{Gsc})_g[scf] = \frac{(\omega_{Gg} M_g)}{\rho_{gsc}} \tag{4.70}$$

where α_c is the volume conversion factor (5.615 ft^3/bbl) and $\omega_{Oo} + \omega_{Go} = \omega_{Og} + \omega_{Gg} = 1$ when water solubility is ignored. From these relationships, it directly follows by substituting them into the appropriate definitions presented in Fig. 4.6:

$$B_o\left[\frac{RB}{STB}\right] = \frac{V_o}{(V_{Osc})_o} = \frac{\rho_{osc}}{\omega_{Oo}\,\rho_o} \tag{4.71}$$

$$B_g\left[\frac{RB}{scf}\right] = \frac{V_g}{(V_{Gsc})_g} = \frac{\rho_{gsc}}{\alpha_c\,\omega_{Gg}\,\rho_g} \tag{4.72}$$

$$R_s\left[\frac{scf}{STB}\right] = \frac{(V_{Gsc})_o}{(V_{Osc})_o} = \alpha_c \frac{\rho_{osc}\,\omega_{Go}}{\rho_{gsc}\,\omega_{Oo}} \tag{4.73}$$

$$R_v\left[\frac{STB}{scf}\right] = \frac{(V_{Osc})_g}{(V_{Gsc})_g} = \frac{1}{\alpha_c}\frac{\rho_{gsc}\,\omega_{Og}}{\rho_{osc}\,\omega_{Gg}} \tag{4.74}$$

from where

$$\frac{R_s}{B_o} = \alpha_c\,\omega_{Go}\frac{\rho_o}{\rho_{gsc}} \tag{4.75}$$

$$\frac{R_v}{B_g} = \omega_{Og}\frac{\rho_g}{\rho_{osc}} \tag{4.76}$$

For water, given that all surface water comes exclusively from the reservoir water phase ($\omega_{Ww} = 1$), one directly writes:

$$B_w = \frac{V_w}{(V_{Wsc})_w} = \frac{\rho_{wsc}}{\rho_w} \tag{4.77}$$

Alternatively, molar concentrations (x_{IJ}'s) could be used for all these calculations. Following Eq (4.3), black-oil *molar* and *mass* compositions are through molecular weight information, as shown:

$$\omega_{IJ} = x_{IJ}\cdot\frac{MW_{Isc}}{MW_J} \tag{4.78}$$

where MW_{Isc} is the molecular weight for the *I-th* surface component at standard conditions (*sc*), MW_J the molecular weight of the *J-th* reservoir phase, and $MW_J = \sum_I^{G,O,W} x_{IJ} MW_{Isc}$. Mass and molar densities (ρ vs. $\hat{\rho}$ respectively) are also related through molecular weight. For reservoir fluid densities one writes, for instance:

$$\hat{\rho}_o = \rho_o / MW_o \tag{4.79}$$

$$\hat{\rho}_g = \rho_g / MW_g \tag{4.80}$$

and for surface or standard densities:

$$\hat{\rho}_{Osc} = \rho_{Osc} / MW_{Osc} \tag{4.81}$$

$$\hat{\rho}_{Gsc} = \rho_{Gsc} / MW_{Gsc} \tag{4.82}$$

Using the corresponding mass to molar density and composition conversions, equations above can be rewritten in terms of molar quantities instead:

$$B_o = \frac{\rho_{osc}}{\omega_{Oo}\,\rho_o} = \frac{\hat{\rho}_{osc}}{x_{Oo}\,\hat{\rho}_o} \tag{4.83}$$

$$B_g = \frac{1}{\alpha_c}\frac{\rho_{gsc}}{\omega_{Gg}\,\rho_g} = \frac{1}{\alpha_c}\frac{\hat{\rho}_{gsc}}{x_{Gg}\,\hat{\rho}_g} \tag{4.84}$$

$$R_s = \alpha_c \frac{\rho_{osc}\,\omega_{Go}}{\rho_{gsc}\,\omega_{Oo}} = \alpha_c \frac{\hat{\rho}_{osc}\,x_{Go}}{\hat{\rho}_{gsc}\,x_{Oo}} \tag{4.85}$$

$$R_v = \frac{1}{\alpha_c}\frac{\rho_{gsc}\,\omega_{Og}}{\rho_{osc}\,\omega_{Gg}} = \frac{\hat{\rho}_{gsc}\,x_{Og}}{\alpha_c\,\hat{\rho}_{osc}\,x_{Gg}} \tag{4.86}$$

$$\frac{R_s}{B_o} = \alpha_c\,x_{Go}\frac{\hat{\rho}_o}{\hat{\rho}_{gsc}} \tag{4.87}$$

$$\frac{R_v}{B_g} = x_{Og}\frac{\hat{\rho}_g}{\hat{\rho}_{osc}} \tag{4.88}$$

For these calculations, *standard oil density* ρ_{osc} may be computed from available values of stock-tank oil gravity (γ_{osc}) as shown:

$$\rho_{osc} = \gamma_{osc}\,\rho_w \tag{4.89}$$

where the reference value ρ_w in this equation is the density of pure water at standard conditions, 62.4 lb/ft^3. For this, stock-tank oil gravity can be estimated if values of initial *API* gravity for the surface oil are available:

$$\gamma_{osc} = \frac{141.5}{API + 131.5} \tag{4.90}$$

Stock-tank molecular weight (MW_{osc}) can also be correlated to API gravities using empirical models (Walsh and Lake, 2003; McCain, 2017); and *molar standard oil densities* ($\hat{\rho}_{osc}$) would follow ($\hat{\rho}_{osc} = \rho_{osc}/MW_{osc}$).

Standard gas density ρ_{gsc} is typically computed from the application of the gas law at standard conditions:

$$\rho_{gsc} = \frac{p_{sc} MW_{gsc}}{RT_{sc}} \tag{4.91}$$

or,

$$\rho_{gsc} = \frac{MW_{gsc}}{379.4\, scf/lbmol} \tag{4.92}$$

for $p_{sc} = 14.7$ *psia* and $T_{sc} = 519.67$ *R* [60°F]. Stock-tank molecular weight (MW_{Gsc}) is computed from values of surface gas gravity (γ_{Gsc}):

$$MW_{gsc} = \gamma_{gsc} MW_{air} \tag{4.93}$$

where the MW_{air} reference value is equal to 28.97 *lb/lbmol*. It should be noted that *molar standard gas density* ($\hat{\rho}_{gsc}$) is always equal to (1/379.4) *lbmol/scf* ($\hat{\rho}_{gsc} = \rho_{gsc}/MW_{gsc}$).

The lower bound of *standard (surface) water density* values is 62.4 lb/ft^3—i.e., density of pure water. Actual values are typically *higher* due to the presence of dissolved solids (salts) in produced waters. Water (brine) densities at standard conditions are thus typically correlated to its salinity or total dissolved solids (*TDS*).

Another concept useful when deploying black-oil fluid models to analyze rich natural gases is that of the *gas-oil equivalency* factor, or R_{GO} (scf/STB) (Walsh and Lake, 2003). R_{GO} values provide the equivalency of each unit standard-volume of surface-oil (*STB*) in terms of surface-gas volume (*scf*). The equivalency relies on the analogy that every mole of surface oil would occupy the same volume as would one mole of surface gas if vaporized. Depending on how dense the surface-oil condensate is, 1 *STB* may occupy anywhere from 600 to 950 *scf* if vaporized at standard conditions. R_{GO} values are directly related to the ratio of *molar standard oil to gas densities* ($\hat{\rho}_{osc}$, $\hat{\rho}_{gsc}$):

$$R_{GO} = \alpha_c \frac{\hat{\rho}_{osc}}{\hat{\rho}_{gsc}} \tag{4.94}$$

where $\hat{\rho}_{gsc}$ equals (1/379.4) *lbmol/scf* and the intrinsic value of $\hat{\rho}_{osc}$ may be calculated from the available *CVD* laboratory test report of the gas condensate fluid. In a typical *CVD* laboratory test report, information about total separator gas contained in the original reservoir fluid is provided from separator tests performed on the initially single-phase gas sample. This information corresponds to the in situ surface-gas reservoir composition, or $(x_{Gg})_{initial}$. From Eq. (4.86), the implicit value of $\hat{\rho}_{osc}$ would then be given by:

$$\hat{\rho}_{osc} = \frac{\hat{\rho}_{gsc}}{\alpha_c R_{vi}} \frac{1 - \left(x_{Gg}\right)_{initial}}{\left(x_{Gg}\right)_{initial}} \tag{4.95}$$

where R_{vi} is the value of R_v at initial conditions.

References

Ahmed, T. *Equations of State and PVT Analysis*, Gulf Publishing Co., Houston, Texas, 2007.

Becker, M.D., M. Zhang, and L.F. Ayala, "On the Pressure-Saturation Path in Infinite-Acting Unconventional Liquid-Rich Gas Reservoirs," *J. Nat. Gas. Sc. & Eng.*, 35:97–113, September 2016.

Danesh, A., *PVT and Phase Behaviour of Petroleum Reservoir Fluids, Developments in Petroleum Science*, Vol. 47, Elsevier, Amsterdam, The Netherlands, 1998.

Firoozabadi, A., *Thermodynamics of Hydrocarbon Reservoirs*, McGraw-Hill, New York, 1999.

McCain, W.D., *The Properties of Petroleum Fluids*, 3rd edition, Penn Well, Tulsa, Oklahoma, 2017.

Michelsen, M. and J. Mollerup, *Thermodynamic Models: Fundamentals & Computational Aspects*, Tie-line Publications, Denmark, 2007.

Okuno, R., R.T. Johns, and K. Sepehrnoori, "A New Algorithm for Rachford-Rice for Multiphase Compositional Simulation," *SPE J.*, 313–325, June 2010.

Peng, D.-Y., and D.B. Robinson, "A New Two-Constant Equation of State," *Ind. Eng. Chem. Fundam.*, 15:59–64, 1976.

Pedersen, K.S., P.L. Christensen, and J.A. Shaikh, *Phase Behavior of Petroleum Reservoir Fluids*, 2nd ed., CRC Press, Boca Raton, FL, 2015.

Rachford, H.H. Jr. and J.D. Rice, "Procedure for Use of Electronic Digital Computers in Calculating Flash Vaporization Hydrocarbon Equilibrium," *J. Pet. Tech.*, 4(10); *Trans. AIME*, 195:327–328, SPE 952327.

Schilthuis, R.J. "Active Oil and Reservoir Energy," *Trans. AIME*, 118: 33,1936.

Singh, M., M. Zhang, H. Emami-Meybodi, and L.F. Ayala, "Use of Rescaled Exponential Models for Boundary-Dominated Liquid-Rich Gas Flow Analysis Under Variable Bottomhole Pressure Conditions," *J. Nat. Gas. Sc. & Eng.*, 46:793–816, 2017.

Soave, G., "Equilibrium Constants from a Modified Redlich–Kwong Equation of State," *Chem. Eng. Sci.*, 27:1197–1203, 1972.

Walsh, M.P. and L.W. Lake, *A Generalized Approach to Primary Hydrocarbon Recovery. Handbook of Petroleum Exploration and Production* (Book 4), Elsevier, Amsterdam, 2003.

Warren, J.H. and M.A. Adewumi, "Polynomial Objective Functions for Flash Calculations: Binary, Ternary, and Quaternary Systems," *Ind. Eng. Chem. Res.*, 32(7):1528–1530, 1993.

Whitson, C. and M. Brule, *Phase Behavior*, SPE Monograph, Vol. 20, Henry Doherty Series, Society of Petroleum Engineers, Richardson, Texas, 2000.

Zhang, M. and L.F. Ayala, "A Density-Based Material Balance Equation for the Analysis of Liquid-Rich Natural Gas Systems," *J. Petrol. Explor. Prod. Technol.*, 6:705–718, 2016.

Zhang, M., M.D. Becker, and L.F. Ayala, "A Similarity Method Approach for Early-Transient Multiphase Flow Analysis of Liquid-Rich Unconventional Gas Reservoirs," *J. Nat. Gas. Sc. & Eng.*, 28:572–586, 2016a.

Zhang, M., M. Singh, and L.F. Ayala, "Rate Forecasting during Boundary-Dominated Multiphase Flow: The Rescaled Exponential Model," *J. Petrol. Sci. Eng.*, 143:199–210, 2016b.

Zhang, M. and L.F. Ayala, "Analytical Study of Constant Gas-Oil-Ratio Behavior as an Infinite-Acting Effect in Unconventional Multiphase Reservoir Systems," *SPE J.* SPE-175079-PA, 22(01):289–299, 2017a.

Zhang, M. and L.F. Ayala, "Similarity-based, Semi-Analytical Assessment of Capillary Pressure Effects in Very Tight, Liquid-Rich Gas Plays during Early-Transient Multiphase Analysis," *J. Nat. Gas. Sc. & Eng.*, 45:189–206, 2017b.

CHAPTER 5

Reservoir Engineering Transport Equations

To predict recovery and performance, reservoir modeling relies on the solution of transport equations describing physical processes taking place in the hydrocarbon system. While reservoir analytical models (Chap. 6) focus on the analysis of simplified systems, reservoir numerical simulation models (Chap. 7) employ a discretized form of the system under study on which more complex governing fluid flow transport equations can be solved. These governing equations are derived from the fundamental laws of conservation of mass, momentum, and energy, which form the basis of the underlying mathematical model. Among these, conservation of mass is the most fundamental relationship that reservoir engineering deploys for modeling purposes. In these reservoir engineering applications, Darcy's law represents the momentum equation mapping velocity fields to flow potential gradient fields; and the energy equation is not typically used given the isothermal flow conditions considered in this book. In this chapter, we will review the most fundamental mass conservation statements used in reservoir engineering using a compositional perspective. For non-isothermal and enhanced-oil recovery applications, Lake et al. (2014) also review the basic equations for fluid flow in permeable media, where energy and entropy balance equations used in reservoir modeling are included in their review.

5.1 Mass Conservation in Compositional Reservoir Engineering Models

One of the most general and useful mass conservation statement in reservoir engineering is the one written for conservation of each species ($I = 1, 2, \ldots, N_c$) found in the reservoir fluid mixture characterization. As discussed in Chap. 4, this compositional approach of fluid characterization embraces the most general scenario: the one where N_c-chemical species are allowed to coexist within any number of phases ($J = 1, 2, \ldots, N_p$) in the thermodynamic reservoir system. Mass conservation of component "*I*" within a phase "*J*" in the control volume (CV) such as the one depicted in Fig. 5.1 requires that, over a period of time Δt:

$$\left\{\begin{matrix}\text{Mass of } I\text{-}th \text{ comp.}\\ \text{entering CV during } \Delta t\end{matrix}\right\} - \left\{\begin{matrix}\text{Mass of } I\text{-}th \text{ comp.}\\ \text{leaving CV during } \Delta t\end{matrix}\right\} + \left\{\begin{matrix}\text{Sink/Sources}\\ I\text{-}th \text{ comp., } \Delta t\end{matrix}\right\}$$

$$= \left\{\begin{matrix}\text{Accumulation}\\ I\text{-}th \text{ comp., } \Delta t\end{matrix}\right\} \tag{5.1}$$

… written of each species $I = 1, 2, \ldots, N_c$ within each phase $J = 1, 2, \ldots, N_p$.

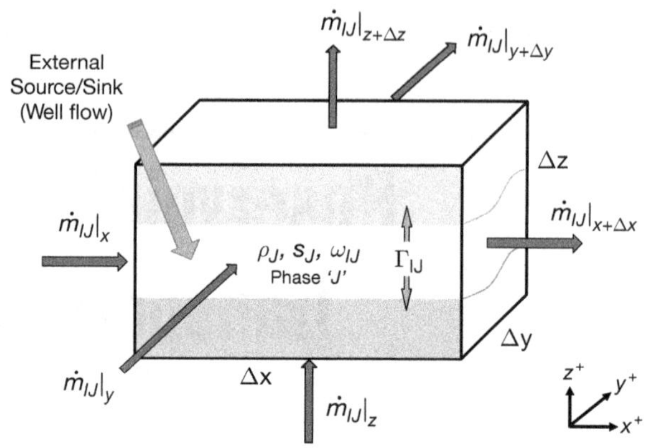

FIGURE 5.1 Control volume (CV) in 3D Cartesian coordinates.

When properties inside the CV are assumed to be uniform (e.g., ϕ, ρ_J, S_J, ω_{IJ}), these terms can be written as follows for the case of 3-D Cartesian flow depicted in Fig. 5.1:

$$\begin{Bmatrix} \text{Mass of } I\text{-}th \text{ comp.} \\ \text{entering CV during } \Delta t \end{Bmatrix} = \left(\dot{m}_{IJ}\big|_x + \dot{m}_{IJ}\big|_y + \dot{m}_{IJ}\big|_z \right)\Delta t \tag{5.2}$$

$$\begin{Bmatrix} \text{Mass of } I\text{-}th \text{ comp.} \\ \text{leaving CV during } \Delta t \end{Bmatrix} = \left(\dot{m}_{IJ}\big|_{x+\Delta x} + \dot{m}_{IJ}\big|_{y+\Delta y} + \dot{m}_{IJ}\big|_{z+\Delta z} \right)\Delta t \tag{5.3}$$

$$\begin{Bmatrix} \text{Accumulation} \\ I\text{-}th \text{ comp., } \Delta t \end{Bmatrix} = M_{IJ}\big|_{t+\Delta t} - M_{IJ}\big|_t \tag{5.4}$$

Using the definitions introduced in Chap. 4, mass flows of the *I-th* component in the *J-th* phase ($\dot{m}_{IJ}$) can be readily written in terms of phase compositions, phase densities and velocities, and flow areas. For incoming flows into the control volume in Fig. 5.1, for example, one can write:

$$\dot{m}_{IJ}\big|_x = (\omega_{IJ}\rho_J v_{xJ} A_x)_x \tag{5.5}$$

$$\dot{m}_{IJ}\big|_y = (\omega_{IJ}\rho_J v_{yJ} A_y)_y \tag{5.6}$$

$$\dot{m}_{IJ}\big|_z = (\omega_{IJ}\rho_J v_{zJ} A_z)_z \tag{5.7}$$

while for the outgoing flows:

$$\dot{m}_{IJ}\big|_{x+\Delta x} = (\omega_{IJ}\rho_J v_{xJ} A_x)_{x+\Delta x} \tag{5.8}$$

$$\dot{m}_{IJ}\big|_{y+\Delta y} = (\omega_{IJ}\rho_J v_{yJ} A_y)_{y+\Delta y} \tag{5.9}$$

$$\dot{m}_{ij}\big|_{z+\Delta z} = (\omega_{IJ}\rho_J v_{zJ} A_z)_{z+\Delta z} \tag{5.10}$$

It should be highlighted that in evaluating mass flow through all control surfaces of Fig. 5.1, as presented in Eqs. (5.5) to (5.10), only advective (velocity-driven) transport into the CV is being considered. In other words, we are explicitly ruling out other mass transfer mechanisms into the CV such as molecular diffusion and mechanical dispersion, which is always the case when Eqs. (5.5) to (5.10) are written in terms of Darcian advective velocities. The implicit assumption is that the flux due to molecular diffusion resulting from species density or concentration gradients would be insignificant compared to this advective flux. This is largely valid for high Peclet number flows in porous media when pore diameters and permeabilities are large enough to result in Darcian-driven fluxes only. The Peclet number is a dimensionless group representing the ratio of mass flow due to fluid bulk motion to mass flow due to molecular diffusion—Pe = (advective transport rate)/(diffusive transport rate).

Sink and sources for component "*I*" in phase "*J*" can be external to the CV, such as the case of having a well of rate q_J in the system, or internal, as it is the case of mass transfer of component "*I*" among phases coexisting in the CV. The latter is captured through the variable Γ_{IJ}, which represents the rate of mass transfer of species "*I*" into phase "*J*" per unit time and bulk volume. This mass transfer can take the form of condensation, vaporization, or any other physical process whose net effect is the exchanging of mass of a species into or out of a phase. Over a period of time Δt, the amount of *I-th* component in the *J-th* phase entering or leaving the control volume through these external and internal sink/sources over a period of time Δt is:

$$\begin{Bmatrix} \text{Sink/Sources} \\ I\text{-}th \text{ comp., } \Delta t \end{Bmatrix} = \omega_{IJ}\rho_J q_J \Delta t + \Gamma_{IJ} V_b \Delta t \tag{5.11}$$

Since the total mass of phase "*J*" inside this control volume is given by $V_b \phi \rho_J S_J$, the amount of mass of the *I-th* component in such *J-th* phase is:

$$M_{IJ} = \omega_{IJ} V_b \phi \, \rho_J S_J \tag{5.12}$$

where bulk volume $V_b = \Delta x \, \Delta y \, \Delta z$ for this CV.

Substituting into Eq. (5.1), one writes:

$$-\begin{bmatrix} (\omega_{IJ}\rho_J v_{xJ} A_x)_{x+\Delta x} - (\omega_{IJ}\rho_J v_{xJ} A_x)_x \\ +(\omega_{IJ}\rho_J v_{yJ} A_y)_{y+\Delta y} - (\omega_{IJ}\rho_J v_{yJ} A_y)_y \\ +(\omega_{IJ}\rho_J v_{zJ} A_z)_{z+\Delta z} - (\omega_{IJ}\rho_J v_{zJ} A_z)_z \end{bmatrix} \Delta t + \omega_{IJ}\rho_J q_J \Delta t$$

$$+\Gamma_{mIJ} V_b \Delta t = (\omega_{IJ} V_b \phi \rho_J S_J)_{t+\Delta t} - (\omega_{IJ} V_b \phi \rho_J S_J)_t \tag{5.13}$$

Recognizing that for the CV shown in Fig. 5.1, flow areas are written as:

$$A_x = \Delta y \, \Delta z, A_y = \Delta x \, \Delta z, A_z = \Delta x \, \Delta y \tag{5.14}$$

and bulk volume is related to flow areas as follows:

$$V_b = A_x \, \Delta x = A_y \, \Delta y = A_z \, \Delta z \tag{5.15}$$

Then, the following alternate expression can be written of Eq. (5.13) by dividing through by Δt:

$$-V_b \cdot \frac{(\omega_{IJ}\rho_J v_{xJ} A_x)_{x+\Delta x} - (\omega_{IJ}\rho_J v_{xJ} A_x)_x}{A_x \Delta x}$$

$$-V_b \cdot \frac{(\omega_{IJ}\rho_J v_{yJ} A_y)_{y+\Delta y} - (\omega_{IJ}\rho_J v_{yJ} A_y)_y}{A_y \Delta y}$$

$$-V_b \cdot \frac{(\omega_{IJ}\rho_J v_{zJ} A_z)_{z+\Delta z} - (\omega_{IJ}\rho_J v_{zJ} A_z)_z}{A_z \Delta z}$$

$$+ \omega_{IJ}\rho_J q_j + \Gamma_{IJ} V_b = \frac{(\omega_{IJ} V_b \phi \rho_J S_J)_{t+\Delta t} - (\omega_{IJ} V_b \phi \rho_J S_J)_t}{\Delta t} \tag{5.16}$$

It is also noted here that the equalities:

$$1 = \frac{V_b}{A_x \Delta x} = \frac{V_b}{A_y \Delta y} = \frac{V_b}{A_z \Delta z} \tag{5.17}$$

are being introduced into each of the spatial differences in Eq. (5.16) to compose the definition of the partial derivative.

For any generic dependent variable F which changes with respect to an independent variable S with all other independent variables kept constant, the first-order partial derivative definition is written as follows:

$$\frac{\partial F}{\partial S} = \lim_{\Delta S \to 0} \frac{F_{S+\Delta S} - F_S}{\Delta S} \tag{5.18}$$

Thus, taking the limits of Eq. (5.16) as $\Delta x, \Delta y, \Delta z, \Delta t \to 0$ and introducing the stated definition of the partial derivative, one will obtain:

$$-\frac{V_b}{A_x}\frac{\partial}{\partial x}(\omega_{IJ}\rho_J v_{xJ} A_x) - \frac{V_b}{A_y}\frac{\partial}{\partial y}(\omega_{IJ}\rho_J v_{yJ} A_y)$$

$$-\frac{V_b}{A_z}\frac{\partial}{\partial z}(\omega_{IJ}\rho_J v_{zJ} A_z) + \omega_{IJ}\rho_J q_J + \Gamma_{IJ} V_b = V_b \frac{\partial}{\partial t}(\omega_{IJ}\phi \rho_J S_J)$$

$$I = 1, 2, \ldots, N_c; J = 1, 2, \ldots, N_p \tag{5.19}$$

where V_b is being assumed invariant with time. Equation (5.19) is a fundamental statement of mass conservation for component "I" in phase "J".

When the statement in Eq. (5.16) considers $(A_x)_x = (A_x)_{x+\Delta x} = A_x$, $(A_y)_y = (A_y)_{y+\Delta y} = A_y$, and $(A_z)_z = (A_z)_{z+\Delta z} = A_z$ while taking the limit as Δx, Δy, $\Delta z \to 0$ the resulting conservation expression ends up written per unit bulk reservoir volume as follows:

$$-\frac{\partial}{\partial x}(\omega_{IJ}\rho_J v_{xJ}) - \frac{\partial}{\partial y}(\omega_{IJ}\rho_J v_{yJ}) - \frac{\partial}{\partial z}(\omega_{IJ}\rho_J v_{zJ}) + \frac{\omega_{IJ}\rho_J q_J}{V_b} + \Gamma_{IJ} = \frac{\partial}{\partial t}(\phi\omega_{IJ}\rho_J S_J)$$

$$I = 1, 2, \ldots, N_c; \quad J = 1, 2, \ldots, N_p \tag{5.20}$$

For the most general case, where there are N_c-species in this system ($I = 1, 2, \ldots, N_c$) distributed among N_p-phases ($J = 1, 2, \ldots, N_p$), Eqs. (5.19) and (5.20) would be written $N_c \cdot N_p$ times. Rather than using Eq. (5.20), the finite-difference work of Ertekin et al. (2001) deliberately carries all *area* terms inside every parenthesis as presented in Eq. (5.19). That choice is made with a numerical purpose in mind, since Eq. (5.19) most straightforwardly allows accommodating changes in formation thickness in the x, y, and z directions (i.e., changes in A_x, A_y, A_z) during the development of finite-difference approximations. Such bulk-volume form of the governing equations [Eq. (5.19), which includes all area terms inside each parenthesis] may be reintroduced later if finite-difference numerical approximations were to be formulated. Alternatively, these area terms would naturally reappear during volume integration of Eq. (5.20) when employing the finite-volume numerical approximations discussed in this work (see Chap. 7).

Equation (5.20) can be more succinctly written in terms of the divergence operator. This allows expressing governing equations in a more compact form that would remain the same regardless of choice of coordinate system:

$$-\nabla \cdot (\omega_{IJ}\rho_J \vec{v_J}) + \frac{\omega_{IJ}\rho_J q_J}{V_b} + \Gamma_{IJ} = \frac{\partial}{\partial t}(\phi\omega_{IJ}\rho_J S_J)$$

$$I = 1, 2, \ldots, N_c; \quad J = 1, 2, \ldots, N_p \tag{5.21}$$

where $\nabla \cdot$ is the divergence operator operating on the mass flux vector of the *I-th* component in the *J-th* phase (i.e., $\omega_{IJ}\rho_J \vec{v_J}$). It is important to recognize that Eq. (5.21) could have been directly derived from the generalized transport equation (GTE) presented in Chap. 2 [Eq. (2.20)]. When GTE is applied to species conservation in porous media, α in Eq. (2.20) corresponds to the total mass amount of component I within the fixed Eulerian control volume per unit CV bulk volume ($\alpha = M_{IJ}/V_b$, with $M_{IJ} = \omega_{IJ}\rho_J V_J$ and $\alpha = \phi S_J \rho_J \omega_{IJ}$). $\vec{V}$ in Eq. (2.20) is the advective velocity of species I, corresponding to the interstitial velocity of the phase that carries it ($\vec{V} = \frac{\vec{v_J}}{\phi S_J}$, with $\vec{v_J}$ being the Darcian superficial velocity). Since molecular diffusion and mechanical dispersion were neglected in this derivation ($\nabla \cdot \vec{J} = 0$), the recognition $\Gamma = \Gamma_{IJ}$ for mass conservation of species in a multiphase environment completes the mapping between the GTE in Eqs. (5.20) and (5.21). See Lake et al. (2014) for another detailed treatment and derivation of mass conservation equations for fluid flow in permeable media.

As also discussed in Chap. 2, the divergence of the mass flux vector ($\nabla \cdot (\omega_{IJ}\rho_J \vec{v_J})$ in Eq. (5.21)) quantifies the amount of mass flux generation per unit volume at each point of the domain. Table 5.1 presents how this divergence operator is expressed for any generic vector $\vec{u}$ and coordinate systems of interest. Example 5.1 illustrates its application to cylindrical coordinates.

Coordinate System	Vector $\vec{u}$ and its divergence $\nabla \cdot \vec{u}$
Cartesian (x, y, z)	$\vec{u} = u_x\hat{\mathbf{i}} + u_y\hat{\mathbf{j}} + u_z\hat{\mathbf{k}}$ $\nabla \cdot \vec{u} = \frac{\partial}{\partial x}(u_x) + \frac{\partial}{\partial y}(u_y) + \frac{\partial}{\partial z}(u_z)$
Cylindrical (r, θ, z)	$\vec{u} = u_r\hat{\mathbf{e}}_r + u_\theta\hat{\mathbf{e}}_\theta + u_z\hat{\mathbf{e}}_z$ $\nabla \cdot \vec{u} = \frac{1}{r}\frac{\partial}{\partial r}(ru_r) + \frac{1}{r}\frac{\partial}{\partial \theta}(u_\theta) + \frac{\partial}{\partial z}(u_z)$
Spherical (r, θ, ϕ)	$\vec{u} = u_r\hat{\mathbf{e}}_r + u_\theta\hat{\mathbf{e}}_\theta + u_\phi\hat{\mathbf{e}}_\phi$ $\nabla \cdot \vec{u} = \frac{1}{r^2}\frac{\partial}{\partial r}(r^2u_r) + \frac{1}{r\sin\theta}\frac{\partial}{\partial \theta}(\sin\theta\, u_\theta) + \frac{1}{r\sin\theta}\frac{\partial}{\partial \phi}(u_\phi)$

TABLE 5.1 Divergence Operator Applied to a Generic Vector $\vec{u}$

Example 5.1: Write Eq. (5.21) for a 3D cylindrical coordinate system using the divergence information provided in Table 5.1.

To expand the divergence operator, we first identify the vector of interest to which the divergence will operate on. For the case of Eq. (5.21), this is the mass flux vector of the I-th component in the J-th phase:

$$\vec{u} = \omega_{IJ}\,\rho_J\,\overrightarrow{v_J} \tag{5.22}$$

which expressed in the coordinates of choice is:

$$\vec{u} = (\omega_{IJ}\,\rho_J\,v_{rJ})\hat{\mathbf{e}}_r + (\omega_{IJ}\,\rho_J\,v_{\theta J})\hat{\mathbf{e}}_\theta + (\omega_{IJ}\,\rho_J\,v_{zJ})\hat{\mathbf{e}}_z \tag{5.23}$$

According to Table 5.1, the corresponding divergence expansion is:

$$\nabla \cdot \vec{u} = \frac{1}{r}\frac{\partial}{\partial r}(ru_r) + \frac{1}{r}\frac{\partial}{\partial \theta}(u_\theta) + \frac{\partial}{\partial z}(u_z) \tag{5.24}$$

where $u_r = \omega_{IJ}\,\rho_J\,v_{rJ}$, $u_\theta = \omega_{IJ}\,\rho_J\,v_{\theta J}$, *and* $u_z = \omega_{IJ}\,\rho_J\,v_{zJ}$, *as per Eq. (5.23). Introducing this information into Eq. (5.21), one writes:*

$$-\frac{1}{r}\frac{\partial}{\partial r}(\omega_{IJ}\,\rho_J\,v_{rJ}) - \frac{1}{r}\frac{\partial}{\partial \theta}(\omega_{IJ}\,\rho_J\,v_{\theta J}) - \frac{\partial}{\partial z}(\omega_{IJ}\,\rho_J\,v_{zJ})$$

$$+\frac{\omega_{IJ}\rho_J q_J}{V_b} + \Gamma_{IJ} = \frac{\partial}{\partial t}(\phi\,\omega_{IJ}\rho_J S_J)$$

$$I = 1, 2, \ldots, N_c; \quad J = 1, 2, \ldots, N_p \tag{5.25}$$

for the 3D cylindrical coordinate system.

5.1.1 Overall (across-all-phases) Compositional Equations

Rather than writing each of the $\boldsymbol{N_c} \cdot \boldsymbol{N_p}$ equations stated in Eq. (5.21), *overall* compositional equations are most often used in hydrocarbon reservoir modeling. To do so, one can combine all equations written for the same component across all phases and end up with a single mass conservation statement for each of the species. In other words, one may add Eq. (5.21) across all phases ($J = 1, 2, \ldots, N_p$) to obtain:

$$-\nabla \cdot \left[\sum_{J=1}^{N_p}\left(\omega_{IJ}\rho_J\vec{v}_J\right)\right] + \frac{1}{V_b}\sum_{J=1}^{N_p}\omega_{IJ}\rho_J q_J + \sum_{J=1}^{N_p}\Gamma_{IJ} = \frac{\partial}{\partial t}\left[\phi\sum_{J=1}^{N_p}S_J\omega_{IJ}\rho_J\right]$$

$$I = 1, 2, \ldots, N_c \tag{5.26}$$

This grouping allows taking advantage of an evident mass transfer constraint across phases:

$$\sum_{J=1}^{N_p}\Gamma_{IJ} = 0 \tag{5.27}$$

which represents a mass conservation statement across phases. Whichever mass a phase gains or loses during interphase mass transfer must be lost or gained by another phase; thus, all Γ_{IJ}'s must add to zero across all phases. When the grouping in Eq. (5.26) is not used, Γ_{IJ} mass transfer models would be required to be postulated in order to achieve mathematical closure of the system of $\boldsymbol{N_c} \cdot \boldsymbol{N_p}$ equations of Eq. (5.21).

Enforcing Eq. (5.27) yields:

$$-\nabla \cdot \left[\sum_{J=1}^{N_p}(\omega_{IJ}\rho_J\vec{v}_J)\right] + \frac{1}{V_b}\sum_{J=1}^{N_p}(\omega_{IJ}\rho_J q_J) = \frac{\partial}{\partial t}\left[\phi\sum_{J=1}^{N_p}S_J\,\omega_{IJ}\,\rho_J\right]$$

$$I = 1, 2, \ldots, N_c \tag{5.28}$$

Invoking **Darcy's law,** reservoir engineering's momentum equation, we can write for the velocity fields of Eq. (5.28), assuming that Darcian advective velocities dominate species transport in porous media:

$$\vec{v}_J = -\frac{\bar{\bar{k}}k_{rJ}}{\mu_J}\nabla\Phi_J \tag{5.29}$$

where $\bar{\bar{k}}$ is the absolute permeability tensor, k_{rJ} is the relative permeability function for phase "J", and Φ_J is the flow potential for each phase. Darcy's law provides the mapping between the fluid velocity vector field $\vec{v}_J$ and the flow potential gradient ($\nabla\Phi_J$). As discussed in Chap. 2, the gradient operator $\boldsymbol{\nabla}$ applied to a pressure or flow potential field produces a *vector* whose magnitude is the rate of change of pressure or potential field, pointing in the direction of the greatest rate of change. Table 5.2 presents the definition of the gradient vector for all coordinate systems of interest. In this chapter, we assume Darcy velocities in Eq. (5.28) to be written in practical field units, or res bbl (RB) per day per ft^2 (RB/day/ft^2), which implies that permeabilities would be expressed in "perms," a unit of convenience that incorporates the suitable field unit conversion (1 Darcy = 1.127 perms) into the equation, flow potential gradients in psi/ft, and viscosities in cp. Note that multiplying Eq. (5.28) times the volume conversion factor α_c (5.615 ft^3/bbl) would express these Darcian velocities in ft/day.

Coordinate System	Gradient vector, ∇p
Cartesian (x, y, z)	$\nabla p = \frac{\partial p}{\partial x}\hat{\boldsymbol{i}} + \frac{\partial p}{\partial y}\hat{\boldsymbol{j}} + \frac{\partial p}{\partial z}\hat{\boldsymbol{k}}$
Cylindrical (r, θ, z)	$\nabla p = \frac{\partial p}{\partial r}\hat{\mathbf{e}}_r + \frac{1}{r}\frac{\partial p}{\partial \theta}\hat{\mathbf{e}}_\theta + \frac{\partial p}{\partial z}\hat{\mathbf{e}}_z$
Spherical (r, θ, ϕ)	$\nabla p = \frac{\partial p}{\partial r}\hat{\mathbf{e}}_r + \frac{1}{r}\frac{\partial p}{\partial \theta}\hat{\mathbf{e}}_\theta + \frac{1}{r\,Sin\theta}\frac{\partial p}{\partial \phi}\hat{\mathbf{e}}_\phi$

Table 5.2 Gradient Operator Applied to a Scalar (e.g., pressure) Field

Substitution of Darcy's law into Eq. (5.28) leads to:

$$\nabla \cdot \left[\sum_{J=1}^{N_p}\left(\omega_{IJ}\rho_J \frac{\overline{\overline{k}}k_{rJ}}{\mu_J}\nabla\Phi_J\right)\right] + \frac{1}{V_b}\sum_{J=1}^{N_p}(\omega_{IJ}\rho_J q_J) = \frac{1}{\alpha_c}\frac{\partial}{\partial t}\left[\phi\sum_{J=1}^{N_p} S_J\,\omega_{IJ}\,\rho_J\right]$$

$$I = 1, 2, \ldots, N_c \tag{5.30}$$

Equation (5.30) is the generalized, overall compositional equation for the analysis of fluid flow in porous media, written for each of the species present in the fluid characterization of interest ($I = 1, 2, \ldots, N_c$). This is the most general material balance formulation used in reservoir engineering modeling, from where a myriad of subsets of flow scenarios may be derived, as will be shown in the following sections. In Eq. (5.30), since practical field units are used in this chapter, densities (ρ_J) are expressed in lb/ft^3, $\underline{\underline{k}}$ in perms, viscosities (μ_J) in cp, flow gradients ($\nabla\Phi_J$) in psi/ft, flow rates (q_J) in res bbl/day (RB/day), and bulk volume (V_b) in ft^3, with all other dependent variables being dimensionless. For this choice of units, the volume conversion factor α_c (5.615 ft^3/bbl) should be placed at the right-hand side on this equation in order to convert accumulation-term volumes from ft^3 to bbl. Example 5.2 summarizes the number of available equations and target unknowns and examines if the compositional problem is solvable as proposed.

Example 5.2: Inventory of Equations and Unknowns in Compositional Simulation

For compositional flow scenarios, flows of N_p-phases and N_c- components are studied. For such problem to be solvable as proposed, the number of target unknowns must match the number of available equations. In other words, one must verify if there are *enough* independent equations available to solve for all the unknowns being sought after by the model. In compositional simulation, the goal is being able to solve for all phase pressures (all N_p-phases), phase saturations (N_p-phases), and phase compositions

(N_p*N_c compositions) at all points of the domain and at every time step. Thus, target unknowns are:

Target Unknowns:

- Phase pressures, p_J ($J = 1, 2, \ldots, N_p$)
- Phase saturations, S_J ($J = 1, 2, \ldots, N_p$)
- Phase compositions, ω_{IJ}(N_p*N_c compositions)

<u>**TOTAL UNKNOWNS**</u> $= 2\,N_p + N_p{*}N_c$

The list of available equations is given below.

Available Equations:

- Mass balance equations [$I = 1, 2, \ldots, N_c$; i.e., Eq. (5.30) written for each fluid component]
- Phase saturation constraint (one equation, all phase saturation must add up to one):

$$\sum_{J=1}^{N_p} S_J = 1 \tag{5.31}$$

- Phase composition constraints ($J = 1, 2, \ldots, N_p$; written for each phase in the system):

$$\sum_{I=1}^{N_c} \omega_{IJ} = 1 \tag{5.32}$$

- Capillary pressure constraints ($J = 1,2, \ldots, N_p - 1$; written for each fluid interface in the system):

$$p_c(S_j) = p_{non-wet} - p_{wet} \tag{5.33}$$

- Equilibrium thermodynamic constraints, in terms of equilibrium ratios or fugacity equilibrium (one set of N_c values for each component; written for each fluid interface in the system, $J = 1,2, \ldots, N_p - 1$) (see Chap. 4):

$$K_{IJ} = \frac{\omega_{IJ}}{\omega_{I,Np}} \tag{5.34}$$

<u>**TOTAL EQUATIONS**</u> $= (N_c) + (1) + (N_p) + (N_p - 1) + (N_p - 1){*}N_c$

$= 2\,N_p + N_p{*}N_c$

Given that the number of unknown exactly match the number of available equations, the compositional flow problem is solvable as proposed.

5.2 Incompressible Fluid (e.g., *water*) Flow Problem: Single-Phase ($N_p = 1$), Single-Component ($N_c = 1$)

We now consider one of the most straightforward flow cases one encounters in reservoir modeling, that of the flow of a single-phase incompressible fluid. This is the typical assumption made for the modeling of water aquifers, for example. For the case of single-phase ($N_p = 1$), single-component ($N_c = 1$) flows, we readily recognize that:

$$\omega_{IJ} = S_J = k_{rJ} = 1 \tag{5.35}$$

By letting J = "l" denote a generic liquid phase, and substituting the insight in Eq. (5.35), the generalized Eq. (5.30) can be simplified to:

$$\nabla \cdot \left[\sum_{J=1}^{1} \left((1)\rho_l \frac{\bar{\bar{k}}(1)}{\mu_l} \nabla \Phi \right) \right] + \frac{1}{V_b} \sum_{J=1}^{1} ((1)\rho_l q_l) = \frac{1}{\alpha_c} \frac{\partial}{\partial t} \left[\phi \sum_{J=1}^{1} (1)(1)\rho_l \right] \tag{5.36}$$

Since ρ_l = *constant* for an incompressible fluid, the right-hand side or accumulation term of this partial differential equation would completely vanish once it is further assumed that system porosity remains constant (Fig. 5.2):

$$\nabla \cdot \left[\frac{\bar{\bar{k}}}{B_l \mu_l} \nabla \Phi \right] + \frac{q_{lsc}}{V_b} = 0 \tag{5.37}$$

Equation (5.37) is the governing equation for incompressible fluid flow in an incompressible porous media. The absence of an accumulation term in its right-hand side indicates that no fluid accumulation is possible in a system because both fluid and rock formation are incompressible. In other words, this system may only operate under steady-state conditions: anything flowing into the system would immediately flow out since it cannot accumulate. Note that in Eq. (5.37), the definition of liquid formation

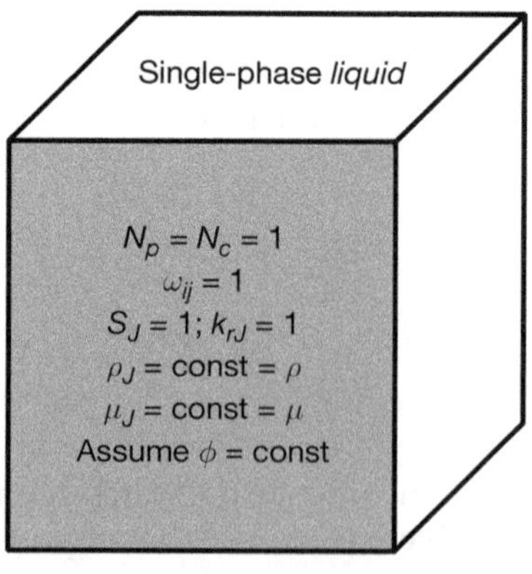

FIGURE 5.2 $N_p = 1$, $N_c = 1$, single-phase incompressible fluid case.

volume factor (B_l) was used in both remaining terms. Following Chap. 4, for the case $\omega_{lJ} = 1$, B_l is always written as:

$$B_l = \frac{\rho_{lsc}}{\rho_l} = \frac{q_l}{q_{lsc}} \tag{5.38}$$

Further simplifications are possible. B_l is obviously constant or pressure-independent for an incompressible fluid, and μ_l is *constant* as well. Therefore,

$$\frac{1}{\mu_l B_l}\nabla \cdot \left(\overline{\overline{k}}\nabla\Phi\right) + \frac{q_{lsc}}{V_b} = 0 \tag{5.39}$$

or

$$\nabla \cdot \left(\overline{\overline{k}}\nabla\Phi\right) + \frac{\mu_l q_l}{V_b} = 0 \tag{5.40}$$

Further, in the study of homogeneous and isotropic media, the permeability tensor $\overline{\overline{k}}$ collapses to a scalar value ($\overline{\overline{k}} = k$, same magnitude in all directions). For such case, the incompressible flow equation simplifies to:

$$\nabla^2\Phi + \frac{\mu_l q_l}{V_b k} = 0 \tag{5.41}$$

If no wells, sources or sinks, are found within the domain of interest:

$$\nabla^2\Phi = 0 \tag{5.42}$$

Equation (5.42) is known as the *Laplace equation*, a second-order partial differential equation which emerges in many fields of science. Laplace-type equations describe behavior of fluid, gravitational, and electric potentials in media with uniform properties. The operator ∇^2 is thus known as the *Laplacian operator* which results from applying the divergence operator to a gradient vector. The Laplacian operator for different coordinate systems is shown in Table 5.3.

Following Table 5.3 and Eq. (5.42), the Laplace equation for incompressible fluid flow in a homogeneous and isotropic medium in 3D Cartesian coordinates is:

$$\frac{\partial^2\Phi}{\partial x^2} + \frac{\partial^2\Phi}{\partial y^2} + \frac{\partial^2\Phi}{\partial z^2} = 0 \tag{5.43}$$

For one-dimensional (1D) flow in a horizontal system with no depth gradients ($\nabla\Phi = \nabla p$), this governing equation becomes:

$$\frac{\partial^2 p}{\partial x^2} = 0 \tag{5.44}$$

which represents one of the simplest governing equations encountered in reservoir modeling.

Coordinate System	Laplacian operator, $\nabla^2 = \nabla \cdot \nabla$
Cartesian (x, y, z)	$\nabla^2 p = \nabla \cdot (\nabla p) = \nabla \cdot \left(\frac{\partial p}{\partial x}\hat{\boldsymbol{i}} + \frac{\partial p}{\partial y}\hat{\boldsymbol{j}} + \frac{\partial p}{\partial z}\hat{\boldsymbol{k}}\right)$ $\nabla^2 p = \frac{\partial^2 p}{\partial x^2} + \frac{\partial^2 p}{\partial y^2} + \frac{\partial^2 p}{\partial z^2}$
Cylindrical (r, θ, z)	$\nabla^2 p = \nabla \cdot (\nabla p) = \nabla \cdot \left(\frac{\partial p}{\partial r}\hat{\mathbf{e}}_r + \frac{1}{r}\frac{\partial p}{\partial \theta}\hat{\mathbf{e}}_\theta + \frac{\partial p}{\partial z}\hat{\mathbf{e}}_z\right)$ $\nabla^2 p = \frac{1}{r}\frac{\partial}{\partial r}\left(r\frac{\partial p}{\partial r}\right) + \frac{1}{r^2}\frac{\partial^2 p}{\partial^2 \theta} + \frac{\partial^2 p}{\partial^2 z}$
Spherical (r, θ, ϕ)	$\nabla^2 p = \nabla \cdot (\nabla p)$ $= \nabla \cdot \left(\frac{\partial p}{\partial r}\hat{\mathbf{e}}_r + \frac{1}{r}\frac{\partial p}{\partial \theta}\hat{\mathbf{e}}_\theta + \frac{1}{r\sin\theta}\frac{\partial p}{\partial \phi}\hat{\mathbf{e}}_\phi\right)$ $\nabla^2 p = \frac{1}{r^2}\frac{\partial}{\partial r}\left(r^2\frac{\partial p}{\partial r}\right) + \frac{1}{r^2 \sin\theta}\frac{\partial}{\partial \theta}\left(\sin\theta\frac{\partial p}{\partial \theta}\right) + \frac{1}{r^2\sin^2\theta}\frac{\partial^2 p}{\partial^2 \phi}$

TABLE 5.3: Laplacian Operator Applied to a Vector Field (e.g., pressure gradient vector)

5.3 Slightly-Compressible Fluid (e.g., oil flow): Single-Phase ($N_p = 1$), Single-Component ($N_c = 1$)

Figure 5.3 shows that for single-phase ($N_p = 1$, J = "o"), single-component ($N_c = 1$, I = "O") flows of slightly compressible liquids, one can readily write:

$$\omega_{Oo} = S_o = k_{ro} = 1, \; N_p = 1 \tag{5.45}$$

Thus, the generalized Eq. (5.30) would simplify to, for this case (I = "O"):

$$\nabla \cdot \left[\sum_{J=1}^{1}\left((1)\rho_o \frac{\bar{\bar{k}}(1)}{\mu_o}\nabla\Phi\right)\right] + \frac{1}{V_b}\sum_{J=1}^{1}((1)\rho_o q_o) = \frac{1}{\alpha_c}\frac{\partial}{\partial t}\left[\phi\sum_{J=1}^{1}(1)(1)\rho_o\right] \tag{5.46}$$

Recalling the definition of oil formation volume factor (Chap. 4) for the case $\omega_{Oo} = 1$,

$$B_o = \frac{\rho_{osc}}{\omega_{Oo}\,\rho_o} = \frac{\rho_{osc}}{\rho_o} \tag{5.47}$$

Equation (5.46) can be rewritten in terms of this B_o-definition by dividing it through by ρ_{osc}, which yields:

$$\nabla \cdot \left[\frac{\bar{\bar{k}}}{\mu_o B_o}\nabla\Phi\right] + \frac{q_o}{V_b B_o} = \frac{1}{\alpha_c}\frac{\partial}{\partial t}(\phi/B_o) \tag{5.48}$$

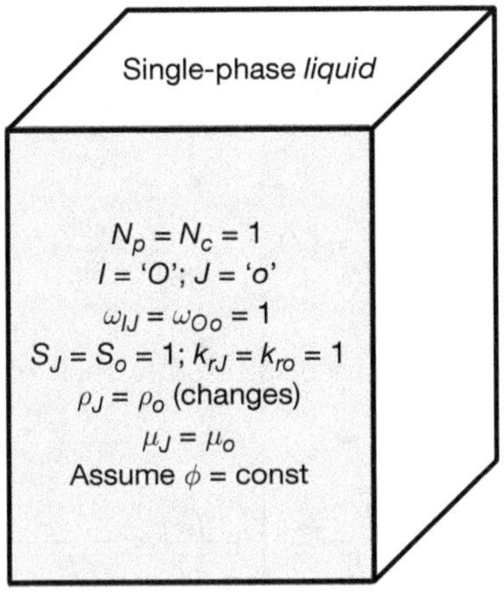

FIGURE 5.3: $N_p = 1$, $N_c = 1$, single-phase slightly compressible fluid case.

It is noted that in Eq. (5.48), oil well flow at reservoir conditions (q_o) can be re-expressed in terms of surface flow as well (see Chap. 4):

$$B_o = \frac{V_o}{(V_{osc})_o} = \frac{q_o}{q_{osc}} \tag{5.49}$$

Thus,

$$\nabla \cdot \left[\frac{\bar{\bar{k}}}{\mu_o B_o} \nabla \Phi \right] + \frac{q_{osc}}{V_b} = \frac{1}{\alpha_c} \frac{\partial}{\partial t} (\phi / B_o) \tag{5.50}$$

The right-hand side in Eq. (5.50) is often re-expressed more directly in terms of pressure. Starting with the definition of isothermal oil compressibility applied to a slightly compressible fluid (Chap. 3):

$$c_o = \frac{1}{\rho_o} \left(\frac{\partial \rho_o}{\partial p} \right)_T \tag{5.51}$$

densities and formation volume factors may be related for single-phase slightly-compressible fluids. For small and constant values of compressibility ($c_o \sim 10^{-6} - 10^{-7}$ psi^{-1}) typical of these slightly compressible fluids, the dependency of density with pressure implied by Eq. (5.51) becomes (Ertekin et al., 2001):

$$\rho_o = \rho_o^{ref} \left[1 + c_o (p - p_{ref}) \right] \tag{5.52}$$

Dividing Eq. (5.52) through ρ_{osc} and using the B_o- definition in Eq. (5.49), one writes:

$$1/B_o = 1/B_o^{ref} \cdot \left[1 + c_o (p - p_{ref}) \right] \tag{5.53}$$

where ρ_o^{ref} and B_o^{ref} are the density and formation volume factor values at the reference conditions ($p = p_{ref}$) of Eqs. (5.52) and (5.53). Equation (5.53) shows that for a slightly compressible fluid,

$$\frac{\partial}{\partial t}(1/B_o) = \frac{c_o}{B_o^{ref}} \cdot \frac{\partial p}{\partial t} \tag{5.54}$$

Therefore, for the case of an incompressible rock (constant ϕ), Eq. (5.50) may be recast as follows:

$$\nabla \cdot \left(\frac{\bar{\bar{k}}}{\mu_o B_o} \nabla \Phi \right) + \frac{q_{osc}}{V_b} = \frac{\phi c_o}{\alpha_c B_o^{ref}} \frac{\partial p}{\partial t} \tag{5.55}$$

Equation (5.55) is the governing equation for slightly compressible fluid flow in an incompressible porous media.

For many practical applications, Eq. (5.48) is often alternatively rewritten in terms of fluid density in both its left- and right-hand sides. This is regularly done for well-test analysis applications. For a single-phase slightly compressible fluid, Eq. (5.48) is written as:

$$\nabla \cdot \left[\rho_o \frac{\bar{\bar{k}}}{\mu_o} \nabla \Phi \right] = \frac{1}{\alpha_c} \frac{\partial}{\partial t}(\phi \rho_o) \tag{5.56}$$

Further, if (a) medium is considered isotropic and homogeneous, (b) porosity changes with pressure, (c) depth gradients are neglected, (d) no wells are considered within the domain, and (e) changes in fluid viscosity with pressure are not considered significant, Eq. (5.56) becomes:

$$\alpha_c \frac{k}{\mu_o} \nabla \cdot [\rho_o \nabla p] = \phi \frac{\partial \rho_o}{\partial t} + \rho_o \frac{\partial \phi}{\partial t} \tag{5.57}$$

Using the definitions of fluid compressibility in Eq. (5.51) and that of pore compressibility (Chap. 3):

$$c_\phi = \frac{1}{\phi} \left(\frac{\partial \phi}{\partial p} \right)_T \tag{5.58}$$

allows writing the gradient equivalencies $d\phi = c_\phi \phi dp$ and $d\rho_o = c_o \rho_o dp$, and Eq. (5.57) becomes:

$$\alpha_c \frac{k}{\mu_o} \left[\rho_o \nabla^2 p + c_o \rho_o (\nabla p)^2 \right] = \phi \rho_o (c_o + c_\phi) \frac{\partial p}{\partial t} \tag{5.59}$$

The second term in the left-hand side of Eq. (5.44) is generally neglected based on the premise that $\nabla^2 p \gg c_o(\nabla p)^2$. This is typically the case when c_o values are very small ($\sim 10^{-6} - 10^{-7}$ psi^{-1}) and when pressure gradients are not large. In such a case:

$$\nabla^2 p = \frac{\phi \mu_o c_t}{\alpha_c k} \frac{\partial p}{\partial t} \tag{5.60}$$

where c_t is the total compressibility ($= c_o + c_\phi$) and $\eta = \alpha_c \dfrac{k}{\phi \mu_o c_t}$ is the hydraulic diffusivity—one of the most important ratios in reservoir analysis. The hydraulic diffusivity characteristic of a reservoir measures the speed of pressure propagation or how fast pressure transients can move through or penetrate the reservoir domain. Dimensional analysis shows it has units of L^2/t, a measure of how much area in a pressure transient is able to cover per unit time. When hydraulic diffusivities (η) are large, the right-hand side of Eq. (5.60) is small and pressure changes propagate fast through the reservoir. When hydraulic diffusivities are small, pressure changes take time to propagate throughout the domain. A well-known subset of Eq. (5.60) is the one used in well-test analysis for the study of pressure propagation in a 1D radial cylindrical domain, which can be obtained from Eq. (5.60) and the application of the corresponding 1D radial cylindrical divergence expression in Table 5.1:

$$\frac{1}{r}\frac{\partial}{\partial r}\left(r\frac{\partial p}{\partial r}\right) = \frac{1}{\eta}\frac{\partial p}{\partial t} \tag{5.61}$$

which is the classical governing equation used in well-test analysis for radial single-phase flow of oil toward a vertical well. This equation will be invoked often in the analytical developments covered in Chap. 6.

5.4 Compressible Fluid (e.g., dry *gas*) Flow Problem: Single-Phase ($N_p = 1$), Single-Component ($N_c = 1$)

For the *compressible*, single-phase ($N_p = 1, J =$ "g"), single-component ($N_c = 1, I =$ "G") case, Fig. 5.4 shows that:

$$\omega_{Gg} = S_g = k_{rg} = 1 \tag{5.62}$$

and thus the generalized Eq. (5.30) becomes, for $I =$ "G":

$$\nabla \cdot \left[\sum_{J=1}^{1}\left((1)\rho_g \frac{\overline{\overline{k}}(1)}{\mu_g}\nabla\Phi\right)\right] + \frac{1}{V_b}\sum_{J=1}^{1}((1)\rho_g q_g) = \frac{1}{\alpha_c}\frac{\partial}{\partial t}\left[\phi\sum_{J=1}^{1}(1)(1)\rho_g\right] \tag{5.63}$$

where gravity effects have been neglected ($\nabla\Phi = \nabla p$) given the relative low density and hydrostatic gradients associated with gases compared to liquids. The definition of gas formation volume factor is (Chap. 4):

$$B_g\left[\frac{\text{RB}}{\text{scf}}\right] = \frac{\rho_{gsc}}{\alpha_c\,\omega_{Gg}\,\rho_g} = \frac{\rho_{gsc}}{\alpha_c\,\rho_g} = \frac{q_g}{q_{gsc}} \tag{5.64}$$

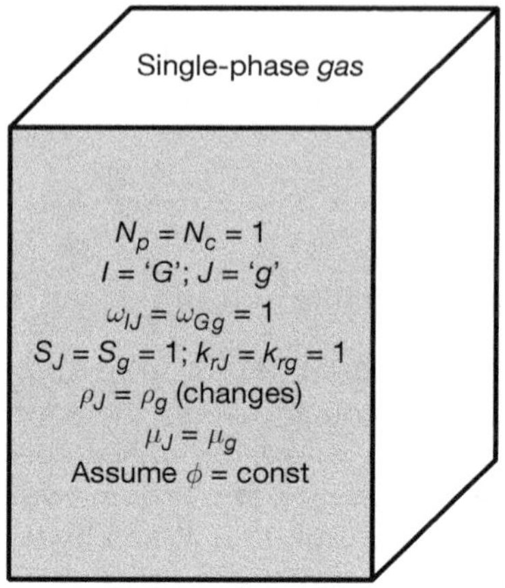

FIGURE 5.4: $N_p = 1$, $N_c = 1$, single-phase compressible (gas) fluid case.

which applies for this dry gas case for which $\omega_{Gg} = 1$ and where α_c is the volume conversion factor (5.615 bbl/ft³ for practical field units).

By multiplying Eq. (5.63) through by α_c/ρ_{gsc} and substituting Eq. (5.64), the B_g-formulation of the compressible (dry-gas) flow governing equation is obtained:

$$\nabla \cdot \left[\frac{\overline{\overline{k}}}{\mu_g B_g} \nabla p \right] + \frac{q_{gsc}}{V_b} = \frac{1}{\alpha_c} \frac{\partial}{\partial t} (\phi / B_g) \tag{5.65}$$

Equation (5.65) is the governing equation for flow of dry natural gases in an incompressible porous media. This governing equation is strongly nonlinear because of the presence of the strong dependency of μ_g and B_g coefficients with pressure inside the divergence operator.

For dry gases ($\omega_{Gg} = 1$), gas formation volume factors can alternatively be calculated as functions of pressure, temperature, and compressibility factors using the real gas law as follows:

$$B_g = \frac{V_g [RB]}{(V_{Gsc})_g [scf]} = \frac{ZnRT/p}{\alpha_c nRT_{sc}/p_{sc}} = \frac{1}{\alpha_c} \frac{p_{sc}}{T_{sc}} \frac{T}{} \frac{Z}{p} \tag{5.66}$$

Therefore, the right-hand side of Eq. (5.65) can be rewritten in terms of pressure as follows:

$$\nabla \cdot \left(\frac{\overline{\overline{k}}}{\mu_g B_g} \nabla p \right) + \frac{q_{gsc}}{V_b} = \phi \frac{T_{sc}}{p_{sc} T} \frac{\partial}{\partial t} (p/Z) \tag{5.67}$$

when the porous medium is considered incompressible (constant ϕ).

For analytical applications, Eq. (5.63) can be alternatively rewritten in terms of *pseudo-pressure* in both its left- and right-hand sides. This is frequently done in well-test applications. *Pseudo-pressure* is a mathematical pressure function that combines changes in gas viscosity, compressibility, and pressure into a single-variable (i.e., *pseudo-pressure*) in order to facilitate analytical treatment of the governing equation for gas flow. One of the most widely deployed pseudo-pressure definitions is that of Al-Hussainy et al. (1966):

$$\Psi(p) = 2\int_0^p \frac{p/Z}{\mu_g}dp \tag{5.68}$$

from where

$$dp = \frac{\mu_g}{2\left(\frac{p}{Z}\right)}d\Psi \tag{5.69}$$

We recall from Eq. (5.63) that for a *compressible,* single-phase ($N_p = 1$), single-component ($N_c = 1$) flow case where $\Delta\Phi = \Delta p$, we write:

$$\nabla \cdot \left[\rho_g \frac{\bar{\bar{k}}}{\mu_g}\nabla p\right] = \frac{1}{\alpha_c}\frac{\partial}{\partial t}(\phi\rho_g) \tag{5.70}$$

Further, if (a) the porous medium is considered isotropic and homogeneous, (b) no wells are considered within the domain, and (c) rock compressibility is considered negligible with respect to gas compressibility, Eq. (5.70) becomes:

$$k\nabla \cdot \left[\frac{\rho_g}{\mu_g}\nabla p\right] = \frac{\phi}{\alpha_c}\frac{\partial \rho_g}{\partial t} \tag{5.71}$$

Using the fluid compressibility definition applied to gases (Chap. 3):

$$c_g = \frac{1}{\rho_g}\left(\frac{\partial \rho_g}{\partial p}\right)_T \tag{5.72}$$

from where $d\rho_g = c_g\rho_g dp$. Substituting Eqs. (5.69) and (5.72) into Eq. (5.71):

$$k\ \nabla \cdot \left[\frac{\rho_g}{\mu_g}\frac{\mu_g}{2(p/Z)}\nabla\Psi\right] = \phi c_g \rho_g \frac{\partial p}{\partial t} = \frac{\phi}{\alpha_c}c_g\rho_g\left(\frac{\mu_g}{2(p/Z)}\frac{\partial \Psi}{\partial t}\right) \tag{5.73}$$

The presence of the $\frac{\rho_g}{(p/Z)}$ can be noted in both the left- and right-hand side. The real gas law prescribes that such ratio, which is equal to:

$$k\frac{\rho_g}{(p/Z)} = \frac{MW}{RT} \tag{5.74}$$

must remain constant and equal to $\frac{MW}{RT}$ (a ratio where all variables are constant for isothermal conditions). Introducing Eq. (5.74) to Eq. (5.73):

$$k\left(\frac{MW}{2RT}\right)\nabla^2\Psi = \frac{\phi}{\alpha_c}c_g\mu_g\left(\frac{MW}{2RT}\right)\frac{\partial\Psi}{\partial t} \tag{5.75}$$

Cancelling the $\left(\frac{MW}{2RT}\right)$ constant coefficient in both sides, the *pseudo-pressure* form of the dry gas governing equation is:

$$\nabla^2\Psi = \frac{\phi\mu_g c_g}{\alpha_c k}\frac{\partial\Psi}{\partial t} \tag{5.76}$$

Equation (5.76) has now been linearized on its left-hand side. However, its right-hand side remains nonlinear given the residual presence of μ_g and c_g, both of which are pressure-dependent properties. For short-term hydrocarbon production with minimal reservoir depletion (such as in early-time well tests of new wells), $\mu_g c_g$ may be approximated to its initial value $\mu_g c_g \cong \mu_{gi} c_{gi}$. However, for longer-term production once extensive reservoir depletion has taken place, the right-hand side must be further linearized using additional mathematical transformation such as pseudo-time (Agarwal, 1979). Ayala and Morgan (2016) present a review of some of these long-term production prediction tools where both the concepts of pseudo-pressure and pseudo-time are needed during natural gas production data analysis. Chapter 6 will discuss how Eq. (5.76) could be rewritten in p^2- or p-forms when further observations about gas behavior are made for certain limiting conditions.

5.5 Immiscible Liquid/Liquid (e.g., *oil/water*) Flow Problem: Two-Phase ($N_p = 2$), Two-Component ($N_c = 2$)

During immiscible multi-phase flow, no mass transfer takes place between flowing phases and thus they do not dissolve in each other. As a result, phase compositions remain invariant with time as depicted in Fig 5.5 for oil and water flow. For oil/water flow, the reservoir oil phase has no water component dissolved in it, thus:

$$\omega_{Wo} = 0 \tag{5.77}$$

$$\omega_{Oo} = 1 \tag{5.78}$$

and the reservoir water phase has no oil component dissolved in it:

$$\omega_{Ow} = 0 \tag{5.79}$$

$$\omega_{Ww} = 1 \tag{5.80}$$

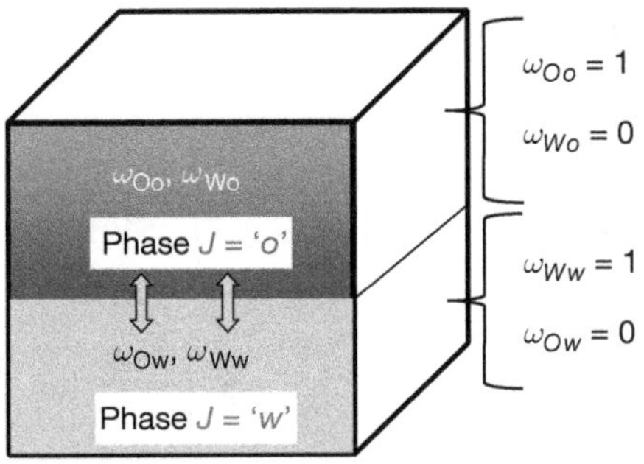

Figure 5.5: $N_p = 2$, $N_c = 2$, immiscible oil/water two-phase flow case.

Since $N_c = 2$, the generalized Eq. (5.30) needs to be written twice: once per component present in the fluid mixture. Therefore, for the oil pseudo-component (I = "O"), Eq. (5.30) becomes:

$$\nabla \cdot \left[\omega_{Oo}\rho_o \frac{\bar{\bar{k}} k_{ro}}{\mu_o} \nabla \Phi_o + \omega_{Ow}\rho_w \frac{\bar{\bar{k}} k_{rw}}{\mu_w} \nabla \Phi_w \right] + \frac{1}{V_b}(\omega_{Oo}\rho_o q_o + \omega_{Ow}\rho_w q_w)$$

$$= \frac{1}{\alpha_c}\frac{\partial}{\partial t}[\phi(S_o\,\omega_{Oo}\,\rho_o + S_w\,\omega_{Ow}\,\rho_w)] \tag{5.81}$$

For the water pseudo-component (I = "W"), Eq. (5.30) is:

$$\nabla \cdot \left[\omega_{Wo}\rho_o \frac{\bar{\bar{k}} k_{ro}}{\mu_o} \nabla \Phi_o + \omega_{Ww}\rho_w \frac{\bar{\bar{k}} k_{rw}}{\mu_w} \nabla \Phi_w \right] + \frac{1}{V_b}(\omega_{Wo}\rho_o q_o + \omega_{Ww}\rho_w q_w)$$

$$= \frac{1}{\alpha_c}\frac{\partial}{\partial t}[\phi(S_o\,\omega_{Wo}\,\rho_o + S_w\,\omega_{Ww}\,\rho_w)] \tag{5.82}$$

Substituting the value of compositions $\omega_{Oo}, \omega_{Ow}, \omega_{Wo}$ and ω_{Ww} of Eqs. (5.77), (5.78), (5.79), and (5.80) into Eqs. (5.81) and (5.82), and introducing the following definitions of oil and water formation volume factors for the case $\omega_{Oo} = \omega_{Ww} = 1$:

$$B_o = \frac{\rho_{osc}}{\omega_{Oo}\rho_o} = \frac{\rho_{osc}}{\rho_o} = \frac{q_o}{q_{osc}} \tag{5.83}$$

$$B_w = \frac{\rho_{wsc}}{\rho_w} = \frac{q_w}{q_{wsc}} \tag{5.84}$$

and further dividing through by ρ_{osc} and ρ_{wsc}, respectively, one obtains:

$$\text{For } I = \text{"O"}: \qquad \nabla \cdot \left(\frac{\bar{\bar{k}} k_{ro}}{\mu_o B_o} \nabla \Phi_o \right) + \frac{q_{osc}}{V_b} = \frac{1}{\alpha_c} \frac{\partial}{\partial t} \left(\phi \frac{S_o}{B_o} \right) \tag{5.85}$$

$$\text{For } I = \text{"W"}: \qquad \nabla \cdot \left(\frac{\bar{\bar{k}} k_{rw}}{\mu_w B_w} \nabla \Phi_w \right) + \frac{q_{wsc}}{V_b} = \frac{1}{\alpha_c} \frac{\partial}{\partial t} \left(\phi \frac{S_w}{B_w} \right) \tag{5.86}$$

Equations (5.68) and (5.69) are the oil pseudo-component and water pseudo-component flow equations, respectively, for immiscible flow conditions. These equations must be solved *simultaneously* by a reservoir simulation model for typical waterflooding simulation problems.

5.6 Standard Black-Oil Formulation for Oil/Gas Flow: Two-Phase ($N_p = 2$), Two-Component ($N_c = 2$)

In standard or classical black-oil fluid models, binary hydrocarbon systems are considered. In these models, binary mixtures of surface-oil and surface-gas components are found coexisting at reservoir conditions. In the standard version of a black-oil fluid model, reservoir phases are allowed to exchange mass but only for *one* of the pseudo-component: the surface-gas component. Surface-gas can transfer between the reservoir oil and gas phases, but the surface-oil cannot transfer to the gas phase (Fig. 5.6).

In the standard black-oil, we thus write for the *reservoir gas* phase:

$$\omega_{Gg} = 1 \ (\textit{surface-gas only}) \tag{5.87}$$

$$\omega_{Og} = 0 \ (\textit{no surface-oil allowed}) \tag{5.88}$$

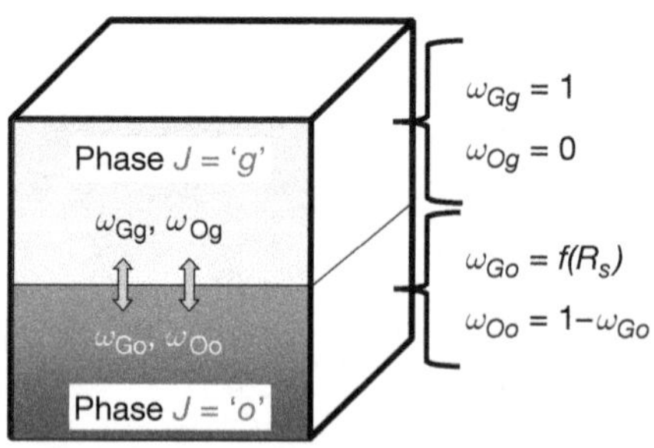

FIGURE 5.6: $N_p = 2$, $N_c = 2$, "standard" black-oil, miscible-oil/gas flow case.

While for *reservoir oil* phase compositions, surface-gas content in the reservoir oil is allowed to vary according to the prevailing R_s values (i.e., solution gas–oil ratio). As derived previously for the black-oil model in Chap. 4:

$$\omega_{Oo} = \frac{\rho_{osc}}{\rho_o\ B_o} \tag{5.89}$$

$$\omega_{Go} = \frac{\rho_{gsc} R_s}{\alpha_c\ \rho_o\ B_o} \tag{5.90}$$

Given the presence of two pseudo-components ($N_c = 2$), one now writes the generalized Eq. (5.30) twice, once per each fluid component. For the surface-oil component ($I =$ "O"), Eq. (5.30) becomes:

$$\nabla \cdot \left[\omega_{Oo}\rho_o \frac{\bar{\bar{k}} k_{ro}}{\mu_o} \nabla\Phi_o + \omega_{Og}\rho_g \frac{\bar{\bar{k}} k_{rg}}{\mu_g} \nabla\Phi_g \right] + \frac{1}{V_b}(\omega_{Oo}\rho_o q_o + \omega_{Og}\rho_g q_g)$$

$$= \frac{1}{\alpha_c}\frac{\partial}{\partial t}\left[\phi(S_o\,\omega_{Oo}\,\rho_o + S_g\,\omega_{Og}\,\rho_g)\right] \tag{5.91}$$

For the surface-gas component ($I =$ "G"), Eq. (5.30) is:

$$\nabla \cdot \left[\omega_{Go}\rho_o \frac{\bar{\bar{k}} k_{ro}}{\mu_o} \nabla\Phi_o + \omega_{Gg}\rho_g \frac{\bar{\bar{k}} k_{rg}}{\mu_g} \nabla\Phi_g \right] + \frac{1}{V_b}(\omega_{Go}\rho_o q_o + \omega_{Gg}\rho_g q_g)$$

$$= \frac{1}{\alpha_c}\frac{\partial}{\partial t}\left[\phi(S_o\,\omega_{Go}\,\rho_o + S_g\,\omega_{Gg}\,\rho_g)\right] \tag{5.92}$$

Substitution of the $\omega_{Oo}, \omega_{Ow}, \omega_{Wo}$ and ω_{Ww} expressions in Eqs. (5.87), (5.88), (5.89), and (5.90) into Eqs. (5.91) and (5.92) leads to:

For $I -$ "O":

$$\nabla \cdot \left[\left(\frac{\rho_{osc}}{\rho_o\ B_o}\right)\rho_o \frac{\bar{\bar{k}} k_{ro}}{\mu_o} \nabla\Phi_o + (0)\rho_g \frac{\bar{\bar{k}} k_{rg}}{\mu_g} \nabla\Phi_g \right] + \frac{1}{V_b}\left(\left(\frac{\rho_{osc}}{\rho_o\ B_o}\right)\rho_o q_o + (0)\rho_g q_g\right)$$

$$= \frac{1}{\alpha_c}\frac{\partial}{\partial t}\left[\phi\left(S_o\left(\frac{\rho_{osc}}{\rho_o B_o}\right)\rho_o + S_g(0)\rho_g\right)\right] \tag{5.93}$$

For $I =$ "G":

$$\nabla \cdot \left[\left(\frac{\rho_{gsc} R_s}{\alpha_c\ \rho_o\ B_o}\right)\rho_o \frac{\bar{\bar{k}} k_{ro}}{\mu_o} \nabla\Phi_o + (1)\rho_g \frac{\bar{\bar{k}} k_{rg}}{\mu_g} \nabla\Phi_g \right] + \frac{1}{V_b}\left(\left(\frac{\rho_{gsc} R_s}{\alpha_c\ \rho_o\ B_o}\right)\rho_o q_o + (1)\rho_g q_g\right)$$

$$= \frac{1}{\alpha_c}\frac{\partial}{\partial t}\left[\phi\left(S_o\left(\frac{\rho_{gsc} R_s}{\alpha_c \rho_o B_o}\right)\frac{\rho_{gsc} R_s}{\alpha_c \rho_o B_o}\rho_o + S_g(1)\rho_g\right)\right] \tag{5.94}$$

Dividing Eqs. (5.93) and (5.94) through by ρ_{osc} and ρ_{gsc}/α_c, respectively, one obtains:

$$\text{For } \boldsymbol{I} = \text{"O"}: \quad \nabla \cdot \left(\frac{\overline{\overline{\boldsymbol{k}}} k_{ro}}{\mu_o B_o} \nabla \Phi_o \right) + \frac{q_o}{B_o V_b} = \frac{1}{\alpha_c} \frac{\partial}{\partial t} \left(\phi \frac{S_o}{B_o} \right) \tag{5.95}$$

$$\text{For } \boldsymbol{I} = \text{"G"}: \quad \nabla \cdot \left(\frac{\overline{\overline{\boldsymbol{k}}} k_{rg}}{\mu_g B_g} \nabla \Phi_g + R_s \frac{\overline{\overline{\boldsymbol{k}}} k_{ro}}{\mu_o B_o} \nabla \Phi_o \right)$$
$$+ \frac{1}{V_b} \left(\frac{q_g}{B_g} + R_s \frac{q_o}{B_o} \right) = \frac{1}{\alpha_c} \frac{\partial}{\partial t} \left[\phi \left(\frac{S_g}{B_g} + R_s \frac{S_o}{B_o} \right) \right] \tag{5.96}$$

where the definition of gas formation volume factor has been used in Eq. (5.96) for the case $\omega_{Gg} = 1$ (Chap. 4):

$$B_g = \frac{\rho_{gsc}}{\alpha_c \omega_{Gg} \rho_g} = \frac{\rho_{gsc}}{\alpha_c \rho_g} \tag{5.97}$$

Since well rates at surface conditions are related to its reservoir counterparts through formation volume factors $[B_o = \frac{q_o}{q_{osc}};\ B_g = \frac{q_g}{q_{gsc}}$, see Chap. 4], Eqs. (5.95) and (5.96) can be alternatively written in terms of surface well rates as follows:

$$\text{For } \boldsymbol{I} = \text{"O"}: \quad \nabla \cdot \left(\frac{\overline{\overline{\boldsymbol{k}}} k_{ro}}{\mu_o B_o} \nabla \Phi_o \right) + \frac{q_{osc}}{V_b} = \frac{1}{\alpha_c} \frac{\partial}{\partial t} \left(\phi \frac{S_o}{B_o} \right) \tag{5.98}$$

$$\text{For } \boldsymbol{I} = \text{"G"}: \quad \nabla \cdot \left(\frac{\overline{\overline{\boldsymbol{k}}} k_{rg}}{\mu_g B_g} \nabla \Phi_g + R_s \frac{\overline{\overline{\boldsymbol{k}}} k_{ro}}{\mu_o B_o} \nabla \Phi_o \right)$$
$$+ \frac{1}{V_b} (q_{gsc} + R_s q_{osc}) = \frac{1}{\alpha_c} \frac{\partial}{\partial t} \left[\phi \left(\frac{S_g}{B_g} + R_s \frac{S_o}{B_o} \right) \right] \tag{5.99}$$

Equations (5.98) and (5.99) are the governing equations for the standard black-oil model for the analysis of two-phase flow of oil and gas at reservoir conditions. It should be emphasized that these governing equations are *compositional* equations, i.e., they have been written for fluid components (namely, surface-oil and surface-gas), and they should not be regarded as equations written for reservoir phases.

5.7 Extended Black-Oil Formulation for Oil/Gas/Water Flow: Three-Phase ($N_p = 3$), Three-Component ($N_c = 3$)

In the "extended" black-oil model, binary hydrocarbon systems (i.e., mixtures of surface-oil and surface-gas) are considered—same as in the classical black-oil model. However, the classical assumption of *no* (volatilized) surface-oil in the gas phase is removed. In the "extended" version, all hydrocarbon reservoir phases are allowed to exchange any of the two pseudo-components among them: *surface-oil* and *surface-gas*.

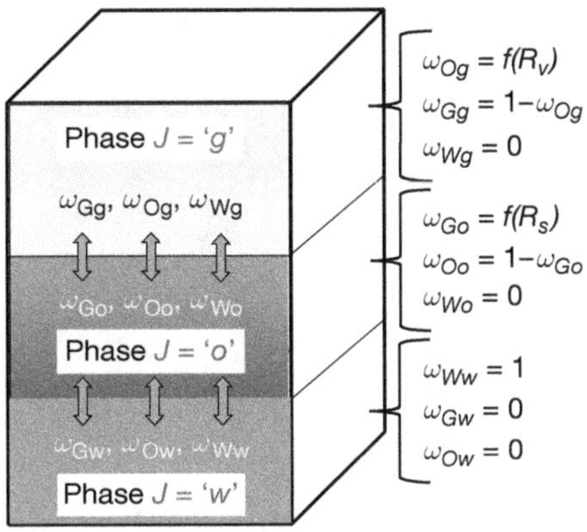

Figure 5.7: $N_p = 3$, $N_c = 3$, "extended" black-oil, miscible-oil/gas flow case with in situ water.

One first recalls how reservoir phase compositions (ω_{IJ}'s) are directly correlated to standard black-oil PVT properties (B_o, B_g, R_s, and R_v), as illustrated in Fig. 5.7. For reservoir oil phase compositions, Chap. 4 demonstrated:

$$\omega_{Oo} = \frac{\rho_{osc}}{\rho_o B_o} \tag{5.100}$$

$$\omega_{Go} = \frac{\rho_{gsc} R_s}{\alpha_c \rho_o B_o} \tag{5.101}$$

$$\omega_{Wo} = 0 \tag{5.102}$$

where it has been assumed that surface-water does not participate in the reservoir oil phase (see Chap. 4).

For reservoir gas phase compositions, Chap. 4 also demonstrated:

$$\omega_{Gg} = \frac{\rho_{gsc}}{\alpha_c \rho_g B_g} \tag{5.103}$$

$$\omega_{Og} = \frac{\rho_{osc} R_v}{\rho_g B_g} \tag{5.104}$$

$$\omega_{Wg} = 0 \tag{5.105}$$

which again considers that surface-water is never part of the reservoir gas phase. For reservoir water phase compositions, hydrocarbon components are assumed insoluble in water and:

$$\omega_{Ww} = 1 \text{ (only surface-water)} \tag{5.106}$$

$$\omega_{Ow} = 0 \text{ (no surface-oil)} \tag{5.107}$$

$$\omega_{Gw} = 0 \text{ (no surface-gas)} \tag{5.108}$$

For this three-component scenario ($N_c = 3$), the generalized Eq. (5.30) must be written three times, once for I = "O," once for I = "G," and once for I = "W".

For the surface-oil component (I = "O"), Eq. (5.30) becomes for $N_p = 3$:

$$\nabla \cdot \left[\omega_{Oo}\rho_o \frac{\bar{\bar{k}} k_{ro}}{\mu_o} \nabla \Phi_o + \omega_{Og}\rho_g \frac{\bar{\bar{k}} k_{rg}}{\mu_g} \nabla \Phi_g + \omega_{Ow}\rho_w \frac{\bar{\bar{k}} k_{rw}}{\mu_w} \nabla \Phi_w \right] + \frac{1}{V_b}(\omega_{Oo}\rho_o q_o + \omega_{Og}\rho_g q_g + \omega_{Ow}\rho_w q_w) = \frac{1}{\alpha_c}\frac{\partial}{\partial t}\left[\phi(S_o \omega_{Oo} \rho_o + S_g \omega_{Og} \rho_g + S_w \omega_{Ow} \rho_w)\right] \tag{5.109}$$

Substituting the expressions for surface-oil compositions ω_{Oo} [Eq. (5.99)], ω_{Og} [Eq. (5.104)], ω_{Ow} [Eq.(5.107)]:

$$\nabla \cdot \left[\left(\frac{\rho_{osc}}{\rho_o B_o}\right)\rho_o \frac{\bar{\bar{k}} k_{ro}}{\mu_o} \nabla \Phi_o + \left(\frac{\rho_{osc} R_v}{\rho_g B_g}\right)\rho_g \frac{\bar{\bar{k}} k_{rg}}{\mu_g} \nabla \Phi_g + (0)\rho_w \frac{\bar{\bar{k}} k_{rw}}{\mu_w} \nabla \Phi_w \right] + \frac{1}{V_b}\left(\left(\frac{\rho_{osc}}{\rho_o B_o}\right)\rho_o q_o + \left(\frac{\rho_{osc} R_v}{\rho_g B_g}\right)\rho_g q_g + (0)\rho_w q_w\right) = \frac{1}{\alpha_c}\frac{\partial}{\partial t}\left[\phi\left(S_o\left(\frac{\rho_{osc}}{\rho_o B_o}\right)\rho_o + S_g\left(\frac{\rho_{osc} R_v}{\rho_g B_g}\right)\rho_g + S_w (0)\rho_w\right)\right] \tag{5.110}$$

Dividing through by ρ_{osc},

$$\nabla \cdot \left[\frac{\bar{\bar{k}} k_{ro}}{\mu_o B_o} \nabla \Phi_o + R_v \frac{\bar{\bar{k}} k_{rg}}{\mu_g B_g} \nabla \Phi_g \right] + \frac{1}{V_b}(q_{osc} + R_v q_{gsc}) = \frac{1}{\alpha_c}\frac{\partial}{\partial t}\left[\phi\left(\frac{S_o}{B_o} + R_v \frac{S_g}{B_g}\right)\right] \tag{5.111}$$

which represents the oil component flow equation for the extended black-oil model. In Eq. (5.111), the relationships $B_o = \frac{q_o}{q_{osc}}$ and $B_g = \frac{q_g}{q_{gsc}}$ were used as before.

For the surface-gas component (I = "G"), Eq. (5.30) becomes:

$$\nabla \cdot \left[\omega_{Go}\rho_o \frac{\bar{\bar{k}} k_{ro}}{\mu_o} \nabla \Phi_o + \omega_{Gg}\rho_g \frac{\bar{\bar{k}} k_{rg}}{\mu_g} \nabla \Phi_g + \omega_{Gw}\rho_w \frac{\bar{\bar{k}} k_{rw}}{\mu_w} \nabla \Phi_w \right] + \frac{1}{V_b}(\omega_{Go}\rho_o q_o + \omega_{Gg}\rho_g q_g + \omega_{Gw}\rho_w q_w) = \frac{1}{\alpha_c}\frac{\partial}{\partial t}\left[\phi(S_o \omega_{Go} \rho_o + S_g \omega_{Gg} \rho_g + S_w \omega_{Gw} \rho_w)\right] \tag{5.112}$$

Substitution of all surface-gas compositions ω_{Go} [Eq. (5.101)], ω_{Gg} [Eq. (5.103)], and ω_{Gw} [Eq. (5.108)] leads to the same gas-component Eq. (5.99) derived for the classical black-oil model, after dividing through by ρ_{gsc}/α_c:

$$\nabla \cdot \left[\frac{\bar{\bar{k}} k_{rg}}{\mu_g B_g} \nabla \Phi_g + R_s \frac{\bar{\bar{k}} k_{ro}}{\mu_o B_o} \nabla \Phi_o \right] + \frac{1}{V_b}(q_{gsc} + R_s q_{osc}) = \frac{1}{\alpha_c} \frac{\partial}{\partial t} \left[\phi \left(\frac{S_g}{B_g} + R_s \frac{S_o}{B_o} \right) \right] \tag{5.113}$$

which represents the gas component flow equation for the extended black-oil model.

Finally, **for the surface-water component** (I = **"W"**), Eq. (5.30) becomes:

$$\nabla \cdot \left[\omega_{Wo} \rho_o \frac{\bar{\bar{k}} k_{ro}}{\mu_o} \nabla \Phi_o + \omega_{Wg} \rho_g \frac{\bar{\bar{k}} k_{rg}}{\mu_g} \nabla \Phi_g + \omega_{Ww} \rho_w \frac{\bar{\bar{k}} k_{rw}}{\mu_w} \nabla \Phi_w \right] + \frac{1}{V_b}(\omega_{Wo} \rho_o q_o + \omega_{Wg} \rho_g q_g + \omega_{Ww} \rho_w q_w) = \frac{1}{\alpha_c} \frac{\partial}{\partial t} \left[\phi (S_o \omega_{Wo} \rho_j + S_g \omega_{Wg} \rho_j + S_w \omega_{Ww} \rho_w) \right] \tag{5.114}$$

After substituting the corresponding expressions for surface-water compositions: ω_{Wo} [Eq. (5.102)], ω_{Wg} [Eq. (5.105)], ω_{Ww} [Eq. (5.106)], one obtains:

$$\nabla \cdot \left[(0)\rho_o \frac{\bar{\bar{k}} k_{ro}}{\mu_o} \nabla \Phi_o + (0)\rho_g \frac{\bar{\bar{k}} k_{rg}}{\mu_g} \nabla \Phi_g + (1)\rho_w \frac{\bar{\bar{k}} k_{rw}}{\mu_w} \nabla \Phi_w \right] + \frac{1}{V_b}((0)\rho_o q_o + (0)\rho_g q_g + (1)\rho_w q_w) = \frac{1}{\alpha_c} \frac{\partial}{\partial t} \left[\phi (S_o (0) \rho_j + S_g (0) \rho_j + S_w (1) \rho_w) \right] \tag{5.115}$$

Dividing through by ρ_{wsc}, and after using the B_w-definition [Eq. (5.84)]:

$$\nabla \cdot \left[\frac{\bar{\bar{k}} k_{rw}}{\mu_w B_w} \nabla \Phi_w \right] + \frac{q_{wsc}}{V_b} = \frac{1}{\alpha_c} \frac{\partial}{\partial t} \left[\phi \left(\frac{S_w}{B_w} \right) \right] \tag{5.116}$$

Equation (5.116) represents the water component flow equation for the extended black-oil model.

In summary, the three-phase, three-component extended black-oil model consists of the following set of three surface-component equations that must be solved simultaneously for oil, gas, and water components, respectively:

$$\nabla \cdot \left[\frac{\bar{\bar{k}} k_{ro}}{\mu_o B_o} \nabla \Phi_o + R_v \frac{\bar{\bar{k}} k_{rg}}{\mu_g B_g} \nabla \Phi_g \right] + \frac{1}{V_b}(q_{osc} + R_v q_{gsc}) = \frac{1}{\alpha_c} \frac{\partial}{\partial t} \left[\phi \left(\frac{S_o}{B_o} + R_v \frac{S_g}{B_g} \right) \right] \tag{5.117}$$

$$\nabla \cdot \left[\frac{\bar{\bar{k}} k_{rg}}{\mu_g B_g} \nabla \Phi_g + R_s \frac{\bar{\bar{k}} k_{ro}}{\mu_o B_o} \nabla \Phi_o \right] + \frac{1}{V_b}(q_{gsc} + R_s q_{osc}) = \frac{1}{\alpha_c} \frac{\partial}{\partial t} \left[\phi \left(\frac{S_g}{B_g} + R_s \frac{S_o}{B_o} \right) \right] \tag{5.118}$$

$$\nabla \cdot \left[\frac{\bar{\bar{k}} k_{rw}}{\mu_w B_w} \nabla \Phi_w \right] + \frac{q_{wsc}}{V_b} = \frac{1}{\alpha_c} \frac{\partial}{\partial t} \left[\phi \left(\frac{S_w}{B_w} \right) \right] \tag{5.119}$$

Example 5.3: Inventory of Equations and Unknowns in Extended Black-Oil Simulation

In the most comprehensive (extended) black-oil scenario, flows of three phases ($N_p = 3$, $J =$ "o," "g," and "w") and three (pseudo-)components ($N_c = 3$, $I =$ "O," "G," and "W") are studied. As described in Example 5.2 for the case of compositional simulation, the number of target unknowns must match the number of available equations if the problem is to be solvable as proposed. Again, one verifies if there are *enough* independent equations available in order to be able to solve for all the unknowns being sought after by the model.

In extended black-oil simulation, the goal is to be able to solve for all phase pressures (p_o, p_g, and p_w), and phase saturations (S_o, S_g, and S_w), at every time step and at all points of the domain. Note that no calculations of phase compositions are made, given that they are no longer independent variables. Phase compositions become uniquely pressure dependent given the thermodynamics of black-oil fluids as discussed in Chap. 4.

In summary, target unknowns and equations are:

Target Unknowns:

- Pressures, p_o, p_g, and p_w
- Saturations, S_o, S_g, and S_w

<u>TOTAL UNKNOWNS</u> = 6

Available Equations:

- Three mass balance equations: *Surface Oil, Gas, Water*
- One phase saturation constraint, $S_o + S_g + S_w = 1$
- One gas/oil capillary pressure constraint, $p_{cgo}(S_g) = p_g - p_o$
- One water/oil capillary pressure constraint, $p_{cow}(S_w) = p_o - p_w$

<u>TOTAL EQUATIONS</u> = 6

Given that the number of unknowns exactly matches the number of available equations, the extended black-oil flow problem is solvable as posed.

Example 5.4: *Write the three-phase "extended" black-oil governing equations in:*
(a) 3D Cartesian coordinates
(b) 1D radial cylindrical coordinates

3D Cartesian coordinates

Let us start with the oil-surface equation in its compact form, Eq. (5.117):

$$\nabla \cdot \left[\frac{\bar{\bar{k}} k_{ro}}{\mu_o B_o} \nabla \Phi_o + R_v \frac{\bar{\bar{k}} k_{rg}}{\mu_g B_g} \nabla \Phi_g \right] + \frac{1}{V_b}(q_{osc} + R_v q_{gsc}) = \frac{1}{\alpha_c} \frac{\partial}{\partial t} \left[\phi \left(\frac{S_o}{B_o} + R_v \frac{S_g}{B_g} \right) \right] \tag{5.120}$$

By inspection, the mass flux vector of interest to which divergence is operating on is:

$$\vec{u} = \frac{\bar{\bar{k}} k_{ro}}{\mu_o B_o} \nabla \Phi_o + R_v \frac{\bar{\bar{k}} k_{rg}}{\mu_g B_g} \nabla \Phi_g \tag{5.121}$$

where, following Table 5.2 for the gradient operator, $\nabla \Phi_o$ and $\nabla \Phi_g$ are written as:

$$\nabla \Phi_o = \frac{\partial \Phi_o}{\partial x} \hat{i} + \frac{\partial \Phi_o}{\partial y} \hat{j} + \frac{\partial \Phi_o}{\partial z} \hat{k} \tag{5.122}$$

$$\nabla \Phi_g = \frac{\partial \Phi_g}{\partial x} \hat{i} + \frac{\partial \Phi_g}{\partial y} \hat{j} + \frac{\partial \Phi_g}{\partial z} \hat{k} \tag{5.123}$$

Therefore, the vectorial equation [Eq. (5.121)] can recast in matrix form as follows, where each of its components is specified:

$$\begin{pmatrix} u_x \\ u_y \\ u_z \end{pmatrix} = \frac{k_{ro}}{\mu_o B_o} \begin{pmatrix} k_x & 0 & 0 \\ 0 & k_y & 0 \\ 0 & 0 & k_z \end{pmatrix} \begin{pmatrix} \dfrac{\partial \Phi_o}{\partial x} \\ \dfrac{\partial \Phi_o}{\partial y} \\ \dfrac{\partial \Phi_o}{\partial z} \end{pmatrix} + R_v \frac{k_{rg}}{\mu_g B_g} \begin{pmatrix} k_x & 0 & 0 \\ 0 & k_y & 0 \\ 0 & 0 & k_z \end{pmatrix} \begin{pmatrix} \dfrac{\partial \Phi_g}{\partial x} \\ \dfrac{\partial \Phi_g}{\partial y} \\ \dfrac{\partial \Phi_g}{\partial z} \end{pmatrix} \tag{5.124}$$

where, as customary, it has been assumed that flow coordinates have been aligned with the principal flow directions in order to write the permeability tensor in terms of its three principal (diagonal) components (k_x, k_y, k_z) only. A similar matrix form for the diagonalized permeability matrix was presented in Chap. 3 for the case of single-phase flow.

By performing the stated matrix multiplication in Eqs. (5.121) and (5.124) for each velocity component, the velocity vector is recognized to be:

$$\vec{u} = \left(\frac{k_x k_{ro}}{\mu_o B_o} \frac{\partial \Phi_o}{\partial x} + R_v \frac{k_x k_{rg}}{\mu_g B_g} \frac{\partial \Phi_g}{\partial x} \right) \hat{i} + \left(\frac{k_y k_{ro}}{\mu_o B_o} \frac{\partial \Phi_o}{\partial y} + R_v \frac{k_y k_{rg}}{\mu_g B_g} \frac{\partial \Phi_g}{\partial y} \right) \hat{j} + \left(\frac{k_z k_{ro}}{\mu_o B_o} \frac{\partial \Phi_o}{\partial z} + R_v \frac{k_z k_{rg}}{\mu_g B_g} \frac{\partial \Phi_g}{\partial z} \right) \hat{k} \tag{5.125}$$

where mass flux vector components are:

$$u_x = \frac{k_x k_{ro}}{\mu_o B_o} \frac{\partial \Phi_o}{\partial x} + R_v \frac{k_x k_{rg}}{\mu_g B_g} \frac{\partial \Phi_g}{\partial x} \tag{5.126}$$

$$u_y = \frac{k_y k_{ro}}{\mu_o B_o} \frac{\partial \Phi_o}{\partial y} + R_v \frac{k_y k_{rg}}{\mu_g B_g} \frac{\partial \Phi_g}{\partial y} \tag{5.127}$$

$$u_z = \frac{k_z k_{ro}}{\mu_o B_o} \frac{\partial \Phi_o}{\partial z} + R_v \frac{k_z k_{rg}}{\mu_g B_g} \frac{\partial \Phi_g}{\partial z} \tag{5.128}$$

by applying the divergence definition for this 3D Cartesian system (Table 5.1):

$$\nabla \cdot \vec{u} = \frac{\partial}{\partial x}(u_x) + \frac{\partial}{\partial y}(u_y) + \frac{\partial}{\partial z}(u_z) \tag{5.129}$$

Then one finally writes:

$$\frac{\partial}{\partial x}\left(\frac{k_x k_{ro}}{\mu_o B_o} \frac{\partial \Phi_o}{\partial x} + R_v \frac{k_x k_{rg}}{\mu_g B_g} \frac{\partial \Phi_g}{\partial x} \right) + \frac{\partial}{\partial y}\left(\frac{k_y k_{ro}}{\mu_o B_o} \frac{\partial \Phi_o}{\partial y} + R_v \frac{k_y k_{rg}}{\mu_g B_g} \frac{\partial \Phi_g}{\partial y} \right) + \frac{\partial}{\partial z}\left(\frac{k_z k_{ro}}{\mu_o B_o} \frac{\partial \Phi_o}{\partial z} + R_v \frac{k_z k_{rg}}{\mu_g B_g} \frac{\partial \Phi_g}{\partial z} \right) + \frac{1}{V_b}(q_{osc} + R_v q_{gsc}) = \frac{1}{\alpha_c} \frac{\partial}{\partial t}\left[\phi \left(\frac{S_o}{B_o} + R_v \frac{S_g}{B_g} \right) \right] \tag{5.130}$$

which is the surface-oil flow equation in three-dimensional (3D) Cartesian coordinates for the extended black-oil formulation. Following the same steps for the surface-gas equation in compact form [Eq. (5.118)], one obtains:

$$\frac{\partial}{\partial x}\left(\frac{k_x k_{rg}}{\mu_g B_g}\frac{\partial \Phi_g}{\partial x} + R_s \frac{k_x k_{ro}}{\mu_o B_o}\frac{\partial \Phi_o}{\partial x}\right)$$

$$+\frac{\partial}{\partial y}\left(\frac{k_y k_{rg}}{\mu_g B_g}\frac{\partial \Phi_g}{\partial y} + R_s \frac{k_y k_{ro}}{\mu_o B_o}\frac{\partial \Phi_o}{\partial y}\right)$$

$$+\frac{\partial}{\partial z}\left(\frac{k_z k_{rg}}{\mu_g B_g}\frac{\partial \Phi_g}{\partial z} + R_s \frac{k_z k_{ro}}{\mu_o B_o}\frac{\partial \Phi_o}{\partial z}\right)$$

$$+\frac{1}{V_b}(q_{gsc} + R_s q_{osc}) = \frac{1}{\alpha_c}\frac{\partial}{\partial t}\left[\phi\left(\frac{S_g}{B_g} + R_s\frac{S_o}{B_o}\right)\right] \tag{5.131}$$

which is the surface-gas flow equation in 3D Cartesian coordinates for the extended black-oil formulation. The analogous expansion of the surface-water equation [Eq. (5.119)] yields for the 3D Cartesian system:

$$\frac{\partial}{\partial x}\left(\frac{k_x k_{rw}}{\mu_w B_w}\frac{\partial \Phi_w}{\partial x}\right) + \frac{\partial}{\partial y}\left(\frac{k_y k_{rw}}{\mu_w B_w}\frac{\partial \Phi_w}{\partial y}\right) + \frac{\partial}{\partial z}\left(\frac{k_z k_{rw}}{\mu_w B_w}\frac{\partial \Phi_w}{\partial z}\right) + \frac{q_{wsc}}{V_b} = \frac{1}{\alpha_c}\frac{\partial}{\partial t}\left[\phi\left(\frac{S_w}{B_w}\right)\right] \tag{5.132}$$

1D radial cylindrical coordinates

Starting with the oil-surface equation in its compact form [Eq. (5.117)] and identifying the mass flux vector of interest:

$$\vec{\boldsymbol{u}} = \frac{\bar{\bar{\boldsymbol{k}}} k_{ro}}{\mu_o B_o}\nabla\Phi_o + R_v \frac{\bar{\bar{\boldsymbol{k}}} k_{rg}}{\mu_g B_g}\nabla\Phi_g \tag{5.133}$$

For 1D-radial flow in *cylindrical* coordinates, $\bar{\bar{\boldsymbol{k}}} = k_r$, $\nabla\Phi_o = \frac{\partial \Phi_o}{\partial r}\hat{\boldsymbol{e}}_r$, $\nabla\Phi_g = \frac{\partial \Phi_g}{\partial r}\hat{\boldsymbol{e}}_r$ (Table 5.2), therefore the only component of the mass flux vector is $\vec{\boldsymbol{u}} = u_r\hat{\boldsymbol{e}}_r$ where

$$u_r = \frac{k_r k_{ro}}{\mu_o B_o}\frac{\partial \Phi_o}{\partial r} + R_v \frac{k_r k_{rg}}{\mu_g B_g}\frac{\partial \Phi_o}{\partial r} \tag{5.134}$$

One now applies the divergence definition for this 1D radial cylindrical system. According to Table 5.1,

$$\nabla \cdot \vec{\boldsymbol{u}} = \frac{1}{r}\frac{\partial}{\partial r}(r u_r) \tag{5.135}$$

Therefore, after expanding Eq. (5.117), one writes:

$$\frac{\partial}{\partial r}\left[r\left(\frac{k_r k_{ro}}{\mu_o B_o}\frac{\partial \Phi_o}{\partial r}+R_v\frac{k_r k_{rg}}{\mu_g B_g}\frac{\partial \Phi_o}{\partial r}\right)\right]+\frac{1}{V_b}(q_{osc}+R_v q_{gsc})=\frac{1}{\alpha_c}\frac{\partial}{\partial t}\left[\phi\left(\frac{S_o}{B_o}+R_v\frac{S_g}{B_g}\right)\right] \tag{5.136}$$

which is the surface-oil flow equation in 1D radial cylindrical coordinates for the extended black-oil formulation. Repeating the same steps for Eqs. (5.118) and (5.119):

$$\frac{\partial}{\partial r}\left[r\left(\frac{k_r k_{rg}}{\mu_g B_g}\frac{\partial \Phi_g}{\partial r}+R_s\frac{k_r k_{ro}}{\mu_o B_o}\frac{\partial \Phi_o}{\partial r}\right)\right]+\frac{1}{V_b}(q_{gsc}+R_s q_{osc})=\frac{1}{\alpha_c}\frac{\partial}{\partial t}\left[\phi\left(\frac{S_g}{B_g}+R_s\frac{S_o}{B_o}\right)\right] \tag{5.137}$$

$$\frac{\partial}{\partial r}\left[r\left(\frac{k_r k_{rw}}{\mu_w B_w}\frac{\partial \Phi_w}{\partial r}\right)\right]+\frac{q_{wsc}}{V_b}=\frac{1}{\alpha_c}\frac{\partial}{\partial t}\left[\phi\left(\frac{S_w}{B_w}\right)\right] \tag{5.138}$$

Example 5.5: ***Rewrite the 3P "extended" black-oil governing equations in 3D Cartesian coordinates for cases of variable A_x, A_y, A_z, in the form used by Ertekin et al. (2001) for the derivation of numerical finite-difference approximations.***

During the derivation of the governing flow equations, we called attention to the presence of A_x, A_y, A_z areas inside of Eq. (5.19), for the so-called bulk-volume form of the equation:

$$-\frac{V_b}{A_x}\frac{\partial}{\partial x}(\omega_{IJ}\rho_J v_{xJ}A_x)-\frac{V_b}{A_y}\frac{\partial}{\partial y}(\omega_{IJ}\rho_J v_{yJ}A_y)-\frac{V_b}{A_z}\frac{\partial}{\partial z}(\omega_{IJ}\rho_J v_{zJ}A_z)$$
$$+\omega_{IJ}\rho_J q_J+\Gamma_{IJ}V_b=V_b\frac{\partial}{\partial t}(\omega_{IJ}\phi\rho_J S_J) \tag{5.139}$$

and showed that this equation may be also rewritten per unit bulk reservoir volume as shown next:

$$-\frac{\partial}{\partial x}(\omega_{IJ}\rho_J v_{xJ})-\frac{\partial}{\partial y}(\omega_{IJ}\rho_J v_{yJ})-\frac{\partial}{\partial z}(\omega_{IJ}\rho_J v_{zJ})+\frac{\omega_{IJ}\rho_J q_J}{V_b}+\Gamma_{IJ}=\frac{\partial}{\partial t}(\omega_{IJ}\phi\rho_J S_J) \tag{5.140}$$

It is noted that in order to transform Eq. (5.140) back into its bulk-volume, variable-flow area form of Eq. (5.139), we multiply it through by bulk-volume (V_b) and leave area terms (A_x, A_y, A_z) explicitly written inside and outside of each of the spatial derivatives, as shown in Eq. (5.139). Such form follows from the 3D Cartesian control-volume derivation of these equations presented earlier.

For the 3D Cartesian *surface-oil equation* shown in Eq. (5.130), this would result in, using practical field units:

$$\frac{V_b}{A_x}\frac{\partial}{\partial x}\left(\frac{k_x A_x k_{ro}}{\mu_o B_o}\frac{\partial \Phi_o}{\partial x}+R_v\frac{k_x A_x k_{rg}}{\mu_g B_g}\frac{\partial \Phi_g}{\partial x}\right)$$
$$+\frac{V_b}{A_y}\frac{\partial}{\partial y}\left(\frac{k_y A_y k_{ro}}{\mu_o B_o}\frac{\partial \Phi_o}{\partial y}+R_v\frac{k_y A_y k_{rg}}{\mu_g B_g}\frac{\partial \Phi_g}{\partial y}\right)$$
$$+\frac{V_b}{A_z}\frac{\partial}{\partial z}\left(\frac{k_z A_z k_{ro}}{\mu_o B_o}\frac{\partial \Phi_o}{\partial z}+R_v\frac{k_z A_z k_{rg}}{\mu_g B_g}\frac{\partial \Phi_g}{\partial z}\right)+q_{osc}$$
$$+R_v q_{gsc}=\frac{V_b}{\alpha_c}\frac{\partial}{\partial t}\left[\phi\left(\frac{S_o}{B_o}+R_v\frac{S_g}{B_g}\right)\right] \tag{5.141}$$

For the *surface-gas equation* in Eq. (5.131), one would write:

$$\frac{V_b}{A_x}\frac{\partial}{\partial x}\left(\frac{k_x A_x k_{rg}}{\mu_g B_g}\frac{\partial \Phi_g}{\partial x}+R_s\frac{k_x A_x k_{ro}}{\mu_o B_o}\frac{\partial \Phi_o}{\partial x}\right)$$
$$+\frac{V_b}{A_y}\frac{\partial}{\partial y}\left(\frac{k_y A_y k_{rg}}{\mu_g B_g}\frac{\partial \Phi_g}{\partial y}+R_s\frac{k_y A_y k_{ro}}{\mu_o B_o}\frac{\partial \Phi_o}{\partial y}\right)$$
$$+\frac{V_b}{A_z}\frac{\partial}{\partial z}\left(\frac{k_z A_z k_{rg}}{\mu_g B_g}\frac{\partial \Phi_g}{\partial z}+R_s\frac{k_z A_z k_{ro}}{\mu_o B_o}\frac{\partial \Phi_o}{\partial z}\right)+q_{gsc}+R_s q_{osc}$$
$$=\frac{V_b}{\alpha_c}\frac{\partial}{\partial t}\left[\phi\left(\frac{S_g}{B_g}+R_s\frac{S_o}{B_o}\right)\right] \tag{5.142}$$

For the *surface-water equation* in Eq. (5.132):

$$\frac{V_b}{A_x}\frac{\partial}{\partial x}\left(\frac{k_x A_x k_{rw}}{\mu_w B_w}\frac{\partial \Phi_w}{\partial x}\right)+\frac{V_b}{A_y}\frac{\partial}{\partial y}\left(\frac{k_y A_y k_{rw}}{\mu_w B_w}\frac{\partial \Phi_w}{\partial y}\right)+\frac{V_b}{A_z}\frac{\partial}{\partial z}\left(\frac{k_z A_z k_{rw}}{\mu_w B_w}\frac{\partial \Phi_w}{\partial z}\right)+q_{wsc}$$
$$=\frac{V_b}{\alpha_c}\frac{\partial}{\partial t}\left[\phi\left(\frac{S_w}{B_w}\right)\right] \tag{5.143}$$

Also, for the original 3D Cartesian control volume presented in Fig. 5.1, and based on Eq. (5.15), one can write for a Cartesian control volume discretization:

$$\frac{V_b}{A_x}=\Delta x \tag{5.144}$$

$$\frac{V_b}{A_y} = \Delta y \tag{5.145}$$

$$\frac{V_b}{A_z} = \Delta z \tag{5.146}$$

Therefore, it is not uncommon to see these governing equations written with Δx, Δy, and Δz coefficients within them in preparation for numerical deployment. Those are the forms of the governing equations used throughout the finite-differences work of Ertekin et al. (2001). If one uses Ertekin et al. (2001)'s preferred form of writing the governing equations, Eqs. (5.141), (5.142), and (5.143) would be rewritten as:

$$\frac{\partial}{\partial x}\left(\frac{k_x A_x k_{ro}}{\mu_o B_o}\frac{\partial \Phi_o}{\partial x} + R_v \frac{k_x A_x k_{rg}}{\mu_g B_g}\frac{\partial \Phi_g}{\partial x}\right)\Delta x + \frac{\partial}{\partial y}\left(\frac{k_y A_y k_{ro}}{\mu_o B_o}\frac{\partial \Phi_o}{\partial y} + R_v \frac{k_y A_y k_{rg}}{\mu_g B_g}\frac{\partial \Phi_g}{\partial y}\right)\Delta y$$
$$+ \frac{\partial}{\partial z}\left(\frac{k_z A_z k_{ro}}{\mu_o B_o}\frac{\partial \Phi_o}{\partial z} + R_v \frac{k_z A_z k_{rg}}{\mu_g B_g}\frac{\partial \Phi_g}{\partial z}\right)\Delta z + q_{osc} + R_v q_{gsc}$$
$$= \frac{V_b}{\alpha_c}\frac{\partial}{\partial t}\left[\phi\left(\frac{S_o}{B_o} + R_v \frac{S_g}{B_g}\right)\right] \tag{5.147}$$

$$\frac{\partial}{\partial x}\left(\frac{k_x A_x k_{rg}}{\mu_g B_g}\frac{\partial \Phi_g}{\partial x} + R_s \frac{k_x A_x k_{ro}}{\mu_o B_o}\frac{\partial \Phi_o}{\partial x}\right)\Delta x + \frac{\partial}{\partial y}\left(\frac{k_y A_y k_{rg}}{\mu_g B_g}\frac{\partial \Phi_g}{\partial y} + R_s \frac{k_y A_y k_{ro}}{\mu_o B_o}\frac{\partial \Phi_o}{\partial y}\right)\Delta y$$
$$+ \frac{\partial}{\partial z}\left(\frac{k_z A_z k_{rg}}{\mu_g B_g}\frac{\partial \Phi_g}{\partial z} + R_s \frac{k_z A_z k_{ro}}{\mu_o B_o}\frac{\partial \Phi_o}{\partial z}\right)\Delta z + q_{gsc} + R_s q_{osc}$$
$$= \frac{V_b}{\alpha_c}\frac{\partial}{\partial t}\left[\phi\left(\frac{S_g}{B_g} + R_s \frac{S_o}{B_o}\right)\right] \tag{5.148}$$

$$\frac{\partial}{\partial x}\left(\frac{k_x A_x k_{rw}}{\mu_w B_w}\frac{\partial \Phi_w}{\partial x}\right)\Delta x + \frac{\partial}{\partial y}\left(\frac{k_y A_y k_{rw}}{\mu_w B_w}\frac{\partial \Phi_w}{\partial y}\right)\Delta y + \frac{\partial}{\partial z}\left(\frac{k_z A_z k_{rw}}{\mu_w B_w}\frac{\partial \Phi_w}{\partial z}\right)\Delta z + q_{wsc}$$
$$= \frac{V_b}{\alpha_c}\frac{\partial}{\partial t}\left[\phi\left(\frac{S_w}{B_w}\right)\right] \tag{5.149}$$

Example 5.6: Rewrite the multiphase, extended black-oil formulation in compact form in terms of total species mobilities (λ_{Isc}) and species accumulation (M_{Isc}). Assume negligible depth gradients. Provide expressions for mass fluxes at reservoir conditions of each pseudo-component in the extended black-oil formulation, as well as GOR and WOR expressions at sandface conditions.

When depth gradients are neglected, the extended black-oil formulation presented in Eqs. (5.117), (5.118), and (5.119) is rewritten as follows:

For the **surface-oil pseudo-component** (I = "O"):

$$\nabla \cdot \left[\frac{\bar{\bar{k}} k_{ro}}{\mu_o B_o} \nabla p_o + R_v \frac{\bar{\bar{k}} k_{rg}}{\mu_g B_g} \nabla p_g \right] = \frac{1}{\alpha_c} \frac{\partial}{\partial t} \left[\frac{\phi S_o}{B_o} + R_v \frac{\phi S_g}{B_g} \right] \tag{5.150}$$

For the **surface-gas pseudo-component** (I = "G"):

$$\nabla \cdot \left[\frac{\bar{\bar{k}} k_{rg}}{\mu_g B_g} \nabla p_g + R_s \frac{\bar{\bar{k}} k_{ro}}{\mu_o B_o} \nabla p_o \right] = \frac{1}{\alpha_c} \frac{\partial}{\partial t} \left[\frac{\phi S_g}{B_g} + R_s \frac{\phi S_o}{B_o} \right] \tag{5.151}$$

For the **surface-water pseudo-component** (I = "W"):

$$\nabla \cdot \left[\frac{\bar{\bar{k}} k_{rw}}{\mu_w B_w} \nabla p_w \right] = \frac{1}{\alpha_c} \frac{\partial}{\partial t} \left[\frac{\phi S_w}{B_w} \right] \tag{5.152}$$

which recast in a more compact form as:

$$\nabla \cdot \left[\bar{\bar{k}} \begin{pmatrix} \lambda_{Osc} \nabla p_o \\ \lambda_{Gsc} \nabla p_g \\ \lambda_{Wsc} \nabla p_w \end{pmatrix} \right] = \frac{1}{\alpha_c} \frac{\partial}{\partial t} \left[\phi \begin{pmatrix} M_{Osc} \\ M_{Gsc} \\ M_{Wsc} \end{pmatrix} \right] \tag{5.153}$$

In here, α_c is the volume conversion factor (5.615 ft^3/bbl for practical field units), λ_{Isc} is the total mobility of the *I-th* component expressed in surface-volume (standard conditions, or sc) units (for I = "O", "G", "W"), and M_{Isc} is the main variable in the accumulation term and represents the total amount of *I-th* component within the reservoir system per unit pore volume and expressed in *surface-volume* units (these *surface-volume* amounts can be readily converted to mass amounts by multiplication with the appropriate *I-th* component surface density). These definitions are:

$$\lambda_{Osc} = \overbrace{\frac{k_{ro}}{\mu_o B_o}}^{\lambda_{Oo}} + R_v \overbrace{\frac{k_{rg}}{\mu_g B_g}}^{\lambda_{Og}} = \lambda_{Oo} + \lambda_{Og} = \lambda_{Oo} + R_v \lambda_{Gg} \tag{5.154}$$

$$\lambda_{Gsc} = \overbrace{\frac{k_{rg}}{\mu_g B_g}}^{\lambda_{Gg}} + R_s \overbrace{\frac{k_{ro}}{\mu_o B_o}}^{\lambda_{Go}} = \lambda_{Gg} + \lambda_{Go} = \lambda_{Gg} + R_s \lambda_{Oo} \tag{5.155}$$

$$\lambda_{Wsc} = \overbrace{\frac{k_{rw}}{\mu_w B_w}}^{\lambda_{Ww}} \tag{5.156}$$

$$M_{Osc} = \overbrace{\frac{S_o}{B_o}}^{M_{Oo}} + R_v \overbrace{\frac{S_g}{B_g}}^{M_{Og}} \tag{5.157}$$

$$M_{Gsc} = \overbrace{\frac{S_g}{B_g}}^{M_{Gg}} + R_s \overbrace{\frac{S_o}{B_o}}^{M_{Go}} \tag{5.158}$$

$$M_{Wsc} = \overbrace{\frac{S_w}{B_w}}^{M_{Ww}} \tag{5.159}$$

where these individual species mobilities are best recognized as:
λ_{Oo} = *Surface-oil mobility in reservoir oil flow*
λ_{Og} = *Surface-oil mobility in reservoir gas flow*
λ_{Gg} = *Surface-gas mobility in reservoir gas flow*
λ_{Go} = *Surface-gas mobility in reservoir oil flow*
λ_{Ww} = *Surface-water mobility in reservoir water flow*

with *surface-volume* amounts of *I-th* component defined as:
M_{Oo} = *Amount of surface-oil in reservoir oil, per unit pore volume*
M_{Og} = *Amount of surface-oil in reservoir gas, per unit pore volume*
M_{Gg} = *Amount of surface-gas in reservoir gas, per unit pore volume*
M_{Go} = *Amount of surface-gas in reservoir oil, per unit pore volume*
M_{Wsc} = *Amount of surface-water in reservoir water, per unit pore volume*

When capillary pressures are neglected ($p_o = p_g = p_w = p$), above equations collapse to:

$$\nabla \circ \left[\bar{\bar{k}} \begin{pmatrix} \lambda_{Osc} \\ \lambda_{Gsc} \\ \lambda_{Wsc} \end{pmatrix} \nabla p \right] = \frac{1}{\alpha_c} \frac{\partial}{\partial t} \left[\phi \begin{pmatrix} M_{Osc} \\ M_{Gsc} \\ M_{Wsc} \end{pmatrix} \right] \tag{5.160}$$

For an isotropic and homogeneous medium with negligible pore compressibility, this can be further simplified to:

$$\nabla \cdot \left[\begin{pmatrix} \lambda_{Osc} \\ \lambda_{Gsc} \\ \lambda_{Wsc} \end{pmatrix} \nabla p \right] = \frac{1}{\alpha_c} \frac{\phi}{k} \frac{\partial}{\partial t} \begin{bmatrix} M_{Osc} \\ M_{Gsc} \\ M_{Wsc} \end{bmatrix} \tag{5.161}$$

which represents the simultaneous system of equations that must be solved for the case of three-phase flow (oil, gas, water) using the extended black-oil model, under the stated assumptions.

The mass flux of each I-component at reservoir conditions is calculated as a function of the mass flux of each phase ($\vec{G}_J = \rho_J \vec{v}_J$; $J = 1, 2, \ldots, N_P$) in the reservoir. That is:

$$\vec{F}_I = \sum_{J=1}^{N_P} \omega_{IJ} \vec{G}_J = \sum_{J=1}^{N_P} \omega_{IJ} \rho_J \vec{v}_J \tag{5.162}$$

For the flow of **surface-oil pseudo-component** (I = "O"), for instance, one writes in terms of surface-volume units after substituting the appropriate definitions for species compositions ω_{Oo}, ω_{Og}:

$$\vec{F}_{Osc} = \frac{\rho_{osc} \vec{q}_{osc}}{A} = \omega_{Oo} \rho_o \vec{v}_o + \omega_{Og} \rho_g \vec{v}_g$$

$$= -\left(\left(\frac{\rho_{osc}}{\rho_o B_o} \right) \rho_o \frac{\bar{\bar{k}} k_{ro}}{\mu_o} \nabla p_o + \left(\frac{\rho_{osc} R_v}{\rho_g B_g} \right) \rho_g \frac{\bar{\bar{k}} k_{rg}}{\mu_g} \nabla p_g \right) \tag{5.163}$$

which for an isotropic and homogenous media can be written as:

$$\vec{q}_{osc} = -kA \left(\lambda_{Oo} \nabla p_o + \lambda_{Og} \nabla p_g \right) \tag{5.164}$$

It should be kept in mind that $q_{Isc} = q_{Jsc}$; $\rho_{Isc} = \rho_{Jsc}$ at surface conditions because $\omega_{IJ} = 1$ for every phase produced at the surface in a black-oil model. When capillary pressures are neglected, the flow of the *surface-oil* pseudo-component at reservoir conditions is given by:

$$\vec{q}_{osc} = -kA \lambda_{Osc} \nabla p \tag{5.165}$$

For the flow of **surface-gas pseudo-component** (I = "G"), after substitution of appropriate definitions for species compositions ω_{Gg}, ω_{Go}:

$$\vec{F}_{Gsc} = \frac{\rho_{gsc}\vec{q}_{gsc}}{A} = \omega_{Gg}\,\rho_g\,\vec{v}_g + \omega_{Go}\,\rho_o\,\vec{v}_o$$
$$= -\left(\left(\frac{\rho_{gsc}}{\alpha_c\,\rho_g\,B_g}\right)\rho_g\,\frac{\bar{\bar{k}}k_{rg}}{\mu_g}\nabla p_g + \left(\frac{\rho_{gsc}R_s}{\alpha_c\,\rho_o\,B_o}\right)\rho_o\,\frac{\bar{\bar{k}}k_{ro}}{\mu_o}\nabla p_o\right) \tag{5.166}$$

which for an isotropic and homogenous media can be written as:

$$\vec{q}_{gsc} = -\frac{kA}{\alpha_c}\left(\lambda_{Gg}\nabla p_g + \lambda_{Go}\nabla p_o\right) \tag{5.167}$$

When capillary pressures are neglected, flow of the *surface-gas* pseudo-component at reservoir conditions is given by:

$$\vec{q}_{gsc} = -\frac{kA}{\alpha_c}\lambda_{Gsc}\,\nabla p \tag{5.168}$$

For the flow of **surface-water pseudo-component** (I = "W"), since $\omega_{Ww} = 1$ and $B_w = \dfrac{\rho_{wsc}}{\rho_w}$:

$$\vec{F}_{Wsc} = \frac{\rho_{wsc}\vec{q}_{wsc}}{A} = \omega_{Ww}\,\rho_w\,\vec{v}_w = -\frac{\rho_{wsc}}{B_w}\frac{\bar{\bar{k}}k_{rw}}{\mu_w}\nabla p_w \tag{5.169}$$

Thus, flow of the *surface-water* pseudo-component at reservoir conditions for an isotropic and homogenous media is given by the expression:

$$\vec{q}_{wsc} = -kA\,\lambda_{Wsc}\,\nabla p_w \tag{5.170}$$

When evaluated at wellbore conditions, these flows and fluxes represent the flow of species in or out of the reservoir. Their ratios will yield producing gas/oil ratios (GOR), producing water/gas ratios (WGR), or producing water/oil ratios (WOR).

For the case of GOR values, one obtains:

$$GOR = \left.\frac{|\vec{q}_{gsc}|}{|\vec{q}_{osc}|}\right|_{sandface} = \left.\frac{\lambda_{Go}|\nabla p_o| + \lambda_{Gg}|\nabla p_g|}{\lambda_{Oo}|\nabla p_o| + \lambda_{Og}|\nabla p_g|}\right|_{sandface}$$
$$= \left.\frac{\lambda_{Gsc} + \lambda_{Gg}\dfrac{|\nabla p_{cgo}|}{|\nabla p_o|}}{\lambda_{Osc} + \lambda_{Og}\dfrac{|\nabla p_{cgo}|}{|\nabla p_o|}}\right|_{sandface} \tag{5.171}$$

where $\nabla p_g = \nabla p_o + \nabla p_{cgo}$. Given that pressure gradients at the sandface are often much more significant than capillary pressure gradients at the sandface location $\left(|\nabla p_o| \gg |\nabla p_{cgo}|\right)$, capillary pressures are neglected in these calculations $\left(\frac{|\nabla p_{cgo}|}{|\nabla p_o|} \sim 0\right)$ and this GOR expression becomes:

$$GOR = \left.\frac{\lambda_{Gsc}}{\lambda_{Osc}}\right|_{sandface} = \left.\frac{\lambda_{Gg} + R_s \lambda_{Oo}}{\lambda_{Oo} + R_v \lambda_{Gg}}\right|_{sandface} \tag{5.172}$$

which corresponds to the ratio of total mobilities of surface-gas to surface-oil pseudo-components evaluated at sandface conditions. Barring any tubing- or well-related production problems, this would be the producing GOR of the reservoir measured at surface conditions.

Following Eq. (5.172), producing GOR is bound to take values in between $\boldsymbol{GOR = R_s}$ (corresponding to the case when the reservoir gas phase is immobile or inexistent, or $k_{rg} = 0, \lambda_{Gg} = 0$) and $\boldsymbol{GOR = 1/R_v}$ (corresponding to the case when the reservoir oil phase is immobile or inexistent, or $k_{ro} = 0, \lambda_{Oo} = 0$). GOR is a time-dependent variable, and its values are bound to change in between these two extremes ($\boldsymbol{R_s} < GOR < 1/\boldsymbol{R_v}$) in response to the evolution of oil and gas mobilities at sandface conditions (λ_{Oo} and λ_{Gg}, respectively).

Finally, for WOR ratios, one writes:

$$WOR = \left.\frac{|\vec{q}_{wsc}|}{|\vec{q}_{osc}|}\right|_{sandface} = \left.\frac{\lambda_{Wsc}|\nabla p_w|}{\lambda_{Oo}|\nabla p_o| + \lambda_{Og}|\nabla p_g|}\right|_{sandface} \tag{5.173}$$

and when capillary pressures are neglected ($\nabla p_w = \nabla p_o = \nabla p_g = \nabla p$),

$$WOR = \left.\frac{|\vec{q}_{wsc}|}{|\vec{q}_{osc}|}\right|_{sandface} = \left.\frac{\lambda_{Wsc}}{\lambda_{Osc}}\right|_{sandface} \tag{5.174}$$

which corresponds to the ratio of total mobilities of surface-water to surface-oil pseudo-components evaluated at sandface conditions. Since in typical black-oil simulation, surface-water is assumed to be only present in the reservoir water phase, $\boldsymbol{WOR = 0}$ if the reservoir water phase is immobile ($k_{rw} = 0, \lambda_{Wsc} = 0$).

References

Agarwal, R.G., "Real Gas Pseudo-Time: A New Function for Pressure Buildup Analysis of MHF Gas Wells", SPE Paper 8279 presented at the 54th Annual SPE Fall Technical Conference and Exhibition, Las Vegas, NV, 1979.

Ayala L.F. and E.C. Morgan, Natural gas production engineering (Chapter 17), In: Riazi, M.R. and E. van Oort, eds., *ASTM Handbook: MNL73 Exploration and Production of Petroleum and Natural Gas Processing*, ASTM International, Conshohocken, PA, 2016.

Al-Hussainy, R., J. Ramey Jr, and P. Crawford, "The Flow of Real Gases Through Porous Media," *J. Petrol. Technol.*18:624–636, 1966.

Ertekin, T., J. Abou-Kassem, and G. King, *Basic Applied Reservoir Simulation*, Society of Petroleum Engineers, SPE Textbook Series Vol. 7, Richardson, TX, 2001.

Lake, L., R. Johns, W. Rossen, and G. Pope, *Fundamentals of Enhanced Oil Recovery*, Society of Petroleum Engineers, Richardson, TX, 2014

CHAPTER 6

Analytical Reservoir Engineering Solutions

In this chapter, analytical solutions to the transport equations in porous media will be examined. In the analytical approach, the solution of the flow problem becomes available at all points in the domain. However, the solution would necessarily capture simplified physics (hopefully, at the least, the most important features of it) given the sometimes significant assumptions and simplifications required to make governing equations amenable to analytical approaches.

Found at the top among the most-often-used reservoir engineering analytical tools are those employed in hydrocarbon production transient analysis (*PTA*) or well test analysis. Well test analysis relies on a myriad of analytical solutions to the flow equations for the purpose of reservoir-well characterization. Thus, most of this analytical chapter will be devoted to the understanding of *PTA*-related analytical solutions. At the end of the chapter, we will review other analytical approaches used to solve problems in porous media.

The overall goal of well testing or *PTA* is to collect information about the flow conditions in the well, around the immediate vicinity of the well, in the virgin portions of the reservoir not influenced by the drilling operations and simulation treatments, and finally about the boundaries of the reservoir. To collect this information, the well flow rate is varied and the resulting pressure transients are measured. The measurement of variation of pressure at the sandface with time provides a pressure transient data which then can be analyzed to determine the formation parameters that characterize the flow conditions that exist in the reservoir-well system. Excellent reviews and discussion of *PTA* methods that can further the information summarized in this chapter can be found in the works of ERCB (1975), Raghavan (1993), Medhat (2009), and Spivey and Lee (2013).

6.1 Analytical Reservoir Solutions and Systems Analysis

PTA or well test analysis can be considered as a "systems analysis" technique. In a systems analysis application, the system under consideration is subjected to a certain kind of perturbation. The system responds to this perturbation and the resulting signals are measured. Figure 6.1 gives a schematic representation of system analysis.

The principles governing the analysis of well tests can be more easily visualized when one considers well test interpretation as a special pattern recognition technique. In a well test, the system "*S*" represents the wellbore and the formation that it is in

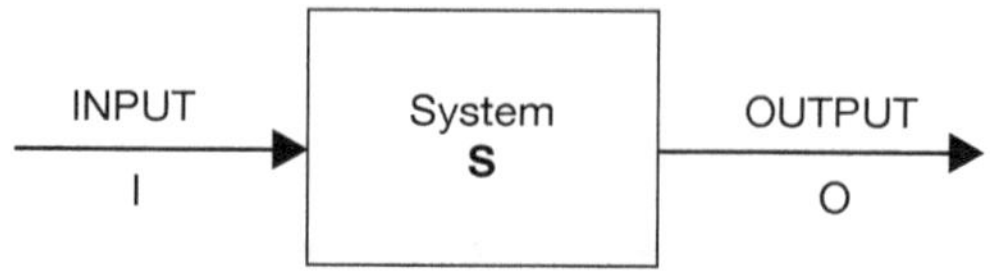

FIGURE 6.1 Schematic representation of interaction of a system with input.

communication with. The input "*I*" typically represents the withdrawal of the reservoir fluid at a constant rate and it can be considered as a forcing function applied to an unknown system "*S*" described earlier. Finally, the response of the system "*O*" represents the change in reservoir pressure and this response is typically measured in the wellbore during the test.

Forward and Inverse Problems

The response of the reservoir can be computed for specific initial and boundary conditions for a reservoir-well system with known properties. This is called the "forward (direct)" problem and is expressed as:

$$[I] \times [S] \rightarrow [O] \tag{6.1}$$

In an "inverse" problem we are attempting to find a well-defined system whose response to the input signal is measured as output. In mathematics, this type of problem is known as the inverse problem and is expressed as:

$$[O]/[I] \rightarrow [S] \tag{6.2}$$

The purpose of well test interpretation is to identify the system, knowing only the input and output signals and possibly some of the system characteristics. It should be recognized that the solution of the inverse problem is not unique. In other words, it is possible to find several systems that would yield similar responses to a given input. However, the number of alternative systems decreases as the number and range of output signal measurements increases.

6.2 Classical Formulations and Hydraulic Diffusivity Concept

The classical well test formulation is tied to a theoretical reservoir /well representation. In this representation, the well is located in the center of an infinitely large reservoir which has isotropic and homogeneous property distributions. The reservoir is 100% saturated with only one fluid (single-phase reservoir). Gravitational forces are ignored. The well fully penetrates the formation and the well in its entirety is exposed to the formation (open-hole completion). Initially, the system has uniform pressure distribution (initial condition). The well is put on production at a constant flow rate (internal boundary condition), and since the reservoir is infinitely large no pressure drop will be experienced at the outer boundary of the system which is located at infinity (outer boundary condition). The classical well test problem can be expressed in one-dimensional radial-cylindrical flow geometry, as shown below. Please note that the flow equation (6.3) was derived in Chap. 5 as the classical governing equation used in well test analysis for radial single-phase flow of a slightly compressible fluid (oil) toward a vertical well.

Flow Equation:

$$\frac{1}{r}\frac{\partial}{\partial r}\left(r\frac{\partial p}{\partial r}\right)=\frac{\phi\mu c}{k}\frac{\partial p}{\partial t} \tag{6.3}$$

Initial Condition:

$$p=p_i \text{ for } r_w \leq r < \infty \text{ and } t=0 \tag{6.4}$$

The initial condition as indicated above states that, initially, the pressure distribution in the reservoir is uniform.

Boundary Conditions:

(i) Inner Boundary Condition:

$$r\frac{\partial p}{\partial r}\bigg|_{r=r_w}=\frac{q\mu B}{2\pi kh}\bigg|_{r=r_w} \quad \text{for all } t>0 \tag{6.5}$$

It should be noted that the inner boundary condition simply states that the well has been producing at a constant rate. Mathematically speaking, constant flowrate specification implies that the gradient of pressure at $r=r_w$ (at the sandface) is fixed.

(ii) Outer Boundary Condition:

$$p \rightarrow p_i \text{ as } r \rightarrow \infty \text{ for all } t \geq 0 \tag{6.6}$$

Finally, the outer boundary condition states that the reservoir is infinitely large so that there is not any pressure change at the outer boundary (at infinity) at any time. It should be noted that no matter how small a reservoir, there will always be a time period during which the reservoir is acting as if it is infinitely large (in other words, during that period of time there will not be a *measurable* pressure drop at the outer boundary).

When the above system of equations is solved, a solution of the following form will be obtained:

$$p=p(r, t, r_w, k, h, ø, \mu, c, B, q, p_i) \tag{6.7}$$

In the forward solution, all of the reservoir and well properties are considered to be available to to the solution expression. The reciprocal of the coefficient that appears in front of the time derivative of Eq. (6.3) is an extremely important group of terms in reservoir engineering and well testing applications:

$$\left(\frac{\phi\mu c}{k}\right)^{-1} \tag{6.8}$$

If one performs a dimensional analysis, as shown in Eq. (6.9), it will be noted that the coefficient is analogous to diffusivity constant:

$$\eta=\frac{k\,[L^2]}{\phi\,[1]\,\mu\left[\frac{M}{Lt}\right]c\left[\frac{Lt^2}{M}\right]}=\frac{[L^2]}{[t]} \tag{6.9}$$

In fluid flow in porous media, the $(k/\phi\mu c)$ group is referred to as the hydraulic diffusivity constant and describes the areal extension rate of a pressure transient over a unit time. It is obvious that in tight formations, the rate of propagation of a pressure transient will be less than the one we would have in permeable systems. This characteristic of the reservoir rock will determine the duration of a well test. In other words, during the implementation of the well test, if it is necessary to communicate with the sections of the reservoir away from the wellbore, then in a tight reservoir it will be necessary to elongate the duration of the test. This is simply because the low permeability of the formation will have a smaller hydraulic diffusivity constant and hence it will take longer for a measurable pressure transient to reach the deeper portions of the reservoir.

6.3 Analytical Models in Well Test Analysis and Purpose of Well Testing

Equations (6.3) through (6.6) provide a mathematical description of the classical well test model. Equation (6.3) is simply known as the diffusivity equation written in one-dimensional radial cylindrical coordinates. This equation with its associated initial boundary and initial conditions culminate in a system of equations that need to be solved simultaneously. The dependent variable to be solved is the pressure as a function of time (t) and position (r). The rock and fluid properties that appear as a group in the form of hydraulic diffusivity constant can be functions of pressure, and as their dependencies on pressure increase, the equation gains a stronger nonlinear character.

The purpose of well testing is to collect pressure transient data which can be utilized after analysis, for decision making in exploration as well as reservoir engineering. In the area of exploration, using the properly analyzed well test data, one can make appropriate decisions toward completing a zone and setting a platform to develop the field. In reservoir engineering, well testing is the most powerful reservoir characterization and parameter estimation tool. Similarly in production engineering, wellbore damage and stimulation treatment effectiveness can be determined with the help of well testing. Figure 6.2 provides a schematic showing the various functions of well test analysis. Analytical models used for the pressure transient analysis are always built in the same manner. The basic models are typically homogeneous. Heterogeneous models have been the subject of many recent developments, and solutions with various inner boundary conditions are now widely used (e.g., double-porosity, multi-layer and composite). In well-test analysis, perhaps the most critical step is the identification of the model that will be used in interpreting the test data. The model identification is achieved using an iterative algorithmic protocol, as summarized in Fig. 6.3.

Practical Field Units Suggested

In this chapter, the standard symbols adopted by the petroleum industry are used. Practical field units are implemented in equations. Also, according to the sign convention adopted for this book, a positive flow rate, $q > 0$, signifies production, while a negative flow rate, $q < 0$, designates injection. Table 6.1 gives a summary of the units used in this chapter. Please note that in this table and in this chapter for well testing purposes,

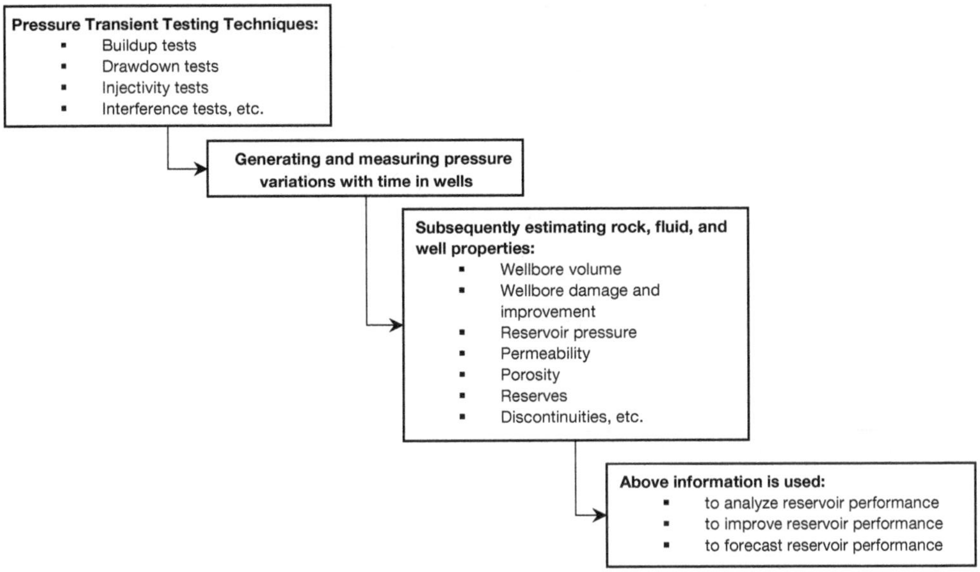

Figure 6.2 Purposes of well testing.

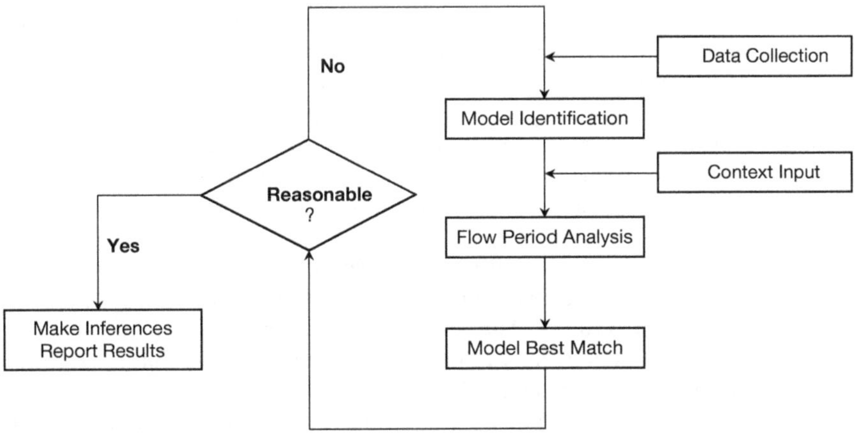

Figure 6.3 Outline of the well test interpretation process.

permeabilities are typically expressed in milidarcy (md), rather than "perms," as used in Chap. 6, and time is written in hours, rather than days, as used in Chap. 6. Thus, conversion factors built in the equations in this chapter will differ slightly when compared to Chap. 6 and this will be reflected in the use of λ and γ unit conversion parameters which will be introduced later.

Parameter	Dimension	Practical field unit
Distance, r	$[L]$	ft
Thickness, h	$[L]$	ft
Permeability, k	$[L^2]$	md
Time, t	$[t]$	hr
Viscosity, μ	$[M/Lt]$	cp
Pressure, p	$[M/Lt^2]$	psi
Volume, V	$[L^3]$	ft^3
Compressibility, c	$[Lt^2/M]$	psi^{-1}
Liquid flow rate, q	$[L^3/t]$	STB/D
Gas flow rate, q	$[L^3/t]$	$MMSCF/D$

TABLE 6.1 Practical Field Units Used in Well Test Analysis Equations

6.4 Derivation of Classical 1-D Radial Cylindrical Flow Equation

We will now derive Eq. (6.3) using differential balances written in cylindrical geometry, and show it to be consistent with the generalized derivations presented in Chap. 5. Equation (6.3) applies to liquid systems, and we will also show how the 1-D radial cylindrical governing equation is written and solved for the case of compressible gas flow.

In well test analysis, one of the most common flow geometries used in practice is the radial-cylindrical flow geometry. This is especially true for the usual case of single-well studies. Most of the applications are in one dimension (r-direction only), even though two- and three-dimensional studies are possible. The one-dimensional representation of the flow geometry forces the assumptions of dealing with homogeneous and isotropic reservoirs.

In this section, we derive a mathematical description for the behavior of a single-well system. This is, how the system (reservoir) reacts and interacts when a perturbation (at the well) is introduced. The idea is that we impose an action on the system, and it will react in a way that may allow us to deduct its own characteristics from the response function (which is measured during the test).

Let us develop the one-dimensional continuity equation in radial-cylindrical coordinates (r-direction only). To do this, let us consider the elemental volume around a single well, as shown in Fig . 6.4.

The volume of this ring-shaped elemental volume can be obtained by subtracting the volume of the inner cylinder from the volume of the outer cylinder, as shown in Eq. (6.10):

$$V_e = \pi(r + \Delta r)^2 h - \pi r^2 h \tag{6.10}$$

by expanding the right-hand-side:

$$V_e = \pi r^2 h + 2\pi r \Delta r h + \pi(\Delta r)^2 h - \pi r^2 h \tag{6.11}$$

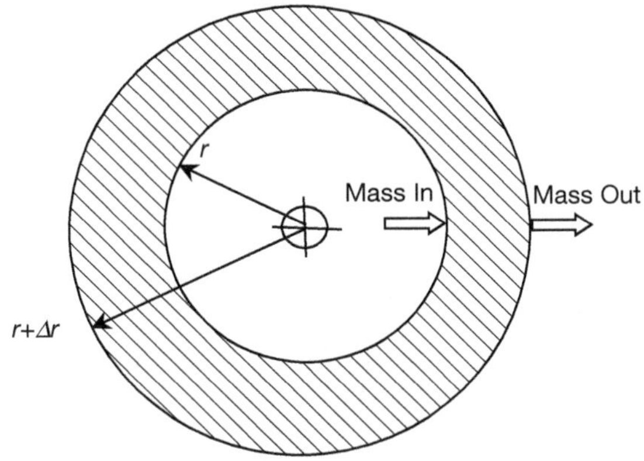

FIGURE 6.4 One-dimensional radial-cylindrical elemental volume.

Above expression can be simplified to Eq. (6.12):

$$V_e = \pi\left[2r\Delta r + (\Delta r)^2\right]h \tag{6.12}$$

To make a mass balance on the fluid that enters and leaves the elemental volume during a period of time Δt, we write the conservation of mass expression (adopting the sign convention as positive outflow and negative inflow):

$$\{Mass\ Out\} - \{Mass\ In\} + \{Mass\ Content\}_{t+\Delta t} - \{Mass\ Content\}_t = 0 \tag{6.13}$$

The entries of Eq. (6.13) are defined as follows:

$$\{Mass\ In\} = A_r(\rho v_r)_r\ \Delta t = 2\pi rh(\rho v_r)_r\Delta t \tag{6.14}$$

$$\{Mass\ Out\} = A_{r+\Delta r}(\rho v_r)_{r+\Delta r}\ \Delta t = 2\pi(r + \Delta r)h(\rho v_r)_{r+\Delta r}\ \Delta t \tag{6.15}$$

$$\{Mass\ Content\}_t = V_e\,(\phi\rho)_t = \pi[2r\Delta r + (\Delta r)^2]h(\phi\rho)_t \tag{6.16}$$

$$\{Mass\ Content\}_{t+\Delta t} = V_e\,(\phi\rho)_{t+\Delta t} = \pi[2r\Delta r + (\Delta r)^2]h(\phi\rho)_{t+\Delta t} \tag{6.17}$$

Introducing these expressions into Eq. (6.13), we get:

$$\begin{aligned}&\{2\pi rh(\rho v_r)_r\Delta t\} - \{2\pi(r + \Delta r)h(\rho v_r)_{r+\Delta r}\Delta t\}\\ &= \pi h\left[2r\Delta r + (\Delta r)^2\right]\left[(\phi\rho)_{t+\Delta t} - (\phi\rho)_t\right]\end{aligned} \tag{6.18}$$

Simplifying and dividing through by $2\pi r(\Delta r)(\Delta t)h$:

$$-\frac{1}{r}\left\{\frac{(r + \Delta r)(\rho v_r)_{r+\Delta r} - r(\rho v_r)_r}{\Delta r}\right\} = \left[1 + \frac{\Delta r}{2r}\right]\frac{\left[(\phi\rho)_{t+\Delta t} - (\phi\rho)_t\right]}{\Delta t} \tag{6.19}$$

or, equivalently,

$$-\frac{1}{r}\left\{\frac{(r\rho v_r)_{r+\Delta r} - (r\rho v_r)_r}{\Delta r}\right\} = \left[1 + \frac{\Delta r}{2r}\right]\frac{\left[(\phi\rho)_{t+\Delta t} - (\phi\rho)_t\right]}{\Delta t} \tag{6.20}$$

Now, taking limits as Δr and $\Delta t \to 0$, and recalling the definition of a derivative:

$$\frac{\partial F}{\partial s} = \lim_{\Delta s \to 0}\left[\frac{F(s+\Delta s) - F(s)}{\Delta s}\right] \tag{6.21}$$

we obtain:

$$-\frac{1}{r}\frac{\partial}{\partial r}(r\rho v) = \frac{\partial}{\partial t}(\phi\rho) \tag{6.22}$$

Equation (6.22) is the 1-D radial cylindrical form of the *continuity equation.* As it stands, it is applicable to any continuum. However, when we substitute the special relationship between pressure gradient and velocity describing flow in porous media (i.e., Darcy's law—the constituent equation), the resulting expression becomes only applicable to *fluid flow dynamics in porous media.* Hence, introducing physics of flow in porous media *via* the Darcy's law:

$$v_r = -\frac{k}{\mu}\frac{\partial p}{\partial r} \tag{6.23}$$

The resulting equation becomes:

$$\frac{1}{r}\frac{\partial}{\partial r}\left(r\rho\frac{k}{\mu}\frac{\partial p}{\partial r}\right) = \frac{\partial}{\partial t}(\phi\rho) \tag{6.24}$$

Equation (6.24) is the continuity equation for porous media or diffusivity equation in radial coordinates, written in one dimension, for single-phase flow. In order for us to be able to integrate it and come up with useful analytical or even numerical solutions, we would require an *equation of state* to account for changes in density with pressure $[\rho = \rho\,(p)]$ for both liquids and gases. It should also be noted that although Eq. (6.24) is known as the diffusivity equation, the physical process is not diffusion but the laminar flow process. It will be sufficient to recall that the Darcian (laminar) flow and diffusional flow problems and their respective formulations are analogous to each other.

Review of Assumptions

Equation (6.24) as we derived has been subjected to a number of assumptions that must be kept in mind. It is always useful to remember those assumptions in order to become more aware of the range of applicability and limitations of the models. The overall assumptions that have been made are as follows:

- *Flow is unidirectional and is in the r-direction.* There is no flow in the θ- and z-directions.
- *Isothermal conditions prevail.* This means that reservoir temperature is to remain constant during the process; otherwise, it would be necessary to include an energy balance equation as well.
- *Flowing fluid is single phase.* This is why neither saturations nor relative permeability functions appear in the development.
- *Homogeneous and isotropic porous medium.* This is because only one-dimensional (radial) flow is considered in the formulation.

- *Gravitational effects are negligible.* We have used pressure (p) as the dependent variable instead of "Φ" or flow potential.
- *Flow is laminar*. This is an inherent assumption of Darcy's law. This may not be a good assumption for the cases of near wellbore gas flow, where the fluid may acquire a relatively high velocity resulting in turbulent flow regime. Nevertheless, it is still a good assumption for liquid and gas flow systems and when the flow equations are implemented at locations in the reservoir away from the wellbore.

6.5 Liquid Flow Analytical Formulation

As discussed earlier, it will be necessary to incorporate a fluid behavior model or equation of state into the differential equation. This will enable us to integrate the diffusivity equation and obtain some workable analytical solutions toward the implementation of well test analysis protocols. In this section, we develop the equation for liquid flow or the continuity equation describing flow of a single-phase liquid in porous media. Additional assumptions are needed to derive this expression. These assumptions are as follows:

- Permeability is independent of pressure.
- Fluid viscosity is constant.
- Fluid compressibility is small and constant.
- Pressure gradients are small.

The assumptions as listed above are reasonable for slightly compressible liquids. We begin with the expression we developed earlier [Eq. (6.24)]:

$$\frac{1}{r}\frac{\partial}{\partial r}\left(r\rho\frac{k}{\mu}\frac{\partial p}{\partial r}\right) = \frac{\partial}{\partial t}(\phi\rho)$$

Expanding the left-hand side, and incorporating the constant viscosity and permeability assumptions, yields:

$$\frac{1}{r}\left(\rho\frac{\partial p}{\partial r} + \frac{\partial \rho}{\partial r}r\frac{\partial p}{\partial r} + \frac{\partial^2 p}{\partial r^2}r\rho\right) = \frac{\mu}{k}\frac{\partial}{\partial t}(\phi\rho) \tag{6.25}$$

This time, expanding the right-hand side and applying the chain rule, one obtains:

$$\frac{1}{r}\rho\frac{\partial p}{\partial r} + \left(\frac{\partial \rho}{\partial p}\frac{\partial p}{\partial r}\right)\frac{\partial p}{\partial r} + \rho\frac{\partial^2 p}{\partial r^2} = \frac{\mu}{k}\left(\rho\frac{\partial \phi}{\partial p}\frac{\partial p}{\partial t} + \phi\frac{\partial \rho}{\partial p}\frac{\partial p}{\partial t}\right) \tag{6.26}$$

Now at this point in the development, it will be necessary to incorporate the model for fluid behavior. As discussed earlier in this chapter, the overall assumption states that liquids are slightly compressible fluids. Fluid compressibility for a liquid may be expressed in terms of density as shown below:

$$c_f = \frac{1}{\rho}\frac{\partial \rho}{\partial p} \quad \text{or} \quad \frac{\partial \rho}{\partial p} = c_f\rho \tag{6.27}$$

Equation (6.27) describes the behavior of density with pressure for slightly compressible fluids. Introducing this expression into Eq. (6.26) results in:

$$\frac{1}{r}\rho\frac{\partial p}{\partial r} + c_f\rho\left(\frac{\partial p}{\partial r}\right)^2 + \rho\frac{\partial^2 p}{\partial r^2} = \frac{\mu}{k}\left(\rho\frac{\partial \phi}{\partial p} + \phi c_f \rho\right)\frac{\partial p}{\partial t} \tag{6.28}$$

Equation (6.28) can be further simplified by crossing out the fluid density coefficients that show up in front of all of the derivatives. Additionally, if one considers that liquid compressibility is a small number ($c_f = 10^{-6}$–10^{-7} psi^{-1}) and together with the assumption that pressure gradients are small, one can comfortably ignore the contribution of the central term in the left-hand side of Eq. (6.28) to obtain:

$$\frac{1}{r}\frac{\partial p}{\partial r} + \frac{\partial^2 p}{\partial r^2} = \frac{\mu}{k}\left(\frac{\partial \phi}{\partial p} + \phi c_f\right)\frac{\partial p}{\partial t} \tag{6.29}$$

Recalling the definition of pore space compressibility,

$$c_\phi = \frac{1}{\phi}\frac{\partial \phi}{\partial p} \text{ or } \frac{\partial \phi}{\partial p} = c\phi \tag{6.30}$$

and substituting Eq. (6.30) into Eq. (6.29), one obtains:

$$\frac{1}{r}\frac{\partial p}{\partial r} + \frac{\partial^2 p}{\partial r^2} = \frac{\mu}{k}\phi(c_\phi + c_f)\frac{\partial p}{\partial t} \tag{6.31}$$

We define the total compressibility (adding up the compressibilities of the rock grains and the formation fluid) as:

$$c_t = c_\phi + c_f \tag{6.32}$$

The final expression becomes:

$$\frac{1}{r}\frac{\partial p}{\partial r} + \frac{\partial^2 p}{\partial r^2} = \frac{\phi\mu c_t}{k}\frac{\partial p}{\partial t} \tag{6.33}$$

which is

$$\frac{1}{r}\frac{\partial}{\partial r}\left(r\frac{\partial p}{\partial r}\right) = \frac{\phi\mu c_t}{k}\frac{\partial p}{\partial t} \tag{6.34}$$

Equation (6.34) is the liquid flow equation or the classical well test model for liquid systems. The most important and influential coefficient in this equation is the *hydraulic diffusivity constant*, defined as:

$$\eta = \frac{k}{\phi\mu c_t} \tag{6.35}$$

As discussed earlier, the hydraulic diffusivity constant controls how fast the pressure transients move away from the wellbore. At this stage of the development, we will simply state that this is the most important single coefficient in well test analysis.

6.6 Compressible Flow Analytical Formulations

In this section, we would like to come up with a governing equation for the flow of a single-phase gas through a porous media. If we look again at the assumptions we had to make to come up with Eq. (6.43), the only one that we could keep for this development is that permeability could be made independent of pressure. This is not always the case, since in tight gas formations, pressure-dependent permeability of the Klinkenberg type may become important (Klinkenberg, 1941).

For compressible gaseous phases, we cannot longer assume that fluid compressibility is small and constant. Fluid viscosity would not be constant either. Pressure gradients for gas flow may be large. We have to account for all these pressure dependencies in the development of the gas flow equation.

Starting expression is again the general continuity equation in radial cylindrical coordinates:

$$\frac{1}{r}\frac{\partial}{\partial r}\left(r\rho\frac{k}{\mu}\frac{\partial p}{\partial r}\right) = \frac{\partial}{\partial t}(\phi\rho) \tag{6.36}$$

The behavior of density with pressure for a real gas can be developed using the real gas law:

$$\rho = \frac{(p)(MW)}{ZRT} \tag{6.37}$$

Substituting Eq. (6.37) into Eq. (6.36) yields:

$$\frac{1}{r}\frac{\partial}{\partial r}\left(r\frac{(p)(MW)}{ZRT}\frac{k}{\mu}\frac{\partial p}{\partial r}\right) = \frac{\partial}{\partial t}\left(\phi\frac{(p)(MW)}{ZRT}\right) \tag{6.38}$$

Dividing both sides of the Eq. (6.38) by (MW/RT) gives:

$$\frac{1}{r}\frac{\partial}{\partial r}\left(r\frac{p}{\mu Z}\frac{\partial p}{\partial r}\right) = \frac{1}{k}\frac{\partial}{\partial t}\left(\phi\frac{p}{Z}\right) \tag{6.39}$$

Note that treating the molecular weight as constant introduces a new assumption into the gas flow formulation, which is "no significant chromatographic separation of gases takes place in the reservoir." This means that the composition of the reservoir gas does not change during production. In reality, composition of reservoir gases may change with production. In fact, it will be correct to state that the reservoir itself acts like a huge chromatograph. When the production starts, first the lighter components are produced and hence the reservoir and well stream gas composition may not be constant during the reservoir life.

The gas flow formulation, in Eq. (6.39), is fundamentally different from that of liquid flow mainly because of a potentially "troublesome" nonlinear group that showed up on the left-hand side of the equation. This group, as shown below, is a strong function of pressure (the dependent variable) and introduces a relatively strong nonlinearity to the formulation:

$$\frac{p}{\mu Z} \rightarrow \text{"}\textit{nonlinear}\text{" group in Eq. (6.39)}$$

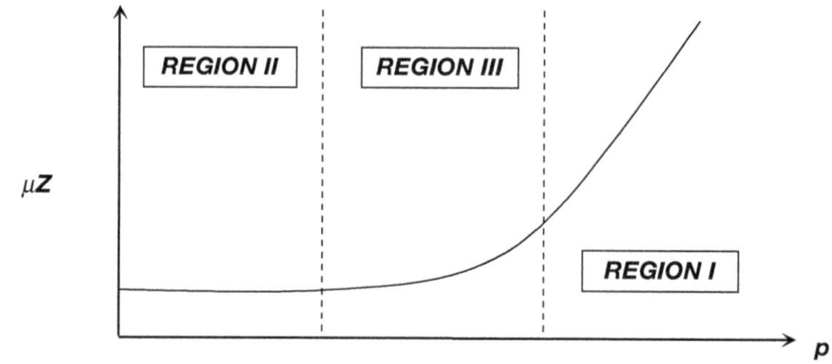

FIGURE 6.5 Behavior of the group (μZ) with pressure (note this qualitative plot is valid only for hydrocarbon gases).

On the right-hand side of Eq. (6.39), $1/Z$ also introduces additional nonlinearities; gas viscosity (μ) and compressibility factor (Z) have a strong dependency on pressure. We would have to know the behavior of the constituents of this group to devise a protocol to linearize the gas-flow equation.

For hydrocarbon gases, the behavior of the denominator (μZ) with pressure is shown in Fig. 6.5.

Three distinct regions are highlighted in Fig. 6.5. The characteristics of each of the regions are summarized below:

REGION I: *HIGH-PRESSURE REGION (typically beyond 3000 psia).* For high pressures, the derivative of $p/\mu Z$ (the nonlinear group) with respect to pressure takes a constant value. This is shown in Region I, as the graph (μZ) vs. p shows a constant slope.

REGION II: *LOW-PRESSURE REGION (typically from 0 psia to 800 psia).* At low pressures, the product (μZ) exhibits a constant behavior.

REGION III: *MID-PRESSURE REGION (typically between 800 and 3000 psia).* In this region, neither μZ nor $p/\mu Z$ takes a constant value.

By capitalizing on the characteristics of each of the three regions discussed above, it will be possible to come up with three different ways of writing the gas-flow equation in a linearized form. These are:

- *REGION I:* The "Pressure Approach (p-approach)" for high reservoir pressures.
- *REGION II:* The "Pressure-Squared Approach (p^2-approach)" for low reservoir pressures.
- *REGION III:* The "Real Gas Potential Approach (ψ-approach or pseudo-pressure approach)" for all pressure conditions.

The most general approach is the one that is going to be derived for Region III, where none of the simplifications observed in Regions I and II apply. In fact, the formulation for Regions I and II will be just subsets of what will be studied for Region III. The approximation used in Region II is often useful for lower pressure gas systems, while the approximation in Region I should seldom be used for gases since its assumptions do not truly hold for the entire depletion life of a gas well.

6.6.1 Compressible Flow Equations: Pressure Approach

If we are working with reservoir gases at high pressures, we can take advantage of the fact that the ratio $p/\mu Z$ could be at times treated to be relatively independent of pressure if reservoir pressures are maintained high. Starting with Eq. (6.39), and bringing out this nonlinear constant coefficient out of the right-hand side, we end up with:

$$\frac{p}{\mu Z}\frac{1}{r}\frac{\partial}{\partial r}\left(r\frac{\partial p}{\partial r}\right) = \frac{1}{k}\frac{\partial}{\partial t}\left(\phi\frac{p}{Z}\right) \tag{6.40}$$

Let us work on the right-hand side of this equation. In the final form of the equation on both sides, we would like to see only the dependent variable p. We start by assuming a constant porosity and applying the chain rule on the right-hand-side of Eq. (6.40):

$$\frac{1}{k}\frac{\partial}{\partial t}\left(\phi\frac{p}{Z}\right) = \frac{\phi}{k}\frac{\partial}{\partial p}\left(\frac{p}{Z}\right)\frac{\partial p}{\partial t} \tag{6.41}$$

We now apply the rule for the derivative of a quotient to get:

$$\frac{1}{k}\frac{\partial}{\partial t}\left(\phi\frac{p}{Z}\right) = \frac{\phi}{k}\left[\frac{Z - p\dfrac{\partial p}{\partial Z}}{Z^2}\right]\frac{\partial p}{\partial t} \tag{6.42}$$

or,

$$\frac{1}{k}\frac{\partial}{\partial t}\left(\phi\frac{p}{Z}\right) = \frac{\phi}{k}\left[\frac{1}{Z} - \frac{p}{Z^2}\frac{\partial p}{\partial Z}\right]\frac{\partial p}{\partial t} = \frac{\phi}{k}\frac{p}{Z}\left[\frac{1}{p} - \frac{1}{Z}\frac{\partial p}{\partial Z}\right]\frac{\partial p}{\partial t} \tag{6.43}$$

Now, we recall the definition for the isothermal gas compressibility in terms of Z-factor:

$$c_g = \frac{1}{p} - \frac{1}{Z}\frac{\partial p}{\partial Z}$$

Introducing this definition into Eq. (6.43), we get a final equivalent expression for the right-hand side of Eq. (6.40), which is:

$$\frac{1}{k}\frac{\partial}{\partial t}\left(\phi\frac{p}{Z}\right) = \frac{\phi}{k}\frac{p}{Z}c_g\frac{\partial p}{\partial t} \tag{6.44}$$

One can now take Eq. (6.44) into Eq. (6.40) to obtain (recalling that the group $p/\mu Z$ is constant):

$$\frac{p}{\mu Z}\frac{1}{r}\frac{\partial}{\partial r}\left(r\frac{\partial p}{\partial r}\right) = \frac{\phi}{k}\frac{p}{Z}c_g\frac{\partial p}{\partial t} \tag{6.45}$$

After dividing both sides of Eq. (6.45) by $(p/\mu Z)$ the final form of the p-equation for gas flow in a high-pressure reservoir is obtained:

$$\frac{1}{r}\frac{\partial}{\partial r}\left(r\frac{\partial p}{\partial r}\right) = \frac{\mu_g \phi c_g}{k}\frac{\partial p}{\partial t} \tag{6.46}$$

Equation (6.46) represents the differential equation for gas flow using the p-approach. The denomination "p-approach" derives from the fact that the dependent variable for both spatial and temporal derivatives is pressure (p). The equation is only applicable for gas reservoirs at high pressures and the gas properties at the right-hand side of the equation must be calculated at the average pressure. Notice that this is, essentially, the same expression that we derived for liquid flow.

There can be significant risk for the reservoir analyst that uses this formulation. Even if the initial gas pressure is high and $p/\mu Z$ is relatively constant around initial reservoir conditions, as soon as the reservoir depletes, $p/\mu Z$ will become a strong nonlinearity (i.e., the system will eventually enter Regions II and III in Fig. 6.5, particularly around near-wellbore conditions). At those conditions, which will eventually occur, the formulation will severely misrepresent the flow problem. Accordingly, the pressure approach is not recommended in general in natural gas engineering practice.

6.6.2 Compressible Flow Equations: p^2-Approach

At low pressures, the product μZ can be treated as a constant in Eq. (6.39). Therefore, Eq. (6.39) can be written as:

$$\frac{1}{\mu Z}\frac{1}{r}\frac{\partial}{\partial r}\left(rp\frac{\partial p}{\partial r}\right) = \frac{1}{k}\frac{\partial}{\partial t}\left(\phi\frac{p}{Z}\right) \tag{6.47}$$

If we recall Eq. (6.44), we can express the right-hand side of this equation as follows:

$$\frac{1}{\mu Z}\frac{1}{r}\frac{\partial}{\partial r}\left(rp\frac{\partial p}{\partial r}\right) = \frac{\phi c_g}{k}\frac{p}{Z}\frac{\partial p}{\partial t} \tag{6.48}$$

Gas compressibility factor cancels out and we see that each pressure derivative is multiplied by pressure. We could bring that pressure coefficient inside of the inner differential by noting that:

$$p\partial p = \frac{1}{2}\partial p^2 \tag{6.49}$$

Therefore, Eq. (6.48) can be recast to the following expression:

$$\frac{1}{r}\frac{\partial}{\partial r}\left(r\frac{\partial p^2}{\partial r}\right) = \frac{\mu_g \phi c_g}{k}\frac{\partial p^2}{\partial t} \tag{6.50}$$

Equation (6.50) represents the differential equation for gas flow using the p^2-approach. The denomination "p^2-approach" derives from the fact that the dependent variable for both spatial and temporal derivatives is now squared pressure (p^2). The equation is only applicable for gases at low pressures and the gas properties at the right-hand side of the equation must be calculated at the average pressure.

6.6.3 Compressible Flow Equations: Real Gas Potential Approach

Figure 6.5 shows that the p- and p^2-linearization strategies are not applicable for the entire pressure range. The most general approach that can be used for the linearization of the gas flow equation is the introduction of a new variable (for the transformation of the dependent variable). For that purpose, the real gas potential (ψ) (also known as pseudo-pressure) function is introduced.

In linearizing the gas flow equation, we introduce the following definition for "ψ" (Al-Hussainy et al., 1966):

$$\psi = \int_0^p \frac{p}{\mu Z} dp \tag{6.51}$$

It is useful to visualize that this pseudo-pressure function has field units of psi^2/cp and represents the area under the $p/\mu Z$ versus p curve. As we would recall, "$p/\mu Z$" was identified as a "problematic" term in Eq. (6.39). Let us now consider Fig. 6.6.

Figure 6.6 shows that for every value of pressure (p), we have a value of real gas potential (ψ) associated with it. A particular value of ψ is calculated as the area under the curve at the corresponding pressure (see the shaded area in Fig. 6.6). In fact, with the definition of real gas potential, we are trying to come up with a new way of expressing the pressure variable that allows the linearization of the gas flow equation. This is why real gas potential is also known as "pseudo-pressure."

Let us work on the derivation of the gas flow equation when the definition given by Eq. (6.51) is implemented. For this, the spatial and temporal derivatives of the real gas potential with respect to those of pressure are needed. Starting with the definition provided by Eq. (6.51) and taking the derivative with respect to pressure, one obtains:

$$\frac{\partial \psi}{\partial p} = \frac{\partial}{\partial p}\left[\int_0^p \frac{p}{\mu Z} dp\right] \tag{6.52}$$

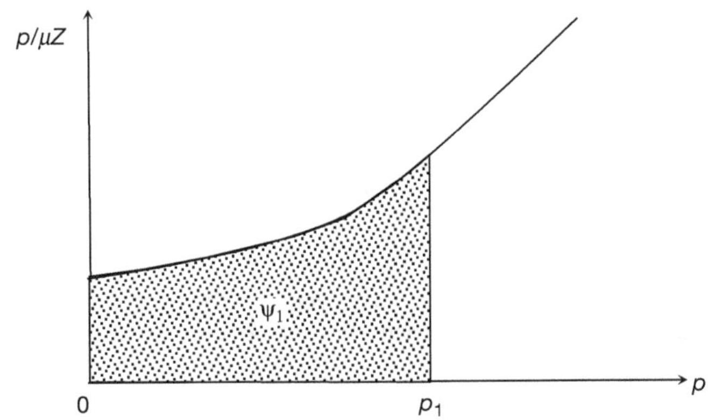

FIGURE 6.6 Real-gas potential definition.

Now, applying the Leibniz's rule, we may interchange the positions of the differential operator and the integral operator. When we do this, the pressure differentials (dp) cancel out and we obtain:

$$\frac{\partial \psi}{\partial p} = \int_0^p \frac{\partial}{\partial p}\left(\frac{p}{\mu Z}\right) dp = \int_0^p d\left(\frac{p}{\mu Z}\right) \tag{6.53}$$

Hence,

$$\partial \psi = \left(\frac{p}{\mu Z}\right) \partial p \tag{6.54}$$

Therefore, the spatial and temporal derivatives of the real gas potential with respect to those of pressure can be written as follows:

$$\frac{\partial \psi}{\partial r} = \frac{p}{\mu Z}\frac{\partial p}{\partial r} \quad \rightarrow \quad \frac{\partial p}{\partial r} = \frac{\mu Z}{p}\frac{\partial \psi}{\partial r} \tag{6.55}$$

$$\frac{\partial \psi}{\partial t} = \frac{p}{\mu Z}\frac{\partial p}{\partial t} \quad \rightarrow \quad \frac{\partial p}{\partial t} = \frac{\mu Z}{p}\frac{\partial \psi}{\partial t} \tag{6.56}$$

Equation (6.48) for compressible flow was written as:

$$\frac{1}{r}\frac{\partial}{\partial r}\left(r\frac{p}{\mu Z}\frac{\partial p}{\partial r}\right) = \frac{1}{k}\frac{\partial}{\partial t}\left(\phi\frac{p}{Z}\right) \tag{6.57}$$

We know from previous sections [Eq. (6.44)] that the right-hand-side of this expression can be written alternatively as:

$$\frac{1}{k}\frac{\partial}{\partial t}\left(\phi\frac{p}{Z}\right) = \frac{\phi}{k}\frac{p}{Z}c_g\frac{\partial p}{\partial t}$$

Therefore, the compressible gas equation becomes:

$$\frac{1}{r}\frac{\partial}{\partial r}\left(r\frac{p}{\mu Z}\frac{\partial p}{\partial r}\right) = \frac{\phi}{k}\frac{p}{Z}c_g\frac{\partial p}{\partial t} \tag{6.58}$$

Substituting Eqs. (6.55) and (6.56) into Eq. (6.58), one obtains:

$$\frac{1}{r}\frac{\partial}{\partial r}\left(r\frac{p}{\mu Z}\frac{\mu Z}{p}\frac{\partial \psi}{\partial r}\right) = \frac{\mu_g \phi c_g}{k}\frac{p}{Z}\frac{Z}{p}\frac{\partial \psi}{\partial t} \tag{6.59}$$

After simplifying Eq. (6.59), it is clearly seen that the troublesome nonlinearities disappear. The real-gas-potential transformation has been instrumental in achieving that.

$$\frac{1}{r}\frac{\partial}{\partial r}\left(r\frac{\partial \psi}{\partial r}\right) = \frac{\mu_g \phi c_g}{k}\frac{\partial \psi}{\partial t} \tag{6.60}$$

To be mathematically correct, Eq. (6.60) is a "quasi-linear" equation because of the dependencies of the μ_g and c_g entries on the right-hand-side on pressure. To make the entire equation linear, one may evaluate the μ_g and c_g terms at initial reservoir pressure (p_i) and neglect the pressure dependency. This could be done only for wells recently placed in production, and for well tests of short duration. As a result, one form the equation for compressible flow using the real-gas-potential approach is:

$$\frac{1}{r}\frac{\partial}{\partial r}\left(r\frac{\partial \psi}{\partial r}\right) = \frac{\mu_{gi}\phi c_{gi}}{k}\frac{\partial \psi}{\partial t} \tag{6.61}$$

Equation (6.61) represents the differential equation for gas flow using the ψ-approach when the $\mu_g c_g$ dependency with pressure has been neglected in the accumulation term. This assumption can and should be relaxed by introducing the concept of pseudo-time (Agarwal, 1979) when the analysis extends to longer-term production data and extensive reservoir depletion has taken place. Pseudo-time, however, does not fully linearize the equation. It is a concept most applicable to boundary-dominated flow conditions and requires modifications when applied during early infinite-acting conditions. In pseudo-time calculations, $\mu_g c_g$ values need to be evaluated at a suitable reference or average pressure defined at every time—typically taken as the recorded wellbore buildup pressure in buildup test analysis or average pressure in the drainage area during boundary-dominated flow analysis.

Nevertheless, the pseudo-pressure equation is more general than the previous two gas formulations (p- and p^2-approaches), and thus it can be collapsed into the other two approaches. Let us consider Region I: high-pressure region (Fig. 6.5) first, for example. For high pressures, the ratio $p/\mu Z$ is taken to be constant. Therefore, for the definition given by Eq. (6.51) we will have:

$$\psi = \int_0^p \frac{p}{\mu Z}dp = \left(\frac{p}{\mu Z}\right)_{const}\int_0^p dp = p\left(\frac{p}{\mu Z}\right)_{const} \tag{6.62}$$

Introducing Eq. (6.62) into Eq. (6.61), we see that:

$$\frac{1}{r}\frac{\partial}{\partial r}\left(r\frac{\partial}{\partial r}p\left(\frac{p}{\mu Z}\right)_{const}\right) = \frac{\mu_{gi}\phi c_{gi}}{k}\frac{\partial}{\partial t}p\left(\frac{p}{\mu Z}\right)_{const} \tag{6.63}$$

Because in this high-pressure approach the group $p/\mu Z$ is being assumed to remain constant, the entire equation can be divided by the same group yielding:

$$\frac{1}{r}\frac{\partial}{\partial r}\left(r\frac{\partial p}{\partial r}\right) = \frac{\mu_{gi}\phi c_{gi}}{k}\frac{\partial p}{\partial t} \tag{6.64}$$

Equation (6.64) is the same expression we obtained as Eq. (6.46) using the p-approach.

Let us now consider Region II (*low-pressure region*). For low pressure, the product μZ has a constant value. Therefore, the pseudo-pressure definition collapses to:

$$\psi = \int_0^p \frac{p}{\mu Z}dp = \left(\frac{1}{\mu Z}\right)_{const}\int_0^p pdp = \frac{1}{2}\left(\frac{1}{\mu Z}\right)_{const}p^2 \tag{6.65}$$

Introducing Eq. (6.65) into Eq. (6.61),

$$\frac{1}{r}\frac{\partial}{\partial r}\left[r\frac{\partial}{\partial r}\left(p^2\frac{1}{2(\mu Z)_{const}}\right)\right]=\frac{\mu_{gi}\phi c_{gi}}{k}\frac{\partial}{\partial t}\left(p^2\frac{1}{2(\mu Z)_{const}}\right) \tag{6.66}$$

Since $1/(2\mu Z)$ is a constant, both sides of Eq. (6.66) can be divided by this constant group to result in:

$$\frac{1}{r}\frac{\partial}{\partial r}\left(r\frac{\partial p^2}{\partial r}\right)=\frac{\mu_{gi}\phi c_{gi}}{k}\frac{\partial p^2}{\partial t} \tag{6.67}$$

Hence, the ψ-approach does collapse to the p^2-approach at low pressures [see Eq. (6.50)].

6.7 Dimensionless Groups and Dimensionless Form of the Governing Equations

Dimensional analysis is a useful tool used in applied mathematics to express functionality in terms of dimensionless variables instead of using dimensional variables. The idea is that the functionality between physical variables should be independent of the unit choice; this is, dimensionally homogeneous. Expressing a given problem and its solution in dimensionless forms provides an avenue for representation of the problem and its solution in a generalized form. The use of dimensionless expressions yields several advantages. It allows reducing the number of variables in a problem while simplifying the governing equations.

In well-test analysis, the representation of the solution in dimensionless form gives an opportunity for deriving a universal solution independent of reservoir properties and choice of units. The first three dimensionless variables that we will find useful for our analysis are:

- *Dimensionless Pressure:*
  The definition of the dimensionless pressure drop is as follows:

$$\Delta p_D=\frac{kh}{\gamma q_{sc}\mu B}\Delta p \tag{6.68}$$

- *Dimensionless Time:*
  The definition of the dimensionless time is as follows:

$$t_D=\frac{\lambda k}{\phi\mu c_t r_w^2}t \tag{6.69}$$

- *Dimensionless Flow Rate:*
  The definition of the dimensionless flow rate is as follows:

$$q_D=\frac{\gamma\mu B}{khp_i}q_{sc} \tag{6.70}$$

The values "λ" and "γ" are unit conversion constants and are dependent on the system and unit of choice. For example, for liquid systems, $\lambda = 2.637.10^{-4}$ and $\gamma = 141.2$ in practical field units and $\lambda = 1$ and $\gamma = 1/2\pi$ in Darcy units. The value of the λ parameter here in field units is directly related to the perms to Darcy conversion (1.127) and volume conversion ($\alpha_c = 5.615$ ft^3/bbl) used in Chap. 6, combined with a change of time units (hours used in well testing vs. days used in Chap. 6), so that $\lambda =$ (1.127 perms/Darcy) $\times$ (1 Darcy/10^3 mD) $\times$ (5.615) $\times$ (1 day/24 hr) $= 2.637.10^{-4}$. The derivation of these dimensionless parameters is based on the assumption that rock and fluid properties are all constants and follows from the manipulation of governing equations, and initial and boundary conditions. In the next section, we will review how to build these dimensionless groups using the liquid flow equation with the appropriate boundary and initial conditions.

6.7.1 Derivation of Dimensionless Groups

Let us consider the liquid flow equation:

$$\frac{1}{r}\frac{\partial}{\partial r}\left(r\frac{\partial p}{\partial r}\right) = \frac{\phi \mu c_t}{k}\frac{\partial p}{\partial t} \tag{6.71}$$

Equation (6.71) is a partial differential equation in two independent variables: space (r) and time (t). The integration of this equation requires taking care of the fact that we have a *spatial derivative of second order* and a *temporal derivative of first order*. Therefore, for the complete specification of the problem, we need to specify two boundary conditions (for the spatial derivative) and one initial condition (for the temporal derivative). These two boundary and initial conditions are reviewed below. It should also be noted that when a partial differential equation is converted to a dimensionless form, all of the associated boundary and initial conditions need to be expressed in dimensionless forms.

Now, consider the following boundary and initial conditions:

Internal Boundary Conditions:

This pertains to the specification of the dependent variable (p) at the well location $r = r_w$, where r_w = well radius. In well test analysis equations, it is common to specify a constant flow rate (q_{sc}) at the well location. Although flow rates may vary throughout the life of the reservoir, one can consider it constant for the time period for which the test takes place. When "q_{sc}" is fixed at some point in the reservoir, what we are doing in reality is to specify the pressure gradient (dp/dr) at that location. This follows from Darcy's law and is referred to as the Neumann type boundary condition. This specification can be expressed as:

$$\left.\frac{q}{2\pi rh}\right|_{@well} = \left.\frac{k}{\mu}\frac{\partial p}{\partial r}\right|_{@well} \quad \text{for all } t > 0 \tag{6.72}$$

or,

$$\left. r\frac{\partial P}{\partial r}\right|_{@well} = \frac{q\mu}{2\pi kh} \quad \text{for all } t > 0 \tag{6.73}$$

As a reminder, we should recall that the sign convention here, and the one we will follow for the remainder of this chapter, is that "q" is positive for production ($q > 0$)

and negative for injection ($q < 0$). Equation (6.73) can be rewritten in terms of flow rate at standard conditions through the introduction of the formation volume factor:

$$r\frac{\partial p}{\partial r}\bigg|_{@\,well} = \frac{q_{sc}\mu B}{2\pi kh} \quad \text{for all } t \geq 0 \tag{6.74}$$

As it will be discussed later, this specification assumes that we have some kind of shutting device at bottomhole conditions. However, we usually shut the wells at the surface and this effect is felt at the sandface later on. The current equations for the idealized model do not consider this, but such deviations which are commonly called "wellbore storage effects" will be incorporated to the solution later in this chapter.

External Boundary Conditions:

This pertains to the specification of the dependent variable (p) at the external (outer) boundary away from the wellbore ($r >> r_w$) and such a specification is called the Dirichlet type specification. Here, we have to make an assumption about the size of the reservoir. For simplicity, we will assume that we are dealing with an infinitely large reservoir and hence the pressure transients that will cause pressure changes will never be felt at the outer boundary in terms of a measurable pressure drop or increase. In other words, at the outer boundary, pressure remains the same as the one we had at initial conditions ($t = 0$, $p = p_i$). Mathematically, the outer boundary is expressed as follows:

$$p \rightarrow p_i \text{ as } r \rightarrow \infty \quad \text{for all } t \geq 0 \tag{6.75}$$

Equation (6.75) states that the pressure transients never reach the boundary and the pressure at the outer boundary remains at p_i.

Initial Conditions:

Here we state the condition of pressure (dependent variable) throughout the reservoir at initial time ($t = 0$). We assume that all points of the reservoir have the same initial pressure, p_i, and hence we write:

$$p = p_i \quad @ \quad t = 0 \quad r_w < r < \infty \tag{6.76}$$

At this point, the definition of the problem is completed, and we are ready to solve the partial differential equation that we have worked on. To come up with the dimensionless form of Eq. (6.71), we define the following terms:

$$\text{Dimensionless distance:} \quad r_D = \frac{r}{r_w} \tag{6.77}$$

$$\text{Pressure drop:} \quad \Delta p = p_i - p \tag{6.78}$$

Now, one can define a *temporary* dimensionless pressure drop:

$$\Delta p'_D = \frac{p_i - p}{p_i} \tag{6.79}$$

First, we focus on the inner boundary condition, starting with Eq. (6.74) which is repeated below:

$$r\frac{\partial p}{\partial r}\bigg|_{@\,well} = \frac{q_{sc}\mu B}{2\pi kh} \quad \text{for all } t > 0$$

Let us manipulate this expression in order to introduce the definitions given by Eq. (6.77) and Eq. (6.79), as shown below:

$$\frac{1}{p_i}\frac{r_w}{r_w}r\frac{\partial p}{\partial r}\bigg|_{r_D=1} = \frac{1}{p_i}\frac{q_{sc}\mu B}{2\pi kh} \tag{6.80}$$

$$\frac{r}{r_w}\frac{\partial(p/p_i)}{\partial(r/r_w)}\bigg|_{r_D=1} = \frac{1}{p_i}\frac{q_{sc}\mu B}{2\pi kh} \tag{6.81}$$

From Eq. (6.79), one can recognize that:

$$\partial(p/p_i) = -\partial(\Delta p'_D) \tag{6.82}$$

We also note that Eq. (6.77) indicates that $r_D = 1$ when $r = r_w$. Therefore,

$$r_D\frac{\partial(\Delta p'_D)}{\partial r_D}\bigg|_{r_D=1} = -\frac{1}{p_i}\frac{q_{sc}\mu B}{2\pi kh} \tag{6.83}$$

Left-hand side of Eq. (6.83) has been non-dimensionalized. For dimensional consistency, the right hand of the equation does not have other choice than being non-dimensional as well (as long as all the arithmetical manipulations are done correctly). A dimensional analysis carried on the right-hand-side of Eq. (6.83) will show that. This "new" dimensionless group that we have "created" on the right-hand-side is called the *dimensionless flow rate*, as it was presented in Eq. (6.70) and repeated below:

$$q_D = \frac{\gamma\mu B}{khp_i}q_{sc} \tag{6.84}$$

with $\gamma = 1/2\pi$.Therefore, the inner boundary condition becomes:

$$r_D\frac{\partial(\Delta p'_D)}{\partial r_D}\bigg|_{r_D=1} = -q_D \tag{6.85}$$

We can work this out a little further to come up with a final definition for the dimensionless pressure group. Dividing through by q_D and recalling that q_D is not a function of r_D [Eq. (6.70)], we may write:

$$r_D\frac{\partial(\Delta p'_D/q_D)}{\partial r_D}\bigg|_{r_D=1} = -1 \tag{6.86}$$

Now, we define the dependent variable of this derivative as the "actual" dimensionless pressure group that we will be using from this point on. Recalling the temporary definition in Eq. (6.79), one can define a new dimensionless pressure group as:

$$\Delta p_D = \frac{\Delta p'_D}{q_D} = \left(\frac{\Delta p}{p_i}\right)\left(\frac{2\pi khp_i}{\mu B}\right) = \frac{2\pi kh}{q_{sc}\mu B}\Delta p \tag{6.87}$$

which is the definition we presented in Eq. (6.68) for, $\gamma = 1/2\pi$, the unit-dependent coefficient in Darcy's units.

With these definitions, the dimensionless form of the inner boundary condition is now expressed as follows:

$$r_D \frac{\partial(\Delta p_D)}{\partial r_D}\bigg|_{r_D=1} = -1 \tag{6.88}$$

Now, we go back to the original partial differential equation [Eq. (6.34)] which is repeated below:

$$\frac{1}{r}\frac{\partial}{\partial r}\left(r\frac{\partial p}{\partial r}\right) = \frac{\phi\mu c_t}{k}\frac{\partial p}{\partial t}$$

Now, it is time to work on the left-hand side of Eq. (6.34). By making the left-hand side of the equation dimensionless, the right-hand side should become dimensionless as well. Along these lines, let us write:

$$r_w^2\frac{1}{r}\frac{\partial}{\partial r}\left(\frac{r_w}{r_w}r\frac{1}{q_D p_i}\frac{\partial p}{\partial r}\right) = \frac{\phi\mu c_t}{k}\frac{\partial p}{\partial t}\left(r_w^2\frac{1}{q_D p_i}\right) \tag{6.89}$$

or,

$$\frac{1}{r/r_w}\frac{\partial}{\partial(r/r_w)}\left(\frac{r}{r_w}\frac{1}{q_D}\frac{\partial(p/p_i)}{\partial(r/r_w)}\right) = \frac{\phi\mu c_t r_w^2}{k}\frac{1}{q_D}\frac{\partial(p/p_i)}{\partial t} \tag{6.90}$$

Recalling Eq. (6.82), and dividing it through by the dimensionless flow rate,

$$\frac{1}{q_D}\partial(p/p_i) = -\partial(\Delta p'_D/q_D) = \partial(\Delta p_D) \tag{6.91}$$

Introducing the definition of the dimensionless radius and substituting Eq. (6.91) into Eq. (6.90), one obtains:

$$\frac{1}{r_D}\frac{\partial}{\partial r_D}\left(r_D\frac{\partial(\Delta p_D)}{\partial r_D}\right) = \frac{\phi\mu c_t r_w^2}{k}\frac{\partial(\Delta p_D)}{\partial t} \tag{6.92}$$

Focusing on the right-hand side of Eq. (6.92) and recognizing that all coefficients on the right-hand side are constants, we may bring them inside of the temporal derivative:

$$\frac{1}{r_D}\frac{\partial}{\partial r_D}\left(r_D\frac{\partial(\Delta p_D)}{\partial r_D}\right) = \frac{\partial(\Delta p_D)}{\partial(tk\ /\ \phi\mu c_t r_w^2)} \tag{6.93}$$

Now, we have come across the definition of *dimensionless time group*, as presented in Eq. (6.69) for $\lambda = 1$,

$$t_D = \frac{k}{\phi \mu c_t r_w^2} t \tag{6.94}$$

This completes our set of dimensionless quantities which are:

$$r_D = \frac{r}{r_w}; \quad q_D = \frac{\mu B}{2\pi k h p_i} q_{sc}; \quad \Delta p_D = \frac{\Delta p}{p_i q_D}; \quad t_D = \frac{k}{\phi \mu c_t r_w^2} t$$

and our dimensionless partial differential equation:

$$\frac{1}{r_D} \frac{\partial}{\partial r_D} \left(r_D \frac{\partial(\Delta p_D)}{\partial r_D} \right) = \frac{\partial(\Delta p_D)}{\partial t_D} \tag{6.95}$$

6.7.2 Dimensionless Form of the Well Test Model

Using the dimensionless groups described in the preceding section, we have seen that the dimensionless equivalent of the flow equation becomes Eq. (6.95), which is repeated here for convenience:

$$\frac{1}{r_D} \frac{\partial}{\partial r_D} \left(r_D \frac{\partial(\Delta p_D)}{\partial r_D} \right) = \frac{\partial(\Delta p_D)}{\partial t_D} \tag{6.96}$$

After converting the partial differential equation into a dimensionless form, we have to convert the associated boundary conditions and the initial condition into dimensionless forms as well. The dimensionless version of the inner boundary condition can now be written as:

$$r_D \frac{\partial(\Delta p_D)}{\partial r_D} \bigg|_{r_D=1} = -1 \quad \text{for all } t_D > 0 \tag{6.97}$$

It is easy to see that the dimensionless version of the outer boundary specification becomes:

$$\Delta p_D \to 0 \quad \text{as} \quad r_D \to \infty \quad \text{for all } t_D > 0 \tag{6.98}$$

Finally, for the dimensionless form of the initial condition, we get:

$$\Delta p_D = 0 \quad \text{at} \quad t_D = 0 \quad \text{for} \quad 1 < r_D < \infty \tag{6.99}$$

Equations (6.96) through (6.99) represent the well test model equations expressed in dimensionless variables. It is obvious that in this new dimensionless formulation the functionality of the dependent variable (Δp_D) can be described by using only two variables:

$$\Delta p_D = f(r_D, t_D) \tag{6.100}$$

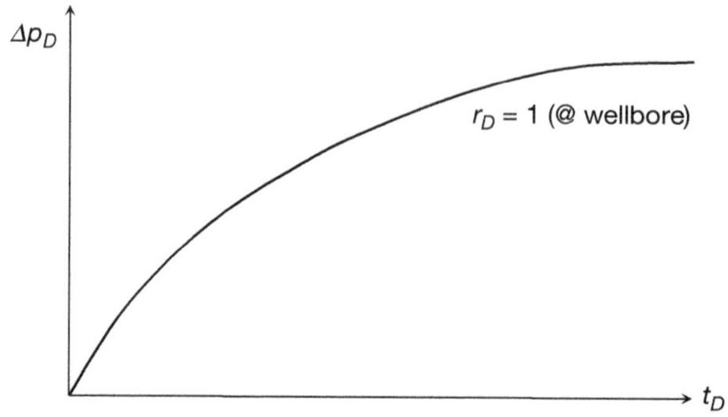

FIGURE 6.7 Dependency of the dimensionless solution $\Delta p_D = f(r_D, t_D)$.

However, it can be seen that for the original dimensional expression [Eqs. (6.71) through (6.74)], the solution of the differential equation will be of the form:

$$p = f(r, t, \phi, \mu, c_t, k, h, q_{sc}, r_w, p_i) \tag{6.101}$$

We have succeeded in effectively reducing the number of independent variables. It is clear that the dimensionless form of the solution is much more compact and more universal (only two principal independent variables are used to define the variation of the dependent variable). We see that for a given r_D, say at wellbore conditions ($r_D = 1$), the dimensionless solution will be simply function of just one variable, t_D. Figure 6.7 shows this unique dependency.

Plots like the one shown in Fig. 6.7 show that it is possible to obtain unique, generalized solutions for all reservoirs for which the assumptions taken during the development hold. In other words, with the help of the dimensionless groups, we have been able to collapse all possible cases into one single curve.

One last word about the dimensionless variables listed in Eqs. (6.68), (6.69), and (6.70); they were derived using Darcy units for the liquid equation. Since we are mostly interested in field units, the appropriate conversion factors must be incorporated into the dimensionless quantities. These conversion factors are "γ" and "λ", and they were incorporated in the original expressions listed in the definitions of dimensionless groups, t_D, Δp_D and q_D. For liquid systems, the unit conversion constants "γ" and "λ" are given in Table 6.2.

	Darcy units	Field units	SI units
λ	1	2.637×10^{-4}	2.637×10^{-4}
γ	$1/2\pi$	141.2	1.842×10^{3}

TABLE 6.2 Numerical Values of "γ" and "λ" Used in the Dimensionless Groups (for Liquid Flow Formulation)

6.8 Reservoir Boundary Idealizations in Classical Analytical Solutions

In the radial flow geometry, there are three fundamental reservoir boundary idealizations that impact the behavior of the solution. The three idealizations have to do with the way the boundary conditions are imposed to the mathematical formulation. The interplay between outer boundary and inner boundary conditions, and the characteristics of the flow equation, control whether or not the reservoir experiences infinite-acting flow (*IAF*), boundary-dominated flow (*BDF*), pseudo-steady-state (*PSS*) flow, and steady-state flow (*SSF*).

Infinite acting reservoir idealization and IAF

The first idealization is the infinite-acting (or infinitely large) reservoir assumption. For an infinitely large reservoir, the following outer boundary condition is imposed to the formulation as discussed earlier in the chapter.

$$\text{As } r \to \infty: \quad p = p_i \quad \text{for all } t \geq 0$$

This idealization considers that the system we are dealing with is, basically, infinitely large. Therefore, any "*measurable*" perturbation introduced to the system through the wellbore will never reach the outer boundary of the reservoir. Therefore, the outer boundary, located at an infinite distance from the wellbore, will never "feel" the presence of wellbore and will maintain a constant pressure "p_i" for all the time. In the implementation of the infinite-acting reservoir concept, it is implied that perturbations to the reservoir are introduced through the wellbore. The infinite-acting reservoir concept is especially useful in field applications when measurable pressure transients have not traveled far enough to reach the closest reservoir boundary (in other words while the reservoir is infinite-acting). Even though one cannot find a "true" infinite reservoir, the infinite acting behavior is not unrealistic and in the implementation of the solution protocol becomes extremely helpful. As mentioned earlier, no matter how small the reservoir is, there is always an infinite-acting period during which no *measurable* pressure change is observed at a point at the actual boundary. This is why it becomes so important to recognize that solution becomes applicable during the time period in which the infinite-acting behavior of the reservoir holds.

Infinite-acting flow (IAF) in petroleum engineering refers to the time period when pressure disturbances generated at the wellbore are continuously spreading toward the outer boundary—but have yet to reach it. Hence, the reservoir is acting as an infinitely large formation would. Since the speed of propagation is controlled by the hydraulic diffusivity, so is the duration of infinite-acting flow period. Large values of the hydraulic diffusivity (e.g., large permeabilities) are associated with short *IAF* periods, and small values (e.g., in tight formations) are associated with extended periods of *IAF*. During *IAF*, only the points in the reservoir domain experiencing depletion, i.e., those behind the pressure propagation or within the radius of propagation, are contributing to fluid production. As a result, drainage areas continuously expand during IAF.

Finite Reservoirs with Closed Boundaries, BDF, and PSS

A second useful reservoir idealization is that of a finite reservoir with closed boundaries. In this idealization, the radial reservoir has a finite size defined by the external radius "r_e" and the boundary is considered closed to flow (a no-flow boundary). From Darcy's law, a no-flow boundary implies that pressure gradient across the boundary (at every point on the boundary) is equal to zero (no-inflow or outflow of fluids across the boundary). Accordingly, the following outer boundary condition is imposed to the well test model that we have developed so far.

$$\text{At } r = r_e \quad \frac{\partial p}{\partial r} = 0 \quad \text{for all } t \geq 0$$

Finite reservoirs with closed boundaries will also experience an early period of *IAF* while the wellbore pressure disturbance propagates outwards toward the location of the external no-flow boundary ($r = r_e$). Once pressure depletion has been experienced by all locations of the no-flow boundary, the reservoir experiences *boundary-dominated flow* (*BDF*). During boundary-dominated flow, all points in the reservoir domain, including the boundary, experience depletion and contribute to fluid production. Drainage areas no longer expanding remain constant during *BDF*. In *BDF*, average reservoir pressure declines continuously with time and every point of the reservoir can experience depletion at a different rate.

Wells produced at constant rate at the inner boundary (wellbore, $r = r_w$) experience a special type of *BDF*. Because of the constant-rate specification, pressure is bound to decline uniformly at every point of the reservoir, i.e., the value of $\partial p/\partial t$ is the same throughout the reservoir. This type of reservoir flow condition has been termed *pseudo-steady state (PSS) flow*, which is a special type of *BDF* where the rate of pressure depletion ($\partial p/\partial t$) remains the same in time and space. *PSS* is also commonly referred to as the "stabilized flow" condition. In *PSS*, average reservoir pressure declines continuously and every point of the reservoir experiences the same $\partial p/\partial t$ depletion rate.

Finite Reservoirs with Constant Pressure Boundaries and SSF

Another idealization in single-well analytical applications is the case of finite reservoirs with constant pressure boundaries. In this idealization, the radial reservoir has a finite size defined by the external radius "r_e". However, the boundary is not closed to flow as it is assumed that the pressure of the boundary remains constant despite of fluid withdrawal or injection through the wellbore. Therefore, the following outer boundary condition is imposed to the mathematical formulation:

$$\text{At } r = r_e, \quad p = p_i \quad \text{for all} \quad t \geq 0$$

In practical applications, reservoirs that are connected to an infinitely large aquifers (reservoirs with strong edge or bottom water drive) or reservoirs associated with a gas cap have constant pressure boundaries where at the interface (between the reservoir and the aquifer or gas cap) the pressure *remains constant*. In these finite reservoirs, when the pressure transient reaches the constant-pressure outer boundary, the pressure distribution will not change with time and true *SSF* conditions will be established. Flow across the boundary will be equal to the flow through the well. In *SSF*, average reservoir pressure does not decline with time ($\partial p/\partial t = 0$) and every point of the reservoir remains at the same pressure.

6.9 Analytical Solutions of the Dimensionless Formulation

6.9.1 Infinite-Acting Analytical Solution

Let us recall the dimensionless form of the flow equations that we derived in earlier.

Flow equation [Eq. (6.95)]:

$$\frac{1}{r_D}\frac{\partial}{\partial r_D}\left(r_D\frac{\partial(\Delta p_D)}{\partial r_D}\right)=\frac{\partial(\Delta p_D)}{\partial t_D}$$

or,

$$\frac{\partial^2(\Delta p_D)}{\partial r_D^2}+\frac{1}{r_D}\frac{\partial(\Delta p_D)}{\partial r_D}=\frac{\partial(\Delta p_D)}{\partial t_D} \quad (6.102)$$

Inner boundary condition [Eq. (6.97)]:

$$r_D\frac{\partial(\Delta p_D)}{\partial r_D}\bigg|_{r_D=1}=-1 \quad \text{for} \quad t_D>0$$

In this development, we represent the well as a "line," where the sandface originally at $r_D = 1$ is now located at the very center ($r_D = 0$). Using the *line-source* representation, we rewrite the inner boundary condition for a modified inner boundary condition for practically the same result as follows:

$$\lim_{r_D\to 0} r_D\frac{\partial(\Delta p_D)}{\partial r_D}=-1 \quad \text{for all} \quad t_D>0 \quad (6.103)$$

Outer boundary condition [Eq. (6.98)]:

$$\Delta p_D \to 0 \quad \text{as} \quad r_D\to\infty \quad \text{for all} \quad t_D>0$$

Initial condition [Eq. (6.99)]:

$$\Delta p_D \to 0 \quad @ \quad t_D=0 \quad \text{for} \quad 1\le r_D<\infty$$

This mathematical formulation considers the case of an *infinitely large reservoir*. The analytical solution of the formulation can be obtained through the introduction of the *Boltzmann's transformation* as we show below. However, the analytical solution of the *finite reservoir* cases is somewhat more involved as it requires the use of *Laplace transformations*.

In the dimensionless formulation, the functionality of the dependent variable (Δp_D) is described through two variables:

$$\Delta p_D = f(r_D, t_D)$$

Fundamental to the analytical solution of the infintely-large reservoir case is the *Boltzmann transformation*. This transformation introduces a change of variables, a new *"dimension"* which considers r_D and t_D at the same time. Thus, the purpose of the transformation is collapsing the dependency:

$$\Delta p_D = f(r_D, t_D)$$

into one single variable, let us say "X", that incorporates both r_D and t_D so that:

$$\Delta p_D = f(X)$$

It should be kept in mind that by applying this transformation, we will ultimately be able to convert the original partial differential equation (*PDE*) into an ordinary differential equation (*ODE*). The proposed change of variable, or the Boltzmann's transformation, is defined as:

$$X = \frac{r_D^2}{4t_D} \tag{6.104}$$

At this point, it will be necessary to express each of the independent variables in the original dimensionless differential equation [Eq. (6.102)]:

$$\frac{\partial^2(\Delta p_D)}{\partial r_D^2} + \frac{1}{r_D}\frac{\partial(\Delta p_D)}{\partial r_D} = \frac{\partial(\Delta p_D)}{\partial t_D}$$

as a function of "X". To build up each of these terms, the application of the "chain rule" becomes instrumental in this process as shown below.

$$\frac{\partial(\Delta p_D)}{\partial t_D} = \frac{\partial(\Delta p_D)}{\partial X}\frac{\partial X}{\partial t_D} \tag{6.105}$$

$$\frac{\partial(\Delta p_D)}{\partial r_D} = \frac{\partial(\Delta p_D)}{\partial X}\frac{\partial X}{\partial r_D} \tag{6.106}$$

and,

$$\frac{\partial^2(\Delta p_D)}{\partial r_D^2} = \frac{\partial}{\partial r_D}\left(\frac{\partial(\Delta p_D)}{\partial X}\frac{\partial X}{\partial r_D}\right) = \frac{\partial}{\partial X}\left(\frac{\partial(\Delta p_D)}{\partial X}\frac{\partial X}{\partial r_D}\right)\frac{\partial X}{\partial r_D} \tag{6.107}$$

As can be seen from the definition of the Boltzmann's transformation [Eq. (6.104)]:

$$\frac{\partial X}{\partial r_D} = \frac{r_D}{2t_D} = \frac{2X}{r_D} \tag{6.108}$$

and,

$$\frac{\partial X}{\partial t_D} = \frac{r_D^2}{4}\left(-\frac{1}{t_D^2}\right) = -\frac{r_D^2}{4t_D^2} = -\frac{4X^2}{r_D^2} \tag{6.109}$$

Substituting into Eqs. (6.105) through (6.107), one obtains:

$$\frac{\partial(\Delta p_D)}{\partial t_D} = \left(-\frac{4X^2}{r_D^2}\right)\frac{\partial(\Delta p_D)}{\partial X} \tag{6.110}$$

$$\frac{\partial(\Delta p_D)}{\partial r_D} = \left(\frac{2X}{r_D}\right)\frac{\partial(\Delta p_D)}{\partial X} \tag{6.111}$$

$$\frac{\partial^2(\Delta p_D)}{\partial r_D^2} = \frac{\partial}{\partial X}\left(\frac{2}{r_D}X\frac{\partial(\Delta p_D)}{\partial X}\right)\frac{2X}{r_D} \tag{6.112}$$

Equation (6.112) needs some further manipulation. For this, let us recall that "r_D" is a function of "X" and also recall that:

$$\frac{\partial}{\partial r}(abc) = ab\frac{\partial c}{\partial r} + ac\frac{\partial b}{\partial r} + bc\frac{\partial a}{\partial r} \tag{6.113}$$

Applying Eq. (6.113) to Eq. (6.112), one gets:

$$\frac{\partial^2(\Delta p_D)}{\partial r_D^2} = \left(\frac{2X}{r_D}\frac{\partial^2(\Delta p_D)}{\partial X^2} + \frac{2}{r_D}\frac{\partial(\Delta p_D)}{\partial X}(1) - X\frac{\partial(\Delta p_D)}{\partial X}\frac{2}{r_D^2}\frac{\partial r_D}{\partial X}\right)\frac{2X}{r_D} \tag{6.114}$$

$$\frac{\partial^2(\Delta p_D)}{\partial r_D^2} = \frac{4X^2}{r_D^2}\frac{\partial^2(\Delta p_D)}{\partial X^2} + \frac{2X}{r_D}\left(\frac{2}{r_D} - \frac{2X}{r_D^2}\frac{r_D}{2X}\right)\frac{\partial(\Delta p_D)}{\partial X} \tag{6.115}$$

$$\frac{\partial^2(\Delta p_D)}{\partial r_D^2} = \frac{4X^2}{r_D^2}\frac{\partial^2(\Delta p_D)}{\partial X^2} + \frac{2X}{r_D^2}\frac{\partial(\Delta p_D)}{\partial X} \tag{6.116}$$

Introducing these new definitions of spatial and temporal derivatives into the original partial differential and dividing through by $4X/r_D^2$ one obtains:

$$X\frac{\partial^2(\Delta p_D)}{\partial X^2} + (1+X)\frac{\partial(\Delta p_D)}{\partial X} = 0 \tag{6.117}$$

Equation (6.117) is the ordinary differential equation we wanted to obtain. This shows that the Boltzmann's transformation has been successful in converting the original partial differential equation into a simpler ordinary differential equation. The accompanying boundary and initial conditions become (using the definition of Boltzmann's transformation):

Initial condition:

$$\Delta p_D = 0 \quad \text{as} \quad X \to \infty \tag{6.118}$$

Outer boundary condition:

$$\Delta p_D \to 0 \quad \text{as} \quad X \to \infty \tag{6.119}$$

Inner boundary condition:

$$\lim_{r_D \to 0} r_D \frac{\partial(\Delta p_D)}{\partial r_D} = -1$$

$$\rightarrow \quad \lim_{r_D \to 0} r_D \frac{\partial(\Delta p_D)}{\partial X}\frac{\partial X}{\partial r_D} = -1$$

$$\rightarrow \quad \lim_{X \to 0} r_D \frac{2X}{r_D}\frac{\partial(\Delta p_D)}{\partial X} = -1$$

$$\rightarrow \quad \lim_{X \to 0} X \frac{\partial(\Delta p_D)}{\partial X} = -\frac{1}{2} \tag{6.120}$$

It should be noted that the initial and the outer boundary conditions collapse to basically the same requirement. The idea is now to solve the ordinary differential equation of Eq. (6.117) subjected to the boundary conditions in Eqs. (6.119) and (6.120). Let us start by reducing the order of the ordinary differential equation [Eq. (6.117)].

$$\text{Let: } \Delta p_D^* = \frac{\partial(\Delta p_D)}{\partial X} \quad \rightarrow \quad \frac{\partial(\Delta p_D^*)}{\partial X} = \frac{\partial^2(\Delta p_D)}{\partial X^2} \tag{6.121}$$

If one substitutes Eq. (6.121) into the ordinary differential equation [Eq. (6.117)], the order of the ordinary differential equation is reduced to first order:

$$X\frac{\partial(\Delta p_D^*)}{\partial X} + (1 + X)\Delta p_D^* = 0 \tag{6.122}$$

The first-order ordinary differential equation can be solved by applying the separation of variables technique:

$$\int \frac{\partial(\Delta p_D^*)}{\Delta p_D^*} = -\int \frac{(1 + X)}{X}\partial X \tag{6.123}$$

$$\rightarrow \quad \ln \Delta p_D^* = -\ln X - X + C \tag{6.124}$$

$$\rightarrow \quad \Delta p_D^* = e^{-\ln X - X + C} = e^{-\ln X}e^{-X}e^{C} \tag{6.125}$$

$$\rightarrow \quad \Delta p_D^* = \frac{c_1}{X}e^{-X} \quad (c_1 = const = e^C) \tag{6.126}$$

Recalling the definition in Eq. (6.121), one obtains that:

$$\frac{\partial(\Delta p_D)}{\partial X} = \frac{c_1}{X}e^{-X} \tag{6.127}$$

By using the boundary condition described by Eq. (6.120) and writing Eq. (6.127) for $X \to \infty$, one can evaluate the integration constant "c_1" as shown below:

$$\lim_{X\to 0} r_D \frac{2X}{r_D}\frac{\partial(\Delta p_D)}{\partial X} = \lim_{X\to 0} 2X\frac{c_1}{X}e^{-X} = -1$$

$$c_1 = -\frac{1}{2} \tag{6.128}$$

Hence, Eq. (6.127) becomes

$$\frac{d(\Delta p_D)}{dX} = -\frac{1}{2}\frac{e^{-X}}{X} \tag{6.129}$$

$$\text{or, } d(\Delta p_D) = -\frac{1}{2}\frac{e^{-X}}{X}dX \tag{6.130}$$

Using the second boundary condition available to us, we can integrate this expression within the domain $X < \infty$ by recalling:

$$\Delta p_D \to 0 \quad \text{as} \quad X \to \infty$$

Therefore, we can set the limits of integration as follows:

$$\int_0^{\Delta p_D} d(\Delta p_D) = -\frac{1}{2}\int_\infty^X \frac{e^{-X}}{X}dX \tag{6.131}$$

The integral that has showed up in the right-hand side of this equation resembles the mathematical function called *Exponential Integral*. The definition of the exponential integral is as follows:

$$E_i(-X) = \int_\infty^X \frac{e^{-X}}{X}dX = -\int_X^\infty \frac{e^{-X}}{X}dX \tag{6.132}$$

$E_i(-X)$ is another mathematical function just like logarithmic, exponential, or circular functions, such as sine, cosine, or tangent functions. Performing the integration in Eq. (6.132) and substituting the definition for the *Exponential Integral*, we get:

$$\Delta p_D = -\frac{1}{2}E_i(-X) \tag{6.133}$$

Now, we recall the definition for the Boltzmann's transformation in Eq. (6.131) to obtain the final form of the solution of the classical well test formulation in dimensionless variables:

$$\Delta p_D = -\frac{1}{2}E_i\left(-\frac{r_D^2}{4t_D}\right) \tag{6.134}$$

Equation (6.134) is the basis of classical well testing analysis protocols. Therefore, it is a good idea to become well acquainted with characteristics of the *Exponential Integral* function.

The Exponential Integral Function, $E_i(-X)$

In mathematical handbooks we can find the exponential integral function expressed in infinite series as shown below:

$$E_i(-X) = \ln(1.781X) - X + \frac{X^2}{2 \cdot 2!} - \frac{X^3}{3 \cdot 3!} + \frac{X^4}{4 \cdot 4!} + \ldots \tag{6.135}$$

$$\text{or,} \quad E_i(-X) = \ln(1.781X) + \sum_{n=1}^{\infty} \frac{(-X)^n}{n \cdot n!} \tag{6.136}$$

It is useful to keep in mind the following simplifications that are available for the exponential integral when its argument becomes either very small or large:

- For large values of the exponential integral argument (typically when $X \geq 5$), the value of the exponential integral is negligible for practical purposes Hence, for practical purposes:

$$E_i(-X) \approx 0 \quad \text{for} \quad X \geq 5 \tag{6.137}$$

From the definition of "X", we may write this cut-off point as:

$$E_i(-X) \approx 0 \quad \text{for} \quad \frac{r_D^2}{4t_D} \geq 5 \tag{6.138}$$

A simple conceptual analysis of Eq. (6.138) shows that when r_D is large and t_D is small the pressure drop is expected to be close to zero away from the wellbore for small times which is completely logical.

- On the other hand, for $X \leq 0.01$ or for small values of the exponential integral argument, the contribution of the terms under the infinite sum in Eq. (6.138) becomes negligible and the exponential integral function can be approximated by:

$$E_i(-X) \approx \ln(1.781X) \quad X \leq 0.01 \tag{6.139}$$

Equivalently, we may rewrite the above equation in terms of common logarithms with base 10.

$$\begin{aligned} E_i(-X) \approx \ln(1.781X) &= \ln(1.781) + \ln(X) \\ &= 0.5771 + 2.303\log_{10}(X) \end{aligned} \tag{6.140}$$

Using the definition of "X", we may write this cut-off point as:

$$E_i(-X) \approx \ln(1.781X) \quad \text{for} \quad \frac{r_D^2}{4t_D} \leq 0.01 \tag{6.141}$$

or,

$$E_i(-X) \approx \ln(1.781X) \quad \text{for} \quad \frac{t_D}{r_D^2} \geq 25 \tag{6.142}$$

A critical question that still needs to be addressed is what happens in between these two "cut-off points" described by Eqs. (6.138) and (6.142)? There are tables and graphs available in the literature where the corresponding values of $E_i(-X)$ can be read off directly for values of the argument X. Table 6.3 gives values of the exponential integral if $0.01 < X < 10$, for example.

As the argument of the exponential integral increases, the value of the exponential integral gets progressively smaller. It should be noted that all values of $E_i(-X)$ are negative. Therefore, all *dimensionless* pressure drop calculated through Eq. (6.134) must be positive. We may introduce above simplifications into Eq. (6.134), the general solution for an infinite-acting reservoir, repeated here:

$$\Delta p_D = -\frac{1}{2}E_i\left(-\frac{r_D^2}{4t_D}\right)$$

For small values of the argument, above equation can be rewritten as:

$$\Delta p_D = -\frac{1}{2}\ln\left(1.781\frac{r_D^2}{4t_D}\right) = \frac{1}{2}\left\{\ln\left[\frac{t_D}{r_D^2}\right] + 0.80907\right\} \quad \text{for} \quad {t_D}/{r_D^2} \geq 25 \tag{6.143}$$

If above expression is written for the wellbore alone ($r_D = 1$),

$$p_t = \Delta p_D\big|_{well} = \frac{1}{2}[\ln t_D + 0.80907] \quad \text{for} \quad t_D \geq 25 \tag{6.144}$$

6.9.2 Finite Reservoir Analytical Solutions

For the case of radial cylindrical flow, constant production rate, and finite circular reservoir with *no flow at outer boundary,* we have the following formulation. It should be noted that only the finite outer boundary condition is different from the original infinite-acting reservoir formulation.

Flow Equation:

$$\frac{1}{r_D}\frac{\partial}{\partial r_D}\left[r_D\frac{\partial(\Delta p_D)}{\partial r_D}\right] = \frac{\partial(\Delta p_D)}{\partial t_D} \tag{6.145}$$

Inner boundary condition:

$$r_D\frac{\partial(\Delta p_D)}{\partial r_D}\bigg|_{r_D=1} = -1 \text{ for all } \quad t_D > 0 \tag{6.146}$$

Outer boundary condition: No-flow (sealed reservoir)

$$\frac{\partial(\Delta p_D)}{\partial r_D}\bigg|_{r_D=r_{eD}} = 0 \text{ where } \quad r_{eD} = \frac{r_e}{r_w} \text{ for all } \quad t_D \geq 0 \tag{6.147}$$

Initial condition:

$$\Delta p_D = 0 \quad \text{at} \quad t_D = 0 \quad \text{for } 1 \leq r_D < \infty \tag{6.148}$$

$-E_i(-X)$, $0.00 < X < 0.209$, interval = 0.001										
X	**0**	**1**	**2**	**3**	**4**	**5**	**6**	**7**	**8**	**9**
0.00	0.00	6.332	5.639	5.235	4.948	4.726	4.545	4.392	4.259	4.142
0.01	4.038	3.944	3.858	3.779	3.705	3.637	3.574	3.514	3.458	3.405
0.02	3.355	3.307	3.261	3.218	3.176	3.137	3.098	3.062	3.026	2.992
0.03	2.959	2.927	2.897	2.867	2.838	2.810	2.783	2.756	2.731	2.706
0.04	2.681	2.658	2.634	2.612	2.590	2.568	2.547	2.527	2.507	2.487
0.05	2.468	2.449	2.431	2.413	2.395	2.377	2.360	2.344	2.327	2.311
0.06	2.295	2.280	2.264	2.249	2.235	2.220	2.206	2.192	2.178	2.164
0.07	2.151	2.138	2.125	2.112	2.099	2.087	2.074	2.062	2.050	2.039
0.08	2.027	2.015	2.004	1.993	1.982	1.971	1.960	1.950	1.939	1.929
0.09	1.919	1.909	1.899	1.889	1.879	1.869	1.860	1.851	1.841	1.832
0.10	1.823	1.814	1.805	1.796	1.788	1.779	1.770	1.762	1.754	1.745
0.11	1.737	1.729	1.721	1.713	1.705	1.697	1.690	1.682	1.674	1.667
0.12	1.660	1.652	1.645	1.638	1.631	1.623	1.616	1.609	1.603	1.596
0.13	1.589	1.582	1.576	1.569	1.562	1.556	1.549	1.543	1.537	1.530
0.14	1.524	1.518	1.512	1.506	1.500	1.494	1.488	1.482	1.476	1.470
0.15	1.464	1.459	1.453	1.447	1.442	1.436	1.431	1.425	1.420	1.415
0.16	1.409	1.404	1.399	1.393	1.388	1.383	1.378	1.373	1.368	1.363
0.17	1.358	1.353	1.348	1.343	1.338	1.333	1.329	1.324	1.319	1.314
0.18	1.310	1.305	1.301	1.296	1.291	1.287	1.282	1.278	1.274	1.269
0.19	1.265	1.261	1.256	1.252	1.248	1.243	1.239	1.235	1.231	1.227
0.20	1.223	1.219	1.215	1.210	1.206	1.202	1.199	1.195	1.191	1.187

TABLE 6.3 Values of Exponential Integral $-E_i(-X)$

0.0	0.000	4.038	3.355	2.959	2.681	2.468	2.295	2.151	2.027	1.919
0.1	1.823	1.737	1.660	1.589	1.524	1.464	1.409	1.358	1.310	1.265
0.2	1.223	1.183	1.145	1.110	1.076	1.044	1.014	0.985	0.957	0.931
0.3	0.906	0.882	0.858	0.836	0.815	0.794	0.774	0.755	0.737	0.719
0.4	0.702	0.686	0.670	0.655	0.640	0.625	0.611	0.598	0.585	0.572
0.5	0.560	0.548	0.536	0.525	0.514	0.503	0.493	0.483	0.473	0.464
0.6	0.454	0.445	0.437	0.428	0.420	0.412	0.404	0.396	0.388	0.381
0.7	0.374	0.367	0.360	0.353	0.347	0.340	0.334	0.328	0.322	0.316
0.8	0.311	0.305	0.300	0.294	0.289	0.284	0.279	0.274	0.269	0.265
0.9	0.260	0.256	0.251	0.247	0.243	0.239	0.235	0.231	0.227	0.223
1.0	0.219	0.216	0.212	0.209	0.205	0.202	0.199	0.195	0.192	0.189
1.1	0.186	0.183	0.180	0.177	0.174	0.172	0.169	0.166	0.164	0.161
1.2	0.158	0.156	0.153	0.151	0.149	0.146	0.144	0.142	0.140	0.138
1.3	0.135	0.133	0.131	0.129	0.127	0.125	0.124	0.122	0.120	0.118
1.4	0.116	0.114	0.113	0.111	0.109	0.108	0.106	0.105	0.103	0.102
1.5	0.1000	0.0985	0.0971	0.0957	0.0943	0.0929	0.0915	0.0902	0.0889	0.0876
1.6	0.0863	0.0851	0.0838	0.0826	0.0814	0.0802	0.0791	0.0780	0.0768	0.0757
1.7	0.0747	0.0736	0.0725	0.0715	0.0705	0.0695	0.0685	0.0675	0.0666	0.0656
1.8	0.0647	0.0638	0.0629	0.0620	0.0612	0.0603	0.0595	0.0586	0.0578	0.0570
1.9	0.0562	0.0554	0.0547	0.0539	0.0531	0.0524	0.0517	0.0510	0.0503	0.0496
2.0	0.0489	0.0482	0.0476	0.0469	0.0463	0.0456	0.0450	0.0444	0.0438	0.0432

Table 6.3 Values of Exponential Integral $-E_i(-X)$ *(Continued)*

$-E_i(-X)$, 2.00 < X < 10.9, interval = 0.1										
X	0	1	2	3	4	5	6	7	8	9
2	0.048901	0.042614	0.037191	0.032502	0.02844	0.024915	0.02185	0.019182	0.016855	0.014824
3	0.013048	0.011494	0.010133	0.008939	0.007891	0.00697	0.00616	0.005448	0.00482	0.004267
4	0.003779	0.003349	0.002969	0.002633	0.002336	0.002073	0.001841	0.001635	0.001453	0.001291
5	0.001148	0.001021	0.000909	0.000809	0.00072	0.000641	0.000571	0.000509	0.000453	0.000404
6	0.00036	0.000321	0.000286	0.000255	0.000228	0.000203	0.000182	0.000162	0.000145	0.000129
7	0.000115	0.000103	9.22×10^{-5}	8.24×10^{-5}	7.36×10^{-5}	6.58×10^{-5}	5.89×10^{-5}	5.26×10^{-5}	4.71×10^{-5}	4.21×10^{-5}
8	3.77×10^{-5}	3.37×10^{-5}	3.02×10^{-5}	2.70×10^{-5}	2.42×10^{-5}	2.16×10^{-5}	1.94×10^{-5}	1.73×10^{-5}	1.55×10^{-5}	1.39×10^{-5}
9	1.24×10^{-5}	1.11×10^{-5}	9.99×10^{-6}	8.95×10^{-6}	8.02×10^{-6}	7.18×10^{-6}	6.44×10^{-6}	5.77×10^{-6}	5.17×10^{-6}	4.64×10^{-6}
10	4.16×10^{-6}	3.73×10^{-6}	3.34×10^{-6}	3.00×10^{-6}	2.69×10^{-6}	2.41×10^{-6}	2.16×10^{-6}	1.94×10^{-6}	1.74×10^{-6}	1.56×10^{-6}

Table 6.3 Values of Exponential Integral $-E_i(-X)$ *(Continued)*
Source: From Nisle (1956).

Solution: In this case, the system does not accept the Boltzmann's transformation due to the lack of an infinite domain. Laplace transforms may be applied to reach a solution and the solution can be expressed in terms of Bessel functions as follows:

$$\Delta p_D = \frac{2}{r_{eD}^2 - 1}\left(\frac{r_D^2}{4} + t_D\right) - \frac{r_{eD}^2 \ln r_D}{r_{eD}^2 - 1} - \frac{\left(3r_{eD}^4 - 4r_{eD}^4 \ln r_{eD} - 2r_{eD}^2 - 1\right)}{4(r_{eD}^2 - 1)^2}$$
$$+ \pi \sum_{n=1}^{\infty} \frac{e^{-\alpha_n^2 t_D} J_1^2(\alpha_n r_{eD})[J_1(\alpha_n)Y_o(\alpha_n r_D) - Y_1(\alpha_n)J_o(\alpha_n r_D)]}{\alpha_n\left[J_1^2(\alpha_n r_{eD}) - J_1^2(\alpha_n)\right]} \tag{6.149}$$

where α_n are the roots of $[J_1(\alpha_n r_{eD})Y_1(\alpha_n) - J_1(\alpha_n)Y_1(\alpha_n r_{eD})] = 0$, J_1 and Y_1 being the Bessel functions of the first and second kind. The most common form of Eq. (6.149) is when it is applied at the wellbore conditions ($r_D = 1$), for which case we have the dimensionless pressure at the well:

$$p_t = \Delta p_D\big|_{well} = \frac{2t_D}{r_{eD}^2} + \ln r_{eD} - \frac{3}{4} + 2\sum_{n=1}^{\infty} \frac{e^{-\alpha_n^2 t_D} J_1^2(\alpha_n r_{eD})}{\alpha_n^2\left[J_1^2(\alpha_n r_{eD}) - J_1^2(\alpha_n)\right]} \tag{6.150}$$

In Eq. (6.150), as "t_D" term increases, terms under the summation term becomes smaller because of the exponential decaying function. Thus, for $r_e >> r_w$, which is usually the case, the dimensionless pressure drop at the well then becomes:

$$p_t = \frac{2t_D}{r_{eD}^2} + \ln r_{eD} - \frac{3}{4} \quad \text{(for large times, } t_D > 0.25 r_{eD}^2\text{)} \tag{6.151}$$

For values of $t_D < 0.25 r_{eD}^2$, outer boundary effects are insignificant and reservoir behaves as if it were infinitely large. Therefore, for the solution at the wellbore, one can confidently use Eq. (6.144) which is repeated here for the sake of completeness:

$$p_t = \frac{1}{2}\{\ln t_D + 0.80907\} \quad \text{for} \quad t_D < 0.25 r_{eD}^2$$

For the case of radial cylindrical flow, constant production rate, and finite circular reservoir with *constant pressure at outer boundary,* we have the following formulation:

Flow equation:

$$\frac{1}{r_D}\frac{\partial}{\partial r_D}\left[r_D \frac{\partial(\Delta p_D)}{\partial r_D}\right] = \frac{\partial(\Delta p_D)}{\partial t_D} \tag{6.152}$$

Inner boundary condition:

$$r_D \frac{\partial(\Delta p_D)}{\partial r_D}\bigg|_{r_D = 1} = -1 \quad \text{for} \quad \text{all} \quad t_D > 0 \tag{6.153}$$

Outer boundary condition: Constant pressure

$$\Delta p_D = 0 \quad \text{at} \quad r_D = r_{eD} \quad \left(r_{eD} = \frac{r_e}{r_w}\right) \quad \text{for all} \quad t_D \geq 0 \tag{6.154}$$

Initial condition:

$$\Delta p_D = 0 \quad \text{at} \quad t_D = 0 \quad \text{for} \quad 1 < r_D < \infty \tag{6.155}$$

Solution: Again, the system does not accept the Boltzmann's transformation. We may apply Laplace transforms to reach a solution, and express the solution in terms of Bessel functions:

$$\Delta p_D = \ln\left(\frac{r_{eD}}{r_D}\right) + \pi\sum_{n=1}^{\infty}\frac{e^{-\beta_n^2 t_D} J_o^2(\beta_n r_{eD})[Y_1(\beta_n)J_o(\beta_n r_D) - J_1(\beta_n)Y_o(\beta_n r_D)]}{\beta_n\left[J_1^2(\beta_n) - J_o^2(\beta_n r_{eD})\right]} \tag{6.156}$$

where β_n are the roots of $[Y_o(\beta_n r_{eD})J_1(\beta_n) - Y_1(\beta_n)J_o(\beta_n r_{eD})] = 0$. When we apply Eq. (6.156) at the wellbore ($r_D = 1$), it simplifies to the form:

$$p_t = \Delta p_D\big|_{well} = \ln r_{eD} - 2\sum_{n=1}^{\infty}\frac{e^{-\beta_n^2 t_D} J_o^2(\beta_n r_{eD})}{\beta_n^2\left[J_1^2(\beta_n) - J_o^2(\beta_n r_{eD})\right]} \tag{6.157}$$

As "t_D" term increases, summation term in above equation becomes negligibly small (again due to the exponential decaying function). Thus, for $r_e >> r_w$, which is usually the case, the pressure at the well is:

$$p_t = \ln r_{eD} \quad (\text{for large times, } t_D > r_{eD}^2) \tag{6.158}$$

For values of $t_D < 0.25r_{eD}^2$, outer boundary effects are insignificant and reservoir behaves as if it were infinitely large. Therefore, for the solution at the wellbore, one can use the infinite-acting solution

$$p_t = \frac{1}{2}\{\ln t_D + 0.80907\} \quad \text{for } t_D < 0.25r_{eD}^2 \tag{6.159}$$

It is important to realize that Eq. (6.158) is simply the *dimensionless* form of Darcy's law. We see that there is no time dependency in this expression; therefore, it should represent the steady-state flow conditions. One can verify these statements by simply inserting the definitions of the dimensionless pressure drop at the wellbore (p_t) and the dimensionless reservoir radius (r_{eD}) into Eq. (6.158).

Aziz and Flock Chart

Aziz and Flock (1963) summarized all the cases we have studied so far in their well-known chart, as shown in Fig. 6.8. Note that the solutions they presented are only valid for dimensionless pressure drops at the wellbore because the chart in Fig. 6.8 represents the equations for the case $r_D = 1$.

Figure 6.8 is a semi-logarithmic plot of "p_t" versus "t_D". The composite graph resembling a "fish-spine" shows the wellbore solution for infinite-acting reservoir, finite reservoir with no-flow external boundary, and finite reservoirs with constant pressure at the external boundary. The "backbone" of the *fish-spine* represents the infinite-acting reservoir solution at the well point ($r_{eD} = \infty$). Notice that the *backbone* shows a behavior of a straight line since it represents the following equation that we have seen several times earlier:

$$p_t = \frac{1}{2}\{\ln t_D + 0.80907\}$$

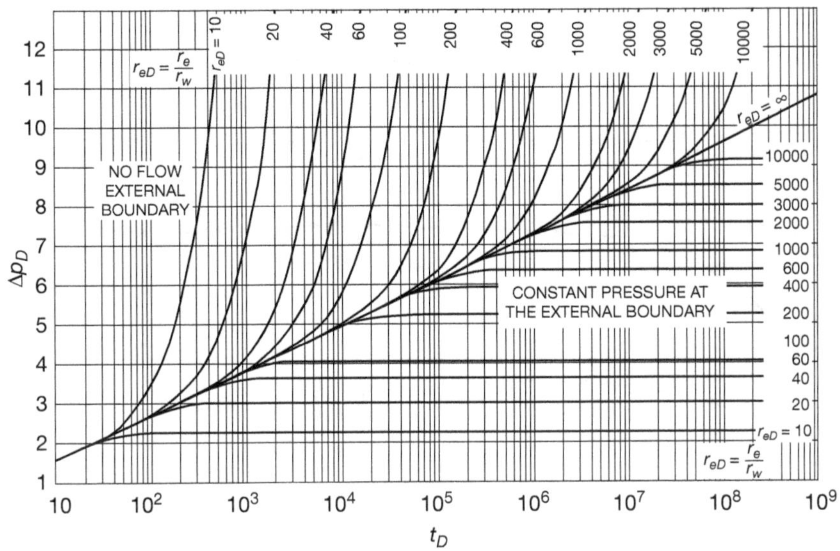

FIGURE 6.8 Aziz and Flock chart for dimensionless pressure drop at wellbore location ($r_D = 1$) in infinite and finite reservoirs (after Aziz and Flock, 1963).

The upper portion of the graph shows solutions for the no-flow external boundary finite reservoir for various values of dimensionless reservoir size (r_{eD}), while the lower portion of it shows solutions for the constant-pressure external boundary finite reservoir for various values of dimensionless reservoir size (r_{eD}).

6.10 Important Well-Test Analysis Concepts

Effective Drainage Radius

The effective drainage concept gives us an idea of the area that is *effectively* being drained by a given well in a reservoir. To better visualize this concept, let us consider a well producing from a finite reservoir with no flow outer boundary for a certain period of time such that some degree of depletion has taken place. If this well is now shut in for a sufficiently long time, the pressure throughout the reservoir will ultimately stabilize to a new uniform value (average reservoir pressure), say "p_R". Because this is a finite reservoir with no flow at the external boundary, this stabilization pressure "p_R" will be smaller than the original reservoir initial pressure "p_i". This situation is illustrated in Fig. 6.9.

Note that the difference ($p_i - p_R$) is an indication of how much depletion has taken place in the system. In Fig. 6.9, the effective drainage radius is marked as the intersection point between the pressure profile before shut-in and the stabilized average reservoir pressure (p_R) in the wellbore. From this figure we can also conclude that if the well is shut in after a moderately short production period, the pressure profile at the shut in time would be considerably higher and the resulting average reservoir pressure at the shut in time would be higher as well. In that case, the intersection of the two pressure curves will occur at a distance closer to the wellbore. Equivalently, it is seen that "r_D" moves away from the wellbore as the well continues to produce for longer times.

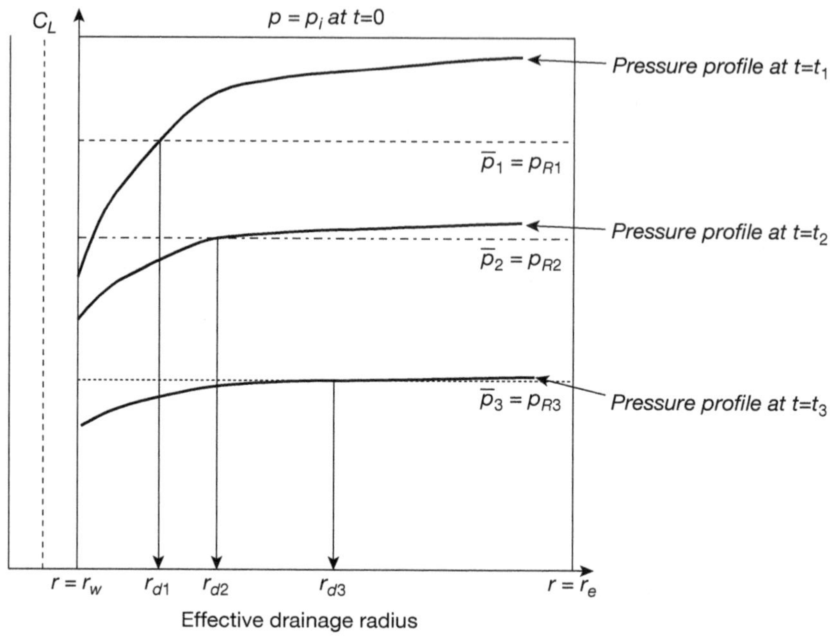

FIGURE 6.9 Effective drainage radius concept.

However, at one point in time, the movement of effective drainage radius comes virtually to a stop. This defines a final value for the effective drainage radius for the well in consideration.

We may want to track down this movement of "r_D" with time. To do this, we will literally force the steady-state equation to describe this transient behavior. This is, by all accounts, only an approximation. For the steady-state profile we have:

$$p_t = \frac{p_i - p_{wf}}{p_i q_D} = \ln\frac{r_e}{r_w} \tag{6.160}$$

Similar to the equation given above, we can write (by replacing p_i with p_R and by replacing r_e with r_d):

$$\frac{p_R - p_{wf}}{p_i q_D} = \ln\frac{r_d}{r_w} \tag{6.161}$$

At early times, not much has been produced and it is reasonable to expect that $p_R \approx p_i$. Additionally, pressure gradients have not reached the boundaries, and the infinite-acting solution is valid:

$$p_t = \Delta p_D\big|_{well} = \frac{1}{2}\{\ln t_D + 0.80907\} \tag{6.162}$$

Combining above equations, one can write:

$$\ln\frac{r_d}{r_w} = \frac{1}{2}\{\ln t_D + 0.80907\} \tag{6.163}$$

Equation (6.163) tracks the movement of "r_d" for early times.

As stated earlier, $\Delta p = (p_i - p_R)$ is an indication of how much depletion has taken place in the reservoir. Now, introducing the definition of compressibility:

$$c = -\frac{1}{V}\left(\frac{\partial V}{\partial p}\right)_T \tag{6.164}$$

It will be possible to come up with a means by which the degree of depletion can be estimated. For this, we rewrite Eq. (6.164) in finite form as follows:

$$\Delta p = p_i - p_R = \frac{\Delta V}{cV} \tag{6.165}$$

Or, in terms of flow rate (q) and reservoir size (r_e), Eq. (6.166) can be expressed as:

$$\Delta p = p_i - p_R = \frac{qt}{c(\pi r_e^2 h\phi)} \tag{6.166}$$

We now will attempt to rewrite this depletion term in terms of dimensionless quantities.

$$q_D = \frac{\mu q}{2\pi k h p_i} \quad \rightarrow \quad q = \frac{2\pi q_D p_i k h}{\mu} \tag{6.167}$$

and,

$$t_D = \frac{kt}{\phi\mu c r_w^2} \quad \rightarrow \quad t = \frac{t_D \phi\mu c r_w^2}{k} \tag{6.168}$$

Substituting Eqs. (6.167) and (6.168) into Eq. (6.166) we obtain:

$$\Delta p = p_i - p_R = \frac{\left(\frac{2\pi q_D p_i k h}{\mu}\right)\left(\frac{t_D \phi\mu c r_w^2}{k}\right)}{c\pi r_e^2 h\phi} \tag{6.169}$$

Simplifying above equation, one obtains:

$$\frac{p_i - p_R}{p_i q_D} = \frac{2t_D}{r_{eD}^2} \tag{6.170}$$

Equation (6.170) represents how depletion changes as a function of time. Let us now recall that for a finite reservoir with no-flow at the outer boundary, the late time approximation was obtained as:

$$\frac{p_i - p_{wf}}{p_i q_D} = \frac{2t_D}{r_{eD}^2} + \ln r_{eD} - \frac{3}{4} \tag{6.171}$$

Recalling that Eq. (6.171) applies for a finite reservoir with no-flow at the outer boundary at large times, when a substantial depletion has taken place. We see that we can subtract Eq. (6.171) from Eq. (6.170) to obtain:

$$\frac{p_R - p_{wf}}{p_i q_D} = \ln r_{eD} - \frac{3}{4} \tag{6.172}$$

Recalling Eq. (6.161) and substituting into Eq. (6.172), one obtains:

$$\ln\frac{r_d}{r_w} = \ln r_{eD} - \frac{3}{4} = \ln\frac{r_e}{r_w} - 0.75 \tag{6.173}$$

From which

$$\ln r_d = \ln\frac{r_e}{2.117} \tag{6.174}$$

or, finally,

$$r_d = 0.472 r_e \tag{6.175}$$

This is a rather interesting result. As we have discussed, at early times "r_d" begins to move toward "r_e" but it stops at "0.472 r_e" at late times. This means that with only one well, we can produce *effectively* only up to 47.2% of the reservoir as the drainage radius stops at that point. It should be noted that we are using the word "effectively" here. In reality the depletion takes throughout the entire reservoir and theoretically, one would be able to deplete the entire system as the total production time approaches to infinity. This discussion supports the general understanding that it is not possible to produce a large reservoir with just a single well drilled in the middle. This is why it is necessary to drill more than one well according to a certain pattern and spacing to develop the reservoir effectively. It should be reiterated that the effective drainage radius is a hypothetical concept which forces the steady-state solution to represent the transient solution. In fact, it does not represent a physical reality, and drainage takes place from the entire reservoir albeit at small rates from the sections of the reservoir beyond the effective drainage radius.

Radius of Investigation

In simple terms, the "radius of investigation" "r_{inv}" is a concept that tracks how far the pressure transient signal has traveled at a given time. In other words, the radius of investigation in a test defines the distance traveled by a measurable pressure transient from the tested well location. Let us consider one of the curves in the Aziz and Flock chart, for a given value of r_{eD}. This situation is illustrated in Fig. 6.10. At early times, the solution follows the infinite-acting behavior, i.e., the line of unit slope. Later on, and once the boundaries are felt, the solution departs from the unit slope line. We can see that we can establish a correspondence between this point of departure and the radius of investigation. When the pressure transient has reached the boundary, we say that at that time, radius of investigation is equal to the distance between the test well and that point of the reservoir (i.e., $r_{inv} = r_e$). If we correlate departure times to the r_{eD} values (or r_{inv} values for this matter), we can come up with the following expression:

$$t_D = 0.25 r_{eD}^2 \tag{6.176}$$

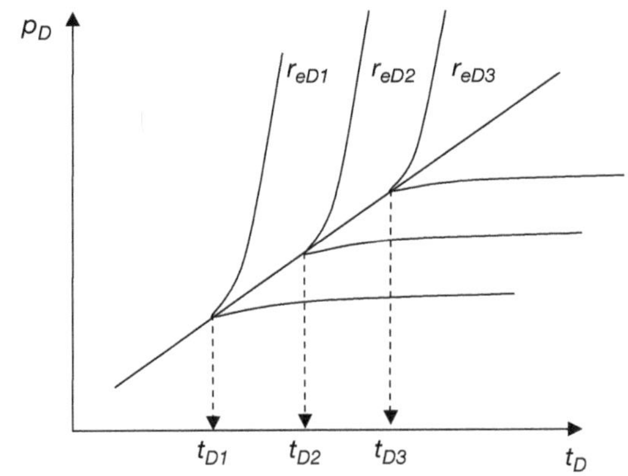

FIGURE 6.10 The concept of radius of investigation.

or, substituting the definitions for the dimensionless variables,

$$\frac{\lambda k}{\phi \mu c r_w^2} t = 0.25 \frac{r_{inv}^2}{r_w^2} \tag{6.177}$$

Solving for the radius of investigation, we obtain the expression:

$$r_{inv} = 2\left(\frac{\lambda k t}{\phi \mu c}\right)^{1/2} \tag{6.178}$$

Equation (6.178) tells us that the radius of investigation is scaled by the hydraulic diffusivity constant:

$$\eta = \frac{k}{\phi \mu c} \tag{6.179}$$

In fact, Eq. (6.178) can be rewritten as:

$$r_{inv} = 2\lambda^{0.5}(\eta t)^{1/2} \tag{6.180}$$

where $r_{inv} < r_e$ because once the pressure transients reach the boundary ($r_{inv} = r_e$), investigation radius stops at the boundary. The radius of investigation depends on the hydraulic diffusivity as the hydraulic diffusivity constant determines how rapid the pressure transients propagate through the reservoir. It is important to see that although pressure transients are introduced into the reservoir by perturbing the flow rate at the well, the speed of the pressure transient does not depend on the magnitude of the flow rate. Flow rate at the well simply scales the magnitude of the pressure change (drop or increase) at a given point and time in the reservoir. The radius of investigation is a useful concept to estimate for how long we need to run a well test for a specific radius of investigation.

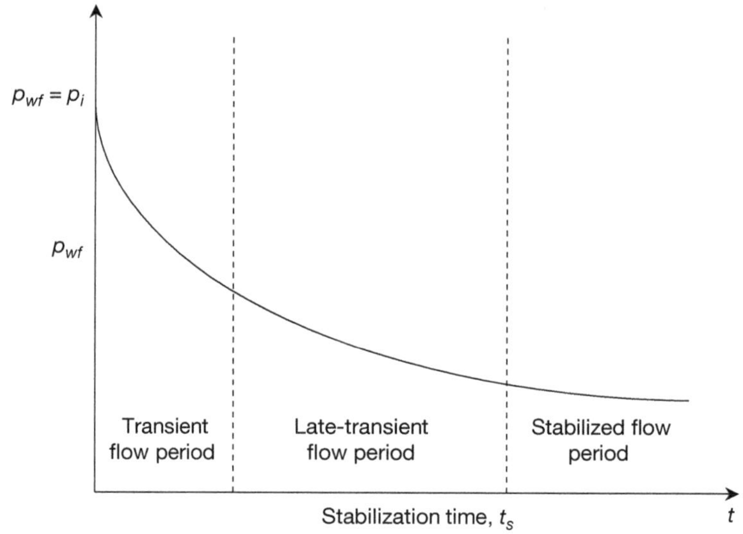

FIGURE 6.11 Flow regimes encountered during a well test (constant q).

Stabilization Time

A well is said to be stabilized when all the points on the outer boundary have felt a measurable pressure transient that has been sent by the well. Then, the time to stabilization is the amount of time that takes for the furthest point in the outer boundary from the well to experience a measurable pressure drop. Once a pressure transient is sent off from the well, it is on its way toward the boundaries. Each point on the outer boundary including the farthest point away from the well will feel the pressure transient given enough time. As soon as this happens, we say that the reservoir-well system has reached stabilization and pressure drop at every point in the reservoir starts to decrease in a linear fashion (pseudo steady-state flow regime).

Figure 6.11 presents the three flow periods that are encountered while well is producing (or while well is being tested):

1. *Infinite-acting (or transient) flow period:* This flow period starts with the implementation of the test and ends when the pressure transient reaches the first (closest) point on the external boundary.
2. *Late-transient flow period:* This flow period starts with the end of the infinite-acting period and continues until the farthest point on the external boundary experiences the pressure transient.
3. *Stabilized or pseudo-steady-state period*: This flow period starts with the end of the late transient period. During this flow period, every point in the reservoir experiences the pressure transient and enters *PSS* flow conditions.

"Time to stabilization" pertains to the duration of both infinite-acting and late-transient flow periods, i.e., the amount of time required for the stabilized flow period to start. For the case of a well in the center of a circular reservoir, all points of the external boundary are felt at the same time. Therefore, time to stabilization is the same as the time when the infinite-acting period ends. In this case, and because there is no late-transient flow period, we may obtain an expression for time to stabilization by calculating the time at which the gradients for the infinite-acting and stabilized flow periods are the same. In other words, $\frac{\partial}{\partial t_D}(p_t \text{ equation for infinite acting}) = \frac{\partial}{\partial t_D}(p_t \text{ equation for finite reservoirs})$

Substituting the appropriate expressions for infinite acting and finite reservoirs,

$$\frac{\partial}{\partial t_D}\left[\frac{1}{2}(\ln t_D + 0.809)\right] = \frac{\partial}{\partial t_D}\left(\frac{2t_D}{r_{eD}^2} + \ln r_{eD} - \frac{3}{4}\right) \tag{6.181}$$

From Eq. (6.181), the following expression for time to stabilization for a well located in the center of a circular reservoir can be extracted

$$t_D = 0.25 r_{eD}^2 \tag{6.182}$$

or, in terms of hydraulic diffusivity

$$t_s = \frac{1}{4}\frac{\phi\mu c r_e^2}{\lambda k} = \frac{1}{4}\left(\frac{r_e^2}{\lambda\eta}\right) \tag{6.183}$$

As you will recall, Eq. (6.182) is the same expression we used in a previous section to switch from the infinite-acting solution to the late-time solution for finite reservoirs with no-flow outer boundary.

6.11 Deviations from Idealized Behavior

Our fundamental, classical well-test analytical model is admittedly highly idealized. It considers a *one*-dimensional, *single*-phase, *infinite*, and *homogeneous* system with *single*-well producing at *constant*-flow rate in the middle of a *circular* drainage area. The analytical solution for the behavior of such a system is the well-known *infinite-acting line-source solution*. In the subsequent sections, we attempt to enhance the capability of this solution by removing some of its restrictions using the techniques of superposition and theory of images, and include other real-life situations which are not yet considered by our idealized model.

6.11.1 Variable Rates and Superposition Principle

Consider the following differential equation:

$$A_2(x)\frac{d^2y}{dx^2} + A_1(x)\frac{dy}{dx} + A_0(x)y = f(x) \tag{6.184}$$

where y is the dependent variable, x the independent variable, A_0, A_1, A_2 the linear coefficients (not dependent on "y"), and $f(x)$ the excitation or perturbation function.

Equation (6.184) describes the response "y" of the system along the x-direction to a perturbation function of $f(x)$. For example, Eq. (6.184) describes the deformation of a simply supported beam upon the action of a force $f(x)$. Let us say that $y_1(x)$ is the solution of Eq. (6.184) when a force $f(x) = F_1(x)$ is applied, so that:

$$A_2(x)\frac{d^2y_1}{dx^2} + A_1(x)\frac{dy_1}{dx} + A_0(x)y_1 = F_1(x) \tag{6.185}$$

Now, let $y_2(x)$ be the solution of Eq. (6.184) when a force $f(x) = F_2(x)$ is applied:

$$A_2(x)\frac{d^2y_2}{dx^2} + A_1(x)\frac{dy_2}{dx} + A_0(x)y_2 = F_2(x) \tag{6.186}$$

Adding Eqs. (6.185) and (6.186) side-by-side, one gets:

$$A_2(x)\frac{d^2(y_2 + y_1)}{dx^2} + A_1(x)\frac{d(y_2 + y_1)}{dx} + A_0(x)(y_2 + y_1) = F_2(x) + F_1 \tag{6.187}$$

By inspecting Eq. (6.187), we see that if we were to apply a force $f(x) = F_1(x) + F_2(x)$ to this system, the response of the system would be "$y_1(x) + y_2(x)$", the sum of the two solutions when the forcing functions $F_1(x)$, $F_2(x)$ acted independently. So we say that we are "superimposing" the solutions. In other words, this means that if we know the response y_1 of the system for a forcing function F_1 and also the response y_2 for a forcing function F_2, we would *not* need to solve the differential equation given by Eq. (6.184) for a new forcing function $F_3 = F_1 + F_2$ because the response of the system will be $y_3 = y_1 + y_2$ (and y_1 and y_2 are already known). Here we assume that the differential equation governing the process is linear and the system characteristics and properties do not change.

In well test analysis, the following analogies can be established:

F_1, F_2	=	Flow rates (perturbations)
y_1, y_2	=	Pressure responses
$A_2(x), A_1(x), A_0(x)$	=	System properties such as k, ϕ, c, μ

Superposition in Time

In well test analysis applications, superposition may be considered to be a problem-solving philosophy in which the pressure behavior at any point at any time is the sum of the histories of each of the effects that may be considered to affect the solution at that point. Superposition in time can be demonstrated by a simple one-well example. This type of superposition pertains to the case of the same well producing at different flow rates in time.

Let us assume a well producing at a constant rate q_1 for a time t_1 and at a rate q_2 until time t_2, and thereafter at a rate q_3. This flow rate history is shown in Fig. 6.12. Three time periods can be considered for the analysis of this problem:

Period $0 < t \le t_1$: Pressure drop at the well is caused by the rate q_1. We can call this pressure drop as Δp_1.

Period $t_1 < t \le t_2$: Additional perturbations are introduced to the reservoir by rate *differences*. During this period, flow rate of the well has been increased by $(q_2 - q_1)$. Because pressure transients cannot be stopped once they are sent, the effects of the original rate "q_1" in the system are still felt. Then, we say that during this period the flow rate "q_1" has been acting on the system creating a pressure drop $(\Delta p_1)'$. The increased flow rate $(q_2 - q_1)$ has also been acting, creating an additional pressure drop of $(\Delta p_2)'$. This is why, at $t = t_2$, the total pressure drop experienced by the system is:

$$\Delta p_2 = (\Delta p_1)' + (\Delta p_2)' \tag{6.188}$$

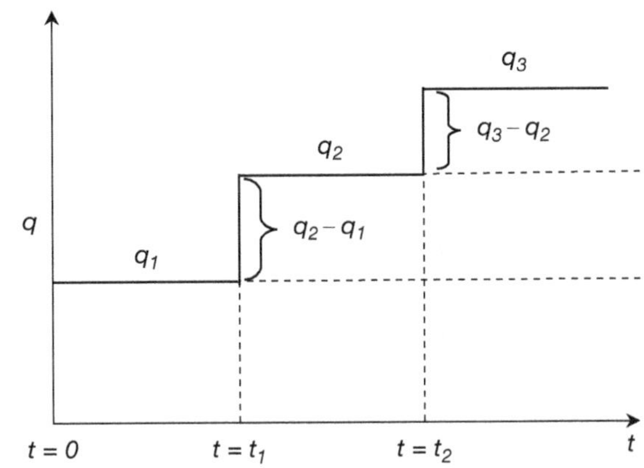

FIGURE 6.12 Flow rate history for a well.

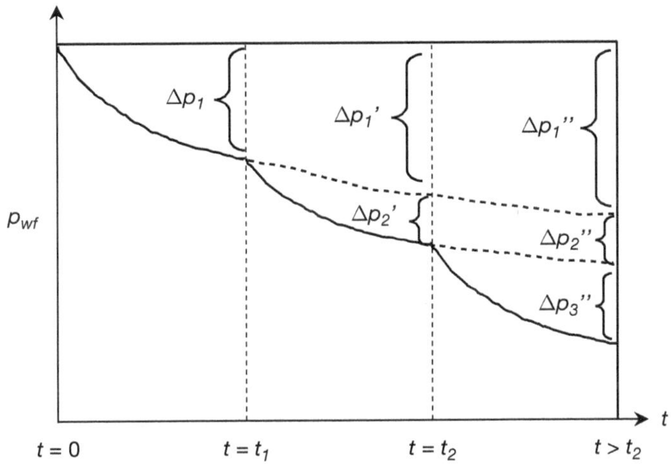

FIGURE 6.13 Superposition in time.

Period $t > t_2$: A new perturbance is introduced to the system by the increased rate $(q_3 - q_2)$. Again, because no past pressure transient can be stopped, the initial rate "q_1" and previous rate perturbance $(q_2 - q_1)$ are creating the pressure drops $(\Delta p_1)''$ and $(\Delta p_2)''$, respectively. An additional pressure drop $(\Delta p_3)''$ is created by the increased rate $(q_3 - q_2)$. The total pressure drop experienced by the well when $t > t_2$ is:

$$\Delta p_3 = (\Delta p_1)'' + (\Delta p_2)'' + (\Delta p_3)'' \tag{6.189}$$

Total pressure drop calculation is schematically shown in Fig. 6.13. In all previous cases, the total pressure drop calculation is obtained by simply superimposing the basic solutions based on the changes in flow rates. This procedure is called superposition in time. It is important to keep in mind that the *first flow rate* will remain active throughout the entire time domain and the new flow rates will exert there influence in the form of flow differentials Δq, starting with the time of their implementation.

In summary, if we wanted to calculate to total pressure drop at the sandface after a time $t = t_3$ where $t_3 > t_2$, we write [see Eq. (6.189)]:

$$\Delta p\big|_{@t=t_3} = (\Delta p_1)'' + (\Delta p_2)'' + (\Delta p_3)'' \tag{6.190}$$

where:

$(\Delta p_1)''$ = Pressure drop that would have been created by the well if it were operating at the constant rate of q_1 during the entire time under consideration

$(\Delta p_2)''$ = Pressure drop created by the flow rate difference $(q_2 - q_1)$ acting between the times $t_1 < t \le t_2$

$(\Delta p_3)''$ = Pressure drop created by the flow rate difference $(q_3 - q_2)$ acting between the times $t_2 < t \le t_3$

We calculate each of these terms using the basic infinite-acting solution. In other words, we assume that pressure transients have not reached any boundaries at the time $t = t_3$. Then, we will have:

$$\Delta p_{D1} = \frac{(\Delta p_1)''}{p_i q_{D1}} = -\frac{1}{2}E_i\left(-\frac{r_D^2}{4t_{D3}}\right) \tag{6.191}$$

$$\Delta p_{D2} = \frac{(\Delta p_2)''}{p_i(q_2 - q_1)_D} = -\frac{1}{2}E_i\left(-\frac{r_D^2}{4(t_3 - t_1)_D}\right) \tag{6.192}$$

$$\Delta p_{D3} = \frac{(\Delta p_3)''}{p_i(q_3 - q_2)_D} = -\frac{1}{2}E_i\left(-\frac{r_D^2}{4(t_3 - t_2)_D}\right) \tag{6.193}$$

Now, because we are evaluating pressure drops at the wellbore ($r_D = 1$), the logarithmic approximation to the exponential integral function as shown earlier could be applied for $t_D > 25$. Accordingly, above equations can be rewritten as:

$$\Delta p_{D1} = \frac{(\Delta p_1)''}{p_i q_{D1}} = \frac{1}{2}[\ln t_{D3} + 0.80907] \tag{6.194}$$

$$\Delta p_{D2} = \frac{(\Delta p_2)''}{p_i(q_2 - q_1)_D} = \frac{1}{2}[\ln(t_3 - t_1)_D + 0.80907] \tag{6.195}$$

$$\Delta p_{D3} = \frac{(\Delta p_3)''}{p_i(q_3 - q_2)_D} = \frac{1}{2}[\ln(t_3 - t_2)_D + 0.80907] \tag{6.196}$$

There is something very important that we must stress here. While one can add up the *dimensional* pressure drops just as we did in Eq. (6.190), a similar operation cannot be done for the *dimensionless* pressure drops. That is, it will not be correct to write the dimensionless version of Eq. (6.190) claiming that you are implementing superposition in time. Keep in mind that *dimensionless* pressure drop group is not only a function of the *dimensional* pressure drop but also a function of the flow rate, which can be a variable while applying superposition in time.

Using above equations, final expression for total pressure drop at $t = t_3$ becomes:

$$\begin{aligned}\Delta p\big|_{@t=t_3} &= -\frac{1}{2}p_i q_{D1} E_i\left(-\frac{r_D^2}{4t_{D3}}\right)\\ &\quad -\frac{1}{2}p_i (q_2 - q_1)_D E_i\left[-\frac{r_D^2}{4(t_3 - t_1)_D}\right]\\ &\quad -\frac{1}{2}p_i (q_3 - q_2)_D E_i\left[-\frac{r_D^2}{4(t_3 - t_2)_D}\right]\end{aligned} \tag{6.197}$$

or, using the logarithmic approximation for pressure drop calculation at the wellbore ($r_D = 1$),

$$\begin{aligned}\Delta p\big|_{@t=t_3} &= \frac{1}{2}p_i q_{D1}\left[\ln t_{D3} + 0.80907\right]\\ &\quad +\frac{1}{2}p_i (q_2 - q_1)_D\left[\ln(t_3 - t_1)_D + 0.80907\right]\\ &\quad +\frac{1}{2}p_i (q_3 - q_2)_D\left[\ln(t_3 - t_2)_D + 0.80907\right]\end{aligned} \tag{6.198}$$

Superposition in time is the basis for the analysis of test such as pressure buildup and multi-rate flows. For example, for the buildup test case, we impose two different flow rates to the system to learn about its characteristics. In summary, the main purpose of applying superposition in time is to remove the constraint of *constant flow rate* that our original well test model has.

Superposition in Space

In the previous section, we devised a way to circumvent the constant flow-rate constraint that the ideal well test model requires. As we go further with the classical well test theory, we will keep trying to get as close as possible to real cases and "loosen up" the original constraints of the ideal well test model. In this section, by means of the concept of *superposition in space*, we will learn how to remove the constraint of *only one well* being drilled in the reservoir.

The principle of superposition in space provides the means to calculating the pressure change at any point of a reservoir where more than one well is perturbing the system. The principle of superposition in space can be demonstrated by considering the case of more than one well acting in a reservoir, as shown in Fig. 6.14.

In Fig. 6.14, the three wells located at points A, B, and C are producing at rates q_A, q_B, and q_C, respectively. Superposition in space tells us that the pressure drop at a given point "P" in this reservoir can be obtained by superimposing the solutions at point "P" due to wells A, B, and C acting alone at the very same point. That is, each of these solutions is independent of the other. Therefore, the total pressure drop at point "P" is:

$$\Delta p\big|_{at\ point\ P} = \Delta p_A\big|_{at\ P\ caused\ by\ well\ A} + \Delta p_B\big|_{at\ P\ caused\ by\ well\ B} + \Delta p_C\big|_{at\ P\ caused\ by\ well\ C} \tag{6.199}$$

If we make point "P" coincide with the locations of any of the wells A, B, or C, we could run an "*interference test.*" In fact, superposition in space forms the basis of interference-type tests used to determine reservoir characteristics such as inter-well

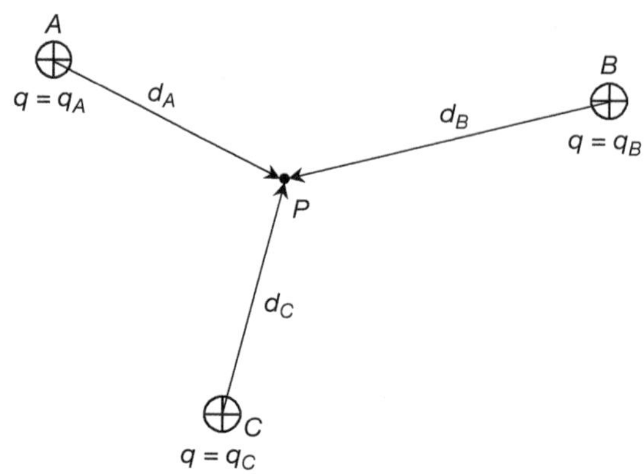

FIGURE 6.14 Superposition in space.

porosity. Interference tests are very useful in determining whether two wells are actually draining the same reservoir (i.e., whether they are communicating or not).

A word of caution is in order here. For the calculation of the individual Δp contributions from each well in Eq. (6.199), the logarithmic approximation that was presented earlier could never be used. The logarithmic approximation is only applicable for pressure changes *at the well location* ($r_D = 1$) and in the case of superposition in space, it is clear that point "*P*" is not located at the wellbore. Therefore, we use the original expressions that utilize the exponential integral functions:

$$\Delta p_A = -\frac{1}{2} p_i q_{DA} E_i\left(-\frac{r_{DA}^2}{4t_{DA}}\right) \tag{6.200}$$

$$\Delta p_B = -\frac{1}{2} p_i q_{DB} E_i\left(-\frac{r_{DB}^2}{4t_{DB}}\right) \tag{6.201}$$

$$\Delta p_C = -\frac{1}{2} p_i q_{DC} E_i\left(-\frac{r_{DC}^2}{4t_{DC}}\right) \tag{6.202}$$

where, from Fig. 6.14, it is seen that:

$$r_{DA} = \frac{d_A}{r_{WA}} \tag{6.203}$$

$$r_{DB} = \frac{d_B}{r_{WB}} \tag{6.204}$$

$$r_{DC} = \frac{d_C}{r_{WC}} \tag{6.205}$$

For none of these cases, $r_D = 1$; therefore, the logarithmic approximation cannot be used.

Superposition in Time and Space

Solutions may be superimposed both in space and in time. For example, in the situation we illustrated in the preceding section shown in Fig. 6.14, for superposition in space, each of the three wells were assumed to be producing at the same constant rate. This does not have to be necessarily the case. Each of the wells (*A*, *B*, and *C*) may experience multi-rates. For example, let us assume that we have the following production history for each of the wells:

Well A

$0 < t < 30$ hr $\quad q_{A1} = 100$ STB/D

$t > 30$ hr $\quad q_{A2} = 200$ STB/D

Well B

$0 < t < 80$ hr $\quad q_{B1} = 200$ STB/D

$t > 80$ hr $\quad q_{B2} = 300$ STB/D

Well C

$0 < t < 100$ hr $\quad q_{C1} = 0$ STB/D

$t > 100$ hr $\quad q_{C2} = 250$ STB/D

Let us say that we would like to calculate the pressure changes at point "*P*" after a week of production ($t = 168$ hr). Since we have more than one well in the reservoir, we apply superposition in space and write:

$$\Delta p\big|_{\text{at point P @}t=168} = \Delta p_A\big|_{\text{at P caused by Well A @}t=168} + \Delta p_B\big|_{\text{at P caused by Well B @}t=168} + \Delta p_C\big|_{\text{at P caused by Well C @}t=168} \tag{6.206}$$

In calculating the pressure drop at point "*P*" in the reservoir, the individual pressure drops caused by each of the wells need to be calculated. However, because we have a multi-rate scenario, for each of these individual calculations, we need to apply superposition in time taking into account their production histories individually as shown below for each well.

$$\begin{aligned}\Delta p_A\big|_{\text{at P caused by Well A at }t=168} &= -\frac{1}{2}p_i q_{DA1} E_i\left[-\frac{r_{DA}^2}{4(t)_{DA}}\right] \\ &\quad -\frac{1}{2}p_i\left(q_{DA2} - q_{DA1}\right)E_i\left[-\frac{r_{DA}^2}{4(t-30)_{DA}}\right]\end{aligned} \tag{6.207}$$

$$\begin{aligned}\Delta p_B\big|_{\text{at P caused by Well B at }t=168} &= -\frac{1}{2}p_i q_{DB1} E_i\left[-\frac{r_{DB}^2}{4(t)_{DB}}\right] \\ &\quad -\frac{1}{2}p_i\left(q_{DB2} - q_{DB1}\right)E_i\left[-\frac{r_{DB}^2}{4(t-80)_{DB}}\right]\end{aligned} \tag{6.208}$$

$$\Delta p_C\Big|_{at\ P\ caused\ by\ Well\ C\ at\ t=168} = -\frac{1}{2}p_i(q_{DC2}-0)E_i\left[-\frac{r_{DC}^2}{4(t-100)_{DC}}\right] \tag{6.209}$$

By calculating the pressure drops caused by each well [Eqs. (6.207) through (6.209)] and bringing the calculated results to Eq. (6.206), solutions are superimposed in both in space and time simultaneously and this is called superposition in time and space. Again, in Eq. (6.206), the logarithmic approximation of the exponential integral cannot be used directly without checking the magnitude of the argument of the exponential integral since $r_{DA} \neq 1$, $r_{DB} \neq 1$, and $r_{DC} \neq 1$.

6.11.2 Noncircular Drainage Areas: Method of Images

In this section, we will study how the classical radial-cylindrical exponential integral solution has been used to investigate flow scenarios where reservoir external boundaries and drainage areas are no longer properly described with the circular geometry assumed by the exponential integral solution.

Let us consider two identical (twin) wells producing at the same rate in an infinitely large reservoir. The wells are spaced at a distance "$2d$" and put on production at the same time. As they continue to produce, their cone of depression in the pressure surface shown in Fig. 6.15 both in radius and height becomes continuously larger with time. At early times, neither of the wells feels the other (i.e., the outer edge of the bases of their cone of depressions have not "touched" each other). When the pressure transients reach each other (i.e., the cone of depressions meet at the middle way "d"), the oil molecule (or the barrel of oil) found on that line becomes stagnant because the pressure gradients are the same toward the both wells. Therefore, a hypothetical no-flow boundary has become structured halfway between them. The situation portrayed here is called an "*even image*" because both wells are of the same type (in this case, both are *producing* at

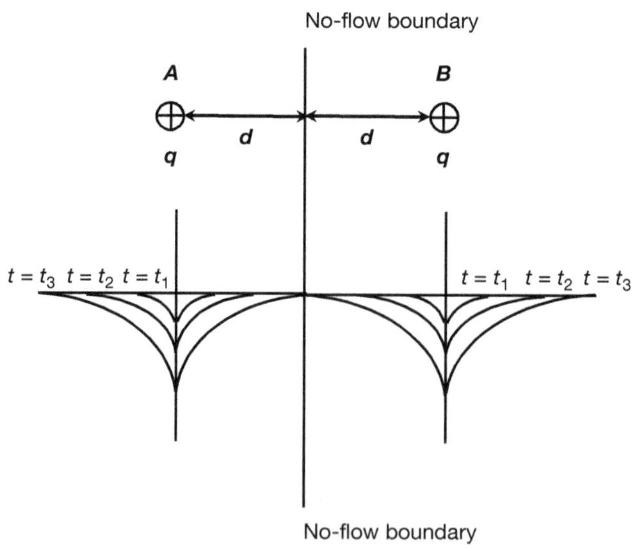

FIGURE 6.15 Cones of depression and formation of a no-flow boundary.

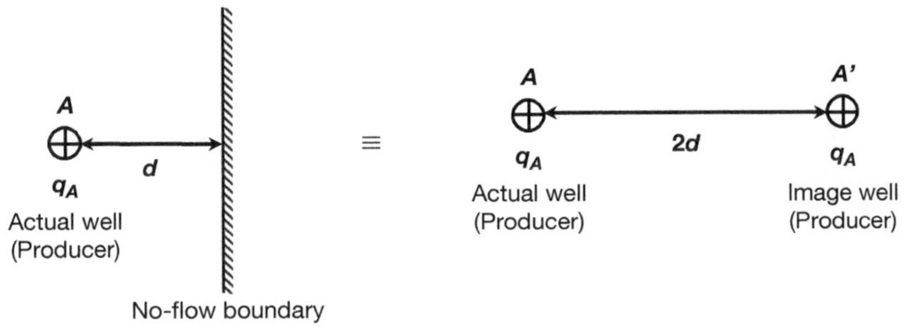

FIGURE 6.16 No-flow boundary and even image application.

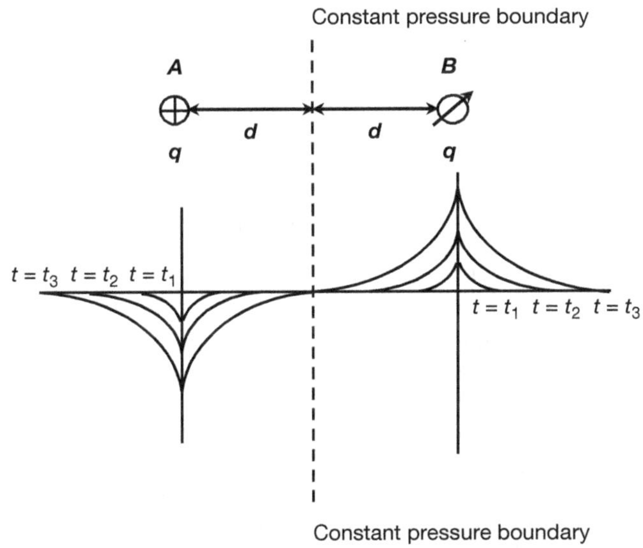

FIGURE 6.17 Cone of depression (producer)/cone of impression (injector).

the same rate). Each of the wells, in this flow geometry, will not be able to drain fluid beyond the midpoint line.

The previous observation suggests that the effect of a no-flow boundary can be represented by suitably replacing an actual no-flow boundary by an even-image well, as shown in Fig. 6.16. "Even image" implies that the actual and image wells are of the same type (producer/producer or injector/injector).

Now, we can consider the case of an *"odd image."* In such a case, one well is producing and the other one is injecting, as illustrated in Fig. 6.17. While well A is causing a pressure drop in the reservoir, well B is creating a pressure increase. The rates of growth of the bases and the heights of each of each cones (cone of depression and cone of impression) will be the same, since both wells are identical in terms of properties and production and injection rates, when the bases of cone of depression and cone of impression meet each other at the halfway location, these effects compensate each other and a point which will stay at constant pressure is created. As the production and

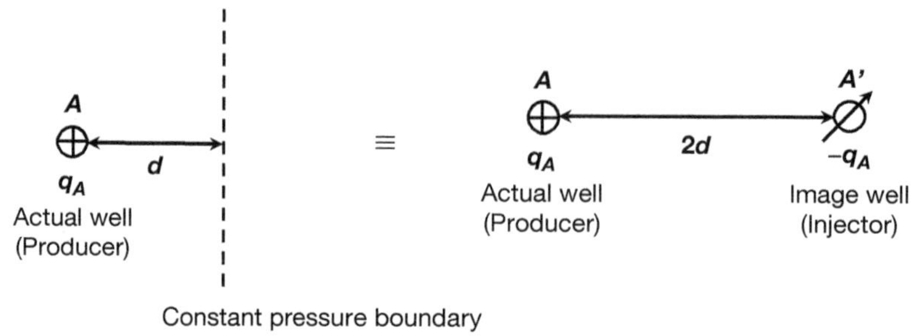

FIGURE 6.18 Constant-pressure boundary and odd image application.

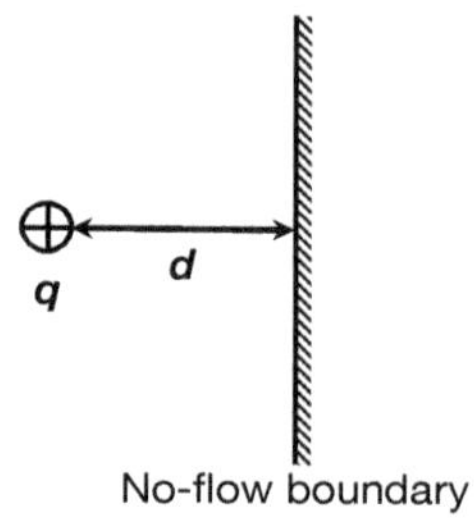

FIGURE 6.19 Well near a barrier.

injection operations continue, this time a hypothetical constant pressure line will be developed. In other words, the line will be located on the inflection points of each pressure surfaces and pressure will not be able to vary at the midpoint line.

The previous analysis suggests that a reservoir system with a constant pressure boundary and a producing well in a semi-infinite reservoir on one side of a constant-pressure boundary can be represented by an equivalent system simply by placing on the other side of the boundary, at an equal distance, an image well injecting at the same flow rate. This identically equivalent reservoir well system is shown in Fig. 6.18. This procedure is known as "odd image" as the actual and image wells are of the opposite kind (producer/injector or injector/producer).

Well Near a Barrier: Image Theory Application

Figure 6.19 presents the situation of a well drilled in a semi-infinite system near permeability (no-flow) barrier. The question that we would like to answer here is: what is the change of pressure at any time and any point in this system? This is a similar situation as the one depicted in Fig. 6.16. Therefore, we can readily provide the following answer:

1. At early times, before the pressure transients hit the barrier or sealing fault, we use the *infinite acting line source* solution. Let us recall that the well does not realize that the fault is there at the very beginning, until the first measurable pressure transient reaches the permeability barrier.
2. For late times, once the boundary has been felt, we can define an exactly equivalent system by locating an "image well" at the other half of the plane producing at the same rate. We are thus eliminating the physical barrier and recreating it artificially with an image well A', as shown in Fig. 6.20.

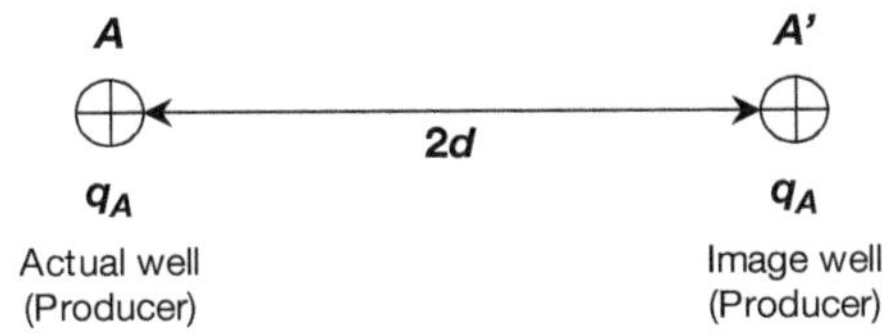

FIGURE 6.20 Equivalent system using the even-image technique.

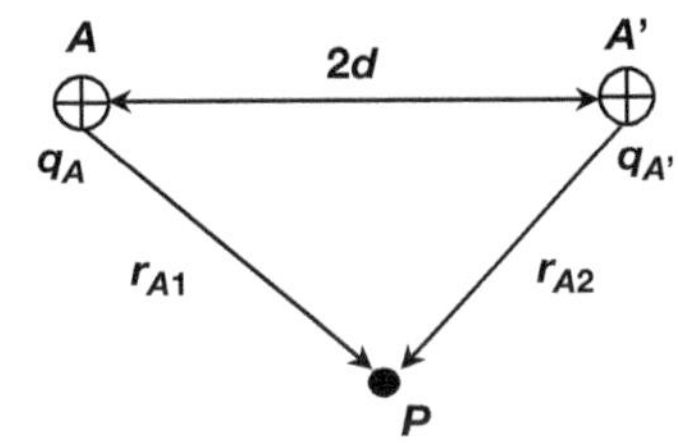

FIGURE 6.21 Pressure drop calculation at point "P".

Let us calculate the change in pressure at some point "P" in this equivalent system (see Fig. 6.21). For the calculation of pressure drop in an infinite-acting reservoir with two wells, we apply *superposition in space*. Therefore, at any point "P" in the reservoir we write:

Then,

$$\Delta p\big|_{at\ point\ P} = \Delta p_A\big|_{at\ P\ caused\ by\ actual\ well\ A} + \Delta p_{A'}\big|_{at\ P\ caused\ by\ image\ well\ A'} \tag{6.210}$$

where

$$\Delta p_A = -\frac{1}{2}p_i(q_A)_D E_i\left(-\frac{r_{DA1}^2}{4t_D}\right) \tag{6.211}$$

$$\Delta p_{A'} = -\frac{1}{2}p_i(q_{A'})_D E_i\left(-\frac{r_{DA2}^2}{4t_D}\right) \tag{6.212}$$

Because well A' is the even image of well A, $(q_A)_D = (q_{A'})_D$. Therefore,

$$\Delta p\big|_{at\ point\ P} = \frac{1}{2}p_i(q_A)_D\left[-E_i\left(-\frac{r_{DA1}^2}{4t_D}\right) - E_i\left(-\frac{r_{DA2}^2}{4t_D}\right)\right] \tag{6.213}$$

Now, if we want to answer the question of how the pressure changes at the sandface of well A as a function of time, Eq. (6.213) still holds. But in such a case, $r_{DA1} = 1$ and $r_{DA2} = 2d/r_w$.

Additionally, because $r_{DA1} = 1$, we could use the logarithmic approximation for the first exponential integral function that appears in Eq. (6.213) Therefore,

$$\Delta p\big|_{at\ well\ A} = \frac{1}{2}p_i(q_A)_D\left[\ln t_D + 0.80907 - E_i\left(-\frac{d^2}{r_w^2 t_D}\right)\right] \tag{6.214}$$

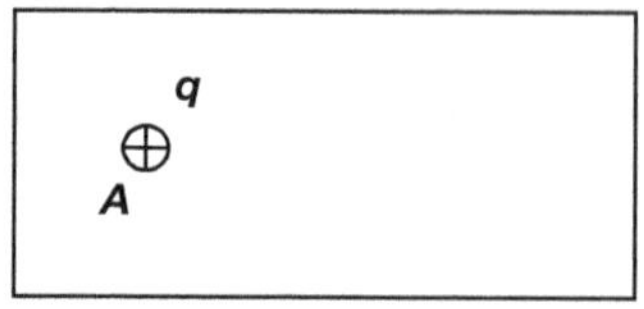

FIGURE 6.22 Well in a rectangular drainage area.

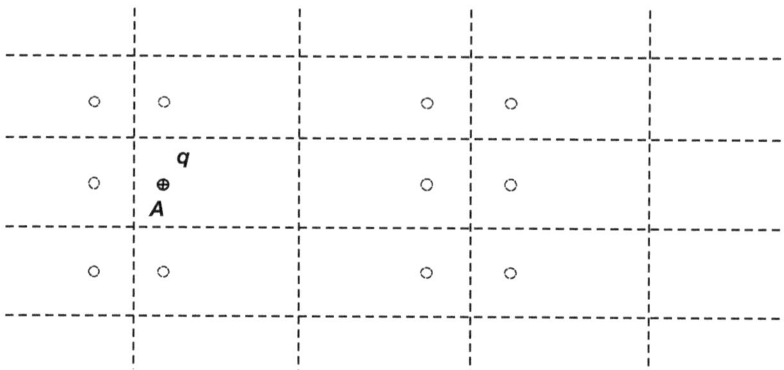

FIGURE 6.23 Image wells for a well inside a rectangular reservoir.

Equation (6.214) provides the pressure drop at the sandface of a well located at a distance "d" of a sealing fault, as it was illustrated in Fig. 6.19.

Well in Closed Rectangular Reservoir: Image Theory Application

Here let us assume that we are presented with the situation portrayed in Fig. 6.22. In this case, we have a well that has been placed within a rectangular drainage area. This is a significant departure from the reservoir-well systems we have seen so far. Nevertheless, the reader should be familiar with what we will attempt to do. Keep in mind that we should always look for ways of converting a given reservoir system into an infinitely large reservoir system; the analytical solution tool that we have is applicable to infinitely large reservoirs.

At a first glance, the practicing engineer may attempt to replace each of the four no-flow boundaries in this rectangle by only four even images. Careful examination of the problem yields the realization that this is not the case. Each of those even images will have its own images as well, and we will actually end up with an infinite number of image wells for the equivalent infinite system, as shown in Fig. 6.23.

Do we need to deal with such an infinite number of wells to calculate pressure drops for this system? Let us keep in mind the following:

1. At early times, the infinite acting solution will be enough to represent the behavior of the system (no image wells would be necessary). This is true until the first measurable pressure transient hits the closest boundary.
2. As time goes by, only a certain number of image wells would feel the pressure transient and we would require adding more image wells to our solution. Let "n" be the number of image wells needed in the solution. It can be easily seen that: As $t \to 0, n \to 0$ and as $t \to \infty, n \to \infty$.

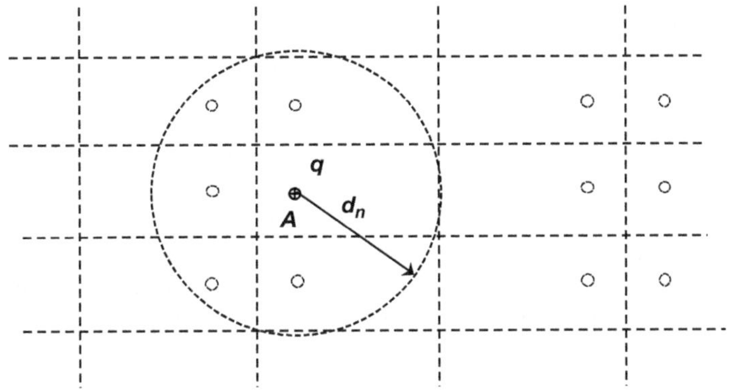

FIGURE 6.24 Number of image wells required for pressure drop calculations at a well located in a rectangular drainage area.

Using superposition in space, and following a similar development as the one presented in the preceding section, the pressure drop at the sandface of actual well A is given by the expression:

$$\Delta p\big|_{\text{at well A}} = \frac{1}{2}p_i(q_A)_D\left[\ln t_D + 0.80907 - \sum_{n=1}^{\infty} E_i\left(-\frac{\phi\mu c d_n^2}{4\lambda kt}\right)\right] \tag{6.215}$$

where "d_n" is the distance from the actual well (A) to the nth image well.

The question now becomes: for a given time "t", how many wells do we have to include in the pressure drop calculations or to what extent we actually have to perform the "infinite" summation in Eq. (6.215)? The answer to this question depends on the contribution that the exponential integral function has in Eq. (6.215). As we recall, when we studied the exponential integral function, we said that:

$$E_i(-X) \approx 0 \quad \text{for} \quad X \geq 5 \tag{6.216}$$

This was first stated in Eq. (6.137). Therefore, we can use this as the criteria that we need to know how many image wells we should consider at a given time. In fact, we may solve for "d_n" by setting the argument of the exponential integral at that particular time equal to 5 as shown below.

$$\frac{\phi\mu c d_n^2}{4\lambda kt} = 5 \tag{6.217}$$

After determining the value of d_n, the number of wells that reside inside the circle of radius "d_n" with center at the location of Well A will be included as the image well at that time. This is illustrated in Fig. 6.24. As stated previously, Eq. (6.215) allows calculating sandface pressure drops at well A. However, in order to calculate pressure changes anywhere else (a point "P") within the rectangular reservoir of Fig. 6.22, superposition in space would yield the following expression:

$$\Delta p\big|_{\text{at point P}} = \frac{1}{2}p_i(q_A)_D\left[-E_i\left(-\frac{\phi\mu c r_1^2}{4\lambda kt}\right) - \sum_{n=1}^{\infty} E_i\left(-\frac{\phi\mu c d_n^2}{4\lambda kt}\right)\right] \tag{6.218}$$

where "r_1" is the distance between point "P" and well A, and d_n is the distance between point "P" and the nth image well.

Well in a Noncircular Drainage Area: Image Theory Application

The method of even images for the representation of no-flow boundaries can be implemented for the study of several other reservoir shapes—in particular reservoirs with shapes resembling regular polygons. In the previous section, we discussed its implementation for a particular reservoir shape (a rectangle).

The concepts of *MBH* dimensionless pressure function (F) and the shape factor (C_A) have been introduced for the generalization of the analysis to different reservoir configurations. The first step of this generalization is the definition of a new *dimensionless time* based on the drainage area A of a particular configuration. In this definition, we replace the square of the well radius ($r_w^{\,2}$) in the denominator of the dimensionless time group by the drainage area A:

$$t_{DA} = \frac{\lambda k}{\phi \mu c A} t \tag{6.219}$$

It is readily seen that our original dimensionless time (t_D) and the dimensionless time based on drainage area (t_{DA}) are related through the expression:

$$t_{DA} = t_D \frac{r_w^2}{A} \tag{6.220}$$

Using these expressions, and for the case of a well in a rectangle, Eq. (6.218) can be written in terms of t_{DA} as follows:

$$\Delta p\Big|_{at\ well\ A} = \frac{1}{2} p_i (q_A)_D \left[\ln\left(\frac{A}{r_w^2} t_{DA}\right) + 0.809 - \sum_{n=1}^{\infty} E_i\left(-\frac{d_n^2}{4t_{DA} A}\right)\right] \tag{6.221}$$

Adding and subtracting the quantity $4\pi t_{DA}$ and noting that $\ln\left(\frac{4}{1.781}\right) = 0.809$:

$$\Delta p\Big|_{at\ well\ A} = \frac{1}{2} p_i (q_A)_D \left[\ln\left(\frac{4A \cdot t_{DA}}{1.781 \cdot r_w^2}\right) + 4\pi t_{DA} - 4\pi t_{DA} - \sum_{n=1}^{\infty} E_i\left(-\frac{d_n^2}{4A \cdot t_{DA}}\right)\right] \tag{6.222}$$

Matthews et al. (1954) defined the following dimensionless pressure function for a rectangular reservoir:

$$F = 4\pi t_{DA} + \sum_{n=1}^{\infty} E_i\left(-\frac{d_n^2}{4A \cdot t_{DA}}\right) \tag{6.223}$$

where F is the Matthews, Brons, and Hazebroek (*MBH*) dimensionless pressure function that allows generalization of the Eq. (6.222) as:

$$\Delta p\Big|_{at\ sandface} = \frac{1}{2} p_i (q_A)_D \left[\ln\left(\frac{4A \cdot t_{DA}}{1.781 \cdot r_w^2}\right) + 4\pi t_{DA} - F\right] \tag{6.224}$$

The values of F have been evaluated by several authors for different reservoir configurations. Equation (6.223) represents the appropriate expression for the evaluation of F for the specific case that the reservoir happens to be rectangular. Note that Eq. (6.224) is only valid at the sandface (well location) and not anywhere else in the reservoir. Therefore, it cannot be used to apply superposition in space if more than one well is located in the reservoir.

Table 6.4 presents the values of F as a function of dimensionless time (t_{DA}) as calculated by Earlougher et al. (1968) for various reservoir-well configurations. To use Table 6.4, it will be necessary to select the column of the reservoir-well configuration of interest and enter with the value of t_{DA} as calculated from Eq. (6.219). The values in the table are given within the range $0.001 < t_{DA} < 10$. For values $t_{DA} < 0.001$, the reservoir is infinite acting and $F = 4\pi t_{DA}$. For large values of t_{DA} ($t_{DA} > 10$), F behaves linearly with the logarithm of dimensionless time; in fact, it is directly proportional to the logarithm of t_{DA}. Thus, a linear extrapolation using the last two values of the column suffices—as long as values of $\log(t_{DA})$ are used in the extrapolation. Figure 6.25 displays the typical behavior of the MBH dimensionless pressure function with dimensionless time.

Figure 6.25 shows that, for large values of t_{DA}, the function F is a straight line with a unit slope. Thus, the late time behavior of F can be represented through the equation:

$$F = \ln t_{DA} + const \tag{6.225}$$

The constant that shows up in Eq. (6.225) is the intercept of the extrapolation of the late-time linear behavior of the curve with the y-axis. This is, the value of F when $t_{DA} = 1$ (or $\ln t_{DA} = 0$) for the linear late-time extrapolation. This late-time behavior represents the pseudo-steady-state flow period, when all the boundaries of the systems have been felt by the pressure transient. The intercept of the straight line portion of the curve with the F-axis is considered to be the "shape factor (C_A)" of the system under consideration. Equation (6.225) can be written as:

$$F = \ln C_A + \ln t_{DA} = \ln(C_A t_{DA}) = \ln\left(\frac{C_A \lambda k t}{\phi \mu c A}\right) \tag{6.226}$$

For small values of dimensionless time t_{DA} ($t_{DA} < 0.001$), $F = 4\pi t_{DA}$ and the generalized Eq. (6.224) becomes:

$$\Delta p\Big|_{at\ sandface} = \frac{1}{2} p_i (q_A)_D \left[\ln\left(\frac{4At_{DA}}{1.781 r_w^2}\right)\right] \tag{6.227}$$

However, for large values of dimensionless time t_{DA} ($t_{DA} > 10$), Eq. (6.226) holds and the generalized Eq. (6.224) becomes

$$\Delta p\Big|_{at\ sandface} = \frac{1}{2} p_i (q_A)_D \left[\ln\left(\frac{4A}{1.781 r_w^2 C_A}\right) + 4\pi t_{DA}\right] \tag{6.228}$$

Dimensionless time t_{DA}	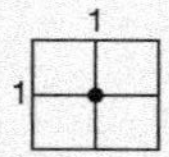	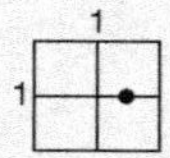	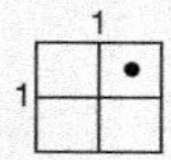	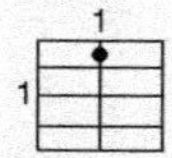	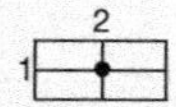	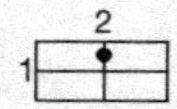	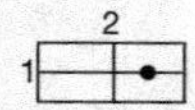	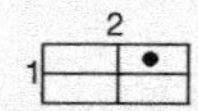
0.001	0.0126	0.0126	0.0126	0.0126	0.0126	0.0126	0.0126	0.0126
0.0015	0.0188	0.0188	0.0188	0.0188	0.0188	0.0188	0.0188	0.0188
0.002	0.0251	0.0251	0.0251	0.0251	0.0251	0.0251	0.0251	0.0251
0.0025	0.0314	0.0314	0.0314	0.0314	0.0314	0.0314	0.0314	0.0314
0.003	0.0377	0.0377	0.0377	0.0368	0.0377	0.0377	0.0377	0.0377
0.004	0.0503	0.0503	0.0503	0.0460	0.0503	0.0502	0.0503	0.0502
0.005	0.0628	0.0628	0.0628	0.0517	0.0628	0.0626	0.0628	0.0626
0.006	0.0754	0.0754	0.0754	0.0537	0.0754	0.0745	0.0754	0.0745
0.007	0.088	0.088	0.0879	0.0524	0.088	0.0858	0.088	0.0858
0.008	0.1005	0.1005	0.1004	0.0483	0.1005	0.0963	0.1005	0.0963
0.009	0.1131	0.113	0.1128	0.0422	0.1131	0.1059	0.1131	0.1059
0.01	0.1257	0.1254	0.1251	0.0345	0.1257	0.1145	0.1257	0.1145
0.015	0.1885	0.1854	0.1823	−0.0162	0.1884	0.1449	0.1884	0.1449
0.02	0.2513	0.2402	0.2287	−0.0700	0.2508	0.1601	0.2505	0.1598
0.025	0.3141	0.2892	0.263	−0.1181	0.3119	0.1678	0.3107	0.1663
0.03	0.3769	0.3333	0.2864	−0.1584	0.3708	0.1723	0.3676	0.1683
0.04	0.5016	0.4108	0.3087	−0.2166	0.4804	0.1812	0.4686	0.1657
0.05	0.6237	0.4791	0.3099	−0.2512	0.5785	0.1955	0.5511	0.1593
0.06	0.7415	0.5413	0.3002	−0.2694	0.6667	0.2169	0.6167	0.1513

TABLE 6.4 MBH Dimensionless Pressure Function Values for Different Reservoir-Well Configurations

0.07	0.8537	0.5991	0.2856	−0.2766	0.7471	0.245	0.6683	0.1428
0.08	0.9597	0.6531	0.27	−0.2765	0.8217	0.2787	0.7085	0.1346
0.09	1.0592	0.7038	0.2553	−0.2716	0.8917	0.3167	0.7398	0.1269
0.1	1.1524	0.7516	0.2427	−0.2633	0.9581	0.3581	0.7642	0.1198
0.15	1.5364	0.9583	0.2226	−0.1951	1.2524	0.5864	0.8261	0.0954
0.2	1.8212	1.1314	0.2637	−0.1027	1.4987	0.8106	0.8476	0.0873
0.25	2.0439	1.2854	0.3412	0.0025	1.7064	1.0104	0.8629	0.0923
0.3	2.2262	1.4257	0.4365	0.1129	1.883	1.1841	0.8829	0.1087
0.4	2.5139	1.672	0.64	0.3300	2.1678	1.4674	0.9447	0.1687
0.5	2.737	1.8797	0.8321	0.5268	2.3905	1.6899	1.0299	0.2537
0.6	2.9193	2.0563	1.0028	0.6994	2.5728	1.8722	1.128	0.3518
0.7	3.0735	2.2083	1.1527	0.8499	2.7269	2.0263	1.2307	0.4545
0.8	3.207	2.3411	1.2847	0.9821	2.8605	2.1599	1.3329	0.5567
0.9	3.3249	2.4586	1.4019	1.0094	2.9782	2.2776	1.4315	0.6552
1	3.4302	2.5638	1.507	1.2045	3.0836	2.383	1.5252	0.7489
2	4.1234	3.2569	2.2	1.8976	3.7768	3.0762	2.2002	1.4239
4	4.8166	3.9501	2.8933	2.5908	4.4701	3.7695	2.8932	2.117
8	5.5099	4.6435	3.5867	3.2842	5.1633	4.4627	3.5866	2.8103
10	5.7331	4.8667	3.8098	3.5073	5.3865	4.6859	3.8098	3.0335

Table 6.4 MBH Dimensionless Pressure Function Values for Different Reservoir-Well Configurations (*Continued*)

Dimensionless time t_{DA}	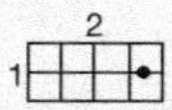		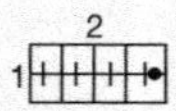	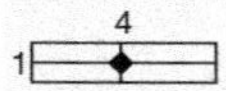	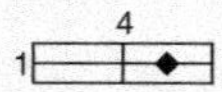	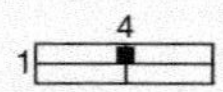	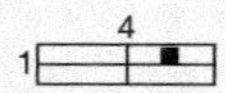	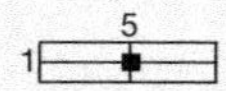
0.001	0.0125	0.0126	0.0126	0.0126	0.0126	0.0126	0.0126	0.0126
0.0015	0.0179	0.0188	0.0179	0.0188	0.0188	0.0188	0.0188	0.0188
0.002	0.0209	0.0251	0.0208	0.0251	0.0251	0.0251	0.0251	0.0251
0.0025	0.0203	0.0314	0.02	0.0314	0.0311	0.0314	0.0311	0.0314
0.003	0.016	0.0377	0.0155	0.0377	0.0368	0.0377	0.0368	0.0377
0.004	−0.0019	0.0502	−0.0027	0.0503	0.0460	0.0503	0.0460	0.0503
0.005	−0.0284	0.0626	−0.0295	0.0628	0.0517	0.0628	0.0517	0.0628
0.006	−0.0596	0.0745	−0.0612	0.0754	0.0537	0.0754	0.0537	0.0753
0.007	−0.0932	0.0858	−0.0951	0.0879	0.0524	0.0879	0.0524	0.0878
0.008	−0.1227	0.0962	−0.1298	0.1004	0.0483	0.1004	0.0483	0.1000
0.009	−0.162	0.1058	−0.1644	0.1129	0.0422	0.1129	0.0422	0.1119
0.01	−0.1957	0.1144	−0.1983	0.1251	0.0345	0.1251	0.0345	0.1234
0.015	−0.3468	0.1445	−0.3502	0.1823	−0.0162	0.1823	−0.0162	0.1713
0.02	−0.467	0.1589	−0.4718	0.2291	−0.0701	0.2291	−0.0701	0.2015
0.025	−0.5615	0.1641	−0.5695	0.2643	−0.1186	0.2643	−0.1187	0.2163
0.03	−0.6357	0.1633	−0.6507	0.2897	−0.1600	0.2897	−0.1600	0.2200
0.04	−0.7395	0.1492	−0.7839	0.3197	−0.2231	0.31	−0.2235	0.2075
0.05	−0.8012	0.1224	−0.8965	0.3332	−0.2662	0.3315	−0.2682	0.1820
0.06	−0.8339	0.0862	−0.9989	0.3385	−0.2957	0.3335	−0.3013	0.1516
0.07	−0.8457	0.0437	−1.0949	0.3399	−0.3158	0.3290	−0.3278	0.1203
0.08	−0.8422	−0.0028	−1.1859	0.3401	−0.3291	0.3199	−0.3510	0.0899

TABLE 6.4 MBH Dimensionless Pressure Function Values for Different Reservoir-Well Configurations (*Continued*)

0.09	−0.8272	−0.0512	−1.2723	0.3403	−0.3375	0.3072	−0.3727	0.0613
0.1	−0.8038	−0.1004	−1.3542	0.3412	−0.3421	0.2915	−0.3942	0.0351
0.15	−0.6223	−0.3322	−1.7021	0.3663	−0.3257	0.1826	−0.5128	−0.0580
0.2	−0.4138	−0.5189	−1.9613	0.4269	−0.2661	0.0468	−0.6499	−0.0935
0.25	−0.2196	−0.658	−2.1508	0.5120	−0.1811	−0.0959	−0.7928	−0.0855
0.3	−0.0479	−0.7555	−2.2854	0.6102	−0.0829	−0.2344	−0.9312	−0.0477
0.4	0.2342	−0.8547	−2.4344	0.8152	0.122	−0.4789	−1.1758	0.0773
0.5	0.4567	−0.8671	−2.4768	1.0075	0.3143	−0.6712	−1.3681	0.2266
0.6	0.6389	−0.8284	−2.4564	1.1783	0.4852	−0.8134	−1.5103	0.3753
0.7	0.7931	−0.762	−2.4011	1.3282	0.6351	−0.9129	−1.6098	0.5143
0.8	0.9267	−0.682	−2.3278	1.4602	0.7670	−0.9775	−1.6744	0.6409
0.9	1.0444	−0.5969	−2.2469	1.5774	0.8843	−1.0145	−1.7114	0.7555
1	1.1497	−0.5115	−2.164	1.6825	0.9894	−1.0301	−1.7270	0.8595
2	1.843	0.1507	−1.5058	2.3755	1.6824	−0.7325	−1.4294	1.5516
4	2.5363	0.8436	−0.8129	3.0688	2.3757	−0.0756	−0.7725	2.2448
8	3.2295	1.537	−0.1195	3.7623	3.0691	0.6173	−0.0796	2.9381
10	3.4527	1.7601	0.1036	3.9854	3.2922	0.8406	0.1438	3.1615

Table 6.4 MBH Dimensionless Pressure Function Values for Different Reservoir-Well Configurations (*Continued*)
Source: Earlougher et al. (1968).

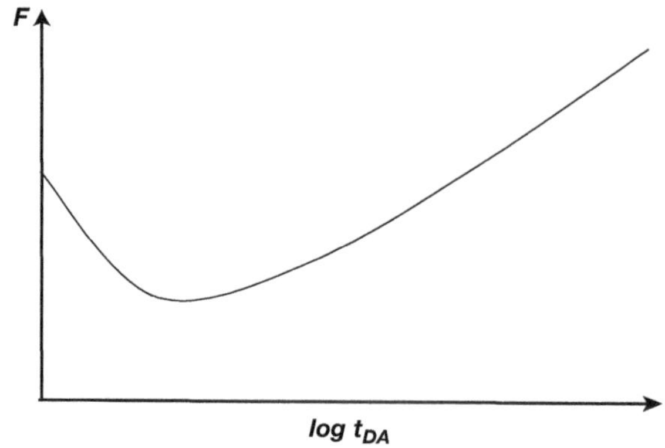

FIGURE 6.25 MBH dimensionless pressure function versus time.

Table 6.5 presents the values of C_A and lnC_A for other various closed single-well drainage areas.

The generalized Eq. (6.224) allows the calculation of the pressure drop at the sandface of a well drilled within a regular polygon. Figure 6.26 shows the three important flow periods that the well experiences throughout its production life and how sandface pressure (P_{wf}) is expected to vary with time. We have had a previous discussion on this, when we introduced the concept of "time to stabilization."

In a reservoir with closed outer boundaries, three flow regimes are observed:

a. *Transient Flow Regime*: This flow regime takes place during the infinite-acting flow period. This flow period is also known as "early transient state." Equation (6.227) describes the pressure changes during this period.

b. *Late Transient Flow Regime:* When presence of the closest boundary is felt at the wellbore, transient flow period ends and the late transient flow period starts. The late transient period ends when the pressure transient moving outwards reaches the farthest point on the boundary. The generalized Eq. (6.224) along Table 6.4 which provides the values of F in Eq. (6.224) applies for this regime.

c. *Pseudo-Steady-State Flow Regime:* The end of the late transient flow period marks the start of the pseudo-steady-state flow. During this flow period, reservoir pressure will decline at the same rate everywhere in the drainage region. The pseudo-steady-state period is also referred to as the "semi-steady state" or "quasi-steady state." Equation (6.228) describes the pressure changes during this period.

Note that the *pseudo*-steady state and the *true*-steady state regimes are completely different concepts. A *true* steady-state flow regime is never achieved in the depletion of hydrocarbon reservoirs with no-flow outer boundaries. For true-steady state to be achieved, the reservoir must have some fluid influx at the boundary that compensates the fluid extraction from the well. Steady-state conditions will then be achieved at late

In Bounded Reservoirs	CA	ln CA	$\frac{1}{2}\ln\left(\frac{2.2458}{C_A}\right)$	Exact for $t_{DA} >$	Less Than 1% Error For $t_{DA} >$	Use Infinite System Solution with Less Than 1% Error For $t_{DA} <$
	31.62	3.4538	−1.3224	0.1	0.06	0.10
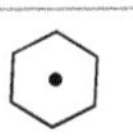	31.6	3.4532	−1.3220	0.1	0.06	0.10
	27.6	3.3178	−1.2544	0.2	0.07	0.09
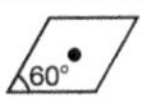	27.1	3.2995	−1.2452	0.2	0.07	0.09
	21.9	3.0865	−1.1387	0.4	0.12	0.08
	0.098	−2.3227	+1.5659	0.9	0.60	0.015
	30.8828	3.4302	−1.3106	0.1	0.05	0.09
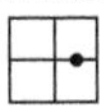	12.9851	2.5638	−0.8774	0.7	0.25	0.03
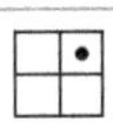	4.5132	1.5070	−0.3490	0.6	0.30	0.025
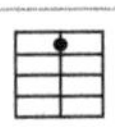	3.3351	1.2045	−0.1977	0.7	0.25	0.01
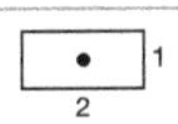	21.8369	3.0836	−1.1373	0.3	0.15	0.025
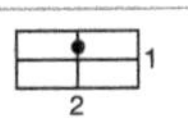	10.8374	2.3830	−0.7870	0.4	0.15	0.025
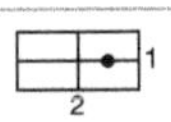	4.5141	1.5072	−0.3491	1.5	0.50	0.06
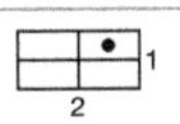	2.0769	0.7309	−0.0391	1.7	0.50	0.02
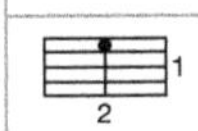	3.1573	1.1497	−0.1703	0.4	0.15	0.005

Table 6.5 Shape Factors for Various Closed Single-Well Drainage Areas
Source: Earlougher (1977).

times when the rate of fluid extraction equals the rate of fluid influx to the reservoir. Accordingly, in reservoirs, with a constant pressure outer boundary, the steady-state flow regime can eventually develop. When true steady-state conditions are established, the pressure profile attained in the reservoir will stay intact as long as the boundary conditions remain the same through the time.

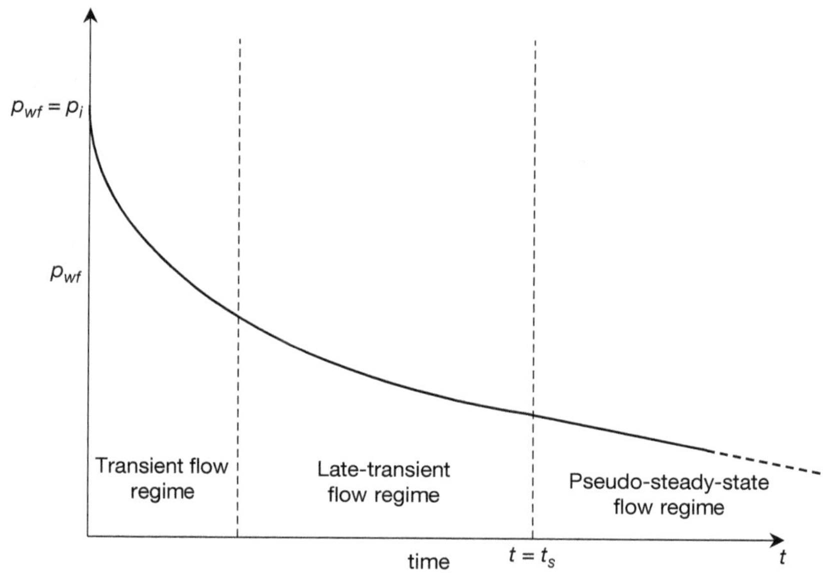

FIGURE 6.26 Flow regimes encountered in a reservoir with closed outer boundaries.

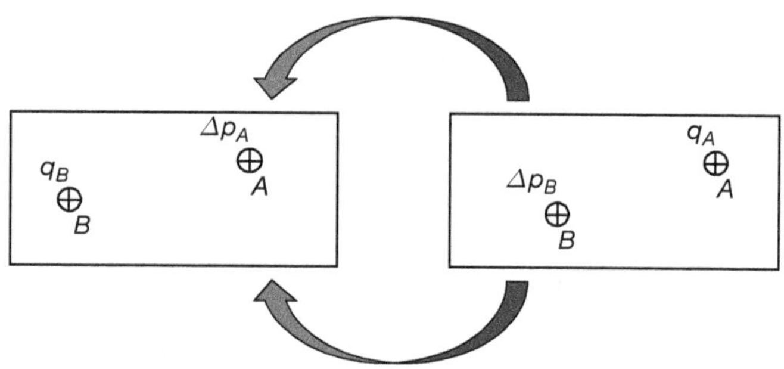

FIGURE 6.27 Schematic representation of reciprocity principle.

The Reciprocity Principle

Reciprocity principle applies to both infinite and finite reservoirs and can be stated as follows:

"The pressure response at an observation Well A caused by production at rate q from Well B is equal to the pressure response at observation Well B caused by the production at rate q from Well A."

This principle is valid if the pressure behavior in the reservoir satisfies the diffusivity equation. Figure 6.27 schematically describes the reciprocity principle. With the help of this principle, it is possible to expand the use of the tabulated solutions for wells located in closed polygons.

6.11.3 Multiphase Flow Effects

Classical well test analytical models are based on single-phase flow assumptions. However, it is not uncommon to find flow of more than one phase (gas, oil, and/or water) while conducting a well test. A rigorous approach to this problem must entail the simultaneous solution of a system of partial differential equations. Nevertheless, there are a number of more simplistic approaches to multiphase flow analysis, although with accuracy levels admittedly questionable. For example, a very common simplification is to come up with a "pseudo"-single phase representing the "bulk" of the behavior of the multiphase fluid flow dynamics. Such a phase is to be compounded of a weighted volumetric average of the actual flowing phases. This simplified multiphase formulation is known as the Perrine–Martin approach, which first proposed and derived the conditions for which single-phase flow equations could be extended to multiphase flow by deployed "pseudo"-single phase properties (Perrine, 1956; Martin, 1959). Their approach assumes that all phases are uniformly distributed in the reservoir and that saturation gradients are negligible.

In this simplified approach, definition of total flow properties is introduced to replace the single-phase flow properties such as mobility and compressibility characteristics. The total mobility of a multiphase flow system is defined as a sum of the mobility of individual phases:

$$\left(\frac{k}{\mu}\right)_{total} = k\left(\frac{k_{ro}}{\mu_o} + \frac{k_{rw}}{\mu_w} + \frac{k_{rg}}{\mu_g}\right) \tag{6.229}$$

The total compressibility is defined in a similar fashion as:

$$(c)_t = (S_o c_{oa}) + (S_w c_{wa}) + (S_g c_{ga}) + c_r \tag{6.230}$$

In the above equation, c_{oa} and c_{wa} are called *apparent* compressibilities for oil and water, respectively because both phases may contain a certain amount of dissolved hydrocarbon gas. Therefore, the definitions of the compressibilities for each phase can be written as:

$$c_{oa} = \frac{B_g}{B_o}\left(\frac{\partial R_{so}}{\partial p}\right) - \frac{1}{B_o}\left(\frac{\partial B_o}{\partial p}\right) \tag{6.231}$$

$$c_{oa} = \frac{B_g}{B_w}\left(\frac{\partial R_{sw}}{\partial p}\right) - \frac{1}{B_w}\left(\frac{\partial B_w}{\partial p}\right) \tag{6.232}$$

$$c_{oa} = -\frac{1}{B_g}\left(\frac{\partial B_g}{\partial p}\right) \tag{6.233}$$

Then, the classical well test equation is used with the following modifications:

$$\Delta p_D = -\frac{1}{2}E_i\left(-\frac{r_D^2}{4t_D}\right) \tag{6.234}$$

with

$$t_D = \frac{\lambda}{\phi r_w^2} \frac{1}{(c)_t} \left(\frac{k}{\mu} \right)_{total} t \tag{6.235}$$

Another methodology that may be taken is to use the values of k, μ, and c_t of the predominant phase in the system. These are, admittedly, rather superficial corrections that can often fail to properly describe multiphase flow effects. For more fundamental approaches, Raghavan (2009) presents a review of multiphase well test analysis approaches that move beyond the pseudo-phase approach of Perrine (1956) and Martin (1959).

6.11.4 Skin and Inertial-Turbulent Effects

As introduced in Chap. 3, the skin factor is a dimensionless quantity that actually represents a dimensionless pressure drop accounting for deviations in the homogeneous reservoir characteristics assumption in the immediate vicinity of the wellbore. Specifically, it mainly accounts for *permeability* heterogeneities in the domain surrounding the wellbore and thus it is a localized phenomenon. Calculation of pressure drop at a point of the reservoir away from the wellbore should be the same regardless of the skin around that wellbore. Other interesting point is that if we were to take an image of a damaged well (say, for the case of a damaged well near a barrier discussed earlier in this chapter), we would have to carry the skin value to the image location as well because both wells must be identical twins.

There are other deviations from idealized well behavior also captured through the skin concept. For example, Darcy's law is strictly applicable for laminar flow conditions, and hence, turbulence effects are not considered by the diffusivity equation in its original form. Inertial-turbulent effects are more likely to be encountered in gas reservoirs because of the low viscosity (high fluidity) of gas. Additionally, as the flow of gas approaches the wellbore, the flow area perpendicular to the flow along the r-direction reduces logarithmically, and hence, the velocity of the gas increases exponentially toward the well.

Turbulent flow effects are felt through higher pressure drops in the wellbore region. In fact, turbulent flow causes a "fictitious" positive skin. This is why turbulence effects are usually accounted for through the introduction of an additional *skin* in our equations. Therefore, for a gas reservoir, the dimensional pressure drop in the wellbore region is written as:

$$(\Delta p_D)_{actual} = (\Delta p_D)_{ideal} + s_{physical} + s_{IT} \tag{6.236}$$

where s_{IT} is the skin due to inertial-turbulence effects. $s_{physical}$ represents the physical skin we have discussed before due to formation stimulation/damage. The skin due to inertial-turbulence effects, s_{IT}, is also a dimensionless pressure drop that represents the additional pressure drop (Δp_{IT}) that the wellbore region experiences due to the existing turbulent flow conditions.

$$s_{IT} = \frac{(\Delta p_{IT})}{p_i q_D} \tag{6.237}$$

For a gas reservoir, this additional dimensionless pressure drop due to turbulence (Δp_{DIT}) is proportional to the gas flow rate:

$$\Delta p_{DIT} = Dq_{sc} \tag{6.238}$$

where D is the inertial-turbulence (IT) flow factor for the system. The composite, total skin of the system (s') will be given by the added contribution of both the physical and IT skins as follows:

$$s' = s_{physical} + s_{IT} = s_{physical} + Dq_{sc} \tag{6.239}$$

Values of composite skin (s') can be determined from field data after conducting well test analysis. If values of composite skin (s') is available for a variety of flow rates (q_{sc}), the physical skin ($s_{physical}$) and IT flow factor (D) can be determined as the intercept and the slope of the plot s' versus q_{sc}, respectively [see Eq. (6.239)].

6.11.5 Skin Due to Restricted Entry

If a well does not penetrate through the formation completely, radial flow regime will be distorted as hemispherical flow geometry is formed. However, this change in the flow geometry in the reservoir takes place only at the immediate vicinity of the wellbore. As one moves away from the wellbore, again radial flow resumes. Distortion of the streamlines (see Fig. 6.28) will cause an extra pressure drop within the flow field, which will create an effect on pressure transient data similar to that of formation damage.

In cased-hole completions, the connection between the wellbore and the formation is established by the perforations. This limited entry to flow also distorts the radial-cylindrical flow geometry and decreases the well's productivity. Around the immediate vicinity of the perforations, almost perfect spherical flow geometry is developed. When pressure transient data from a perforated well is collected, it might indicate the presence of formation damage when in reality there is none, or it might indicate a value larger than the actual physical skin value. Accordingly, this may lead to incorrect decisions such as planning of a well stimulation program when in reality it is

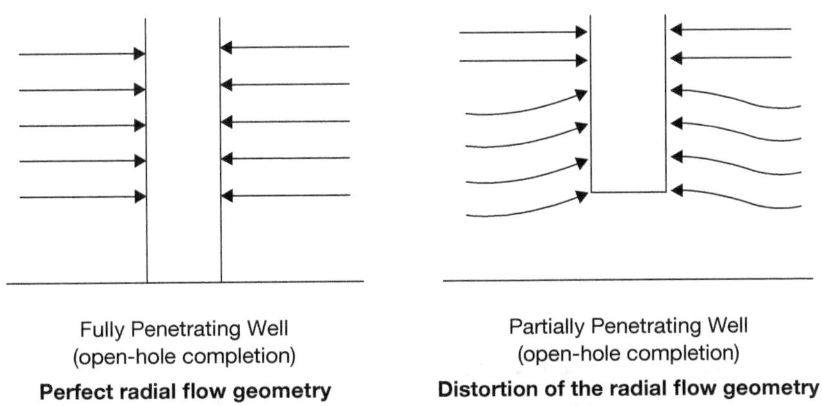

FIGURE 6.28 Distortion of streamlines due to partial penetration.

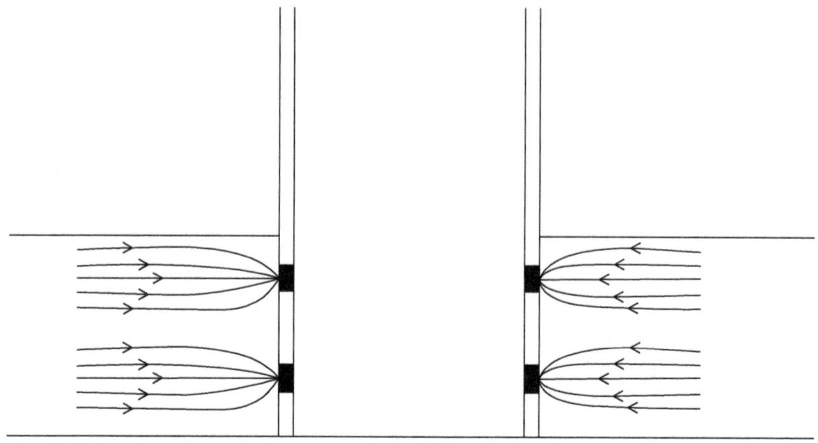

Figure 6.29 Distortion of streamlines for restricted entry due to perforations.

not necessary. Figure 6.29 shows the increased intensity of the streamlines around the perforations.

In the case of restricted entry, the total or composite skin of the wellbore is given as:

$$s' = s_{physical} + s_a \tag{6.240}$$

In Eq. (6.240), s_a is the apparent skin due to restricted entry. If we are working on a gas reservoir, we might need to include an additional term for inertial-turbulence effects (s_{IT}) in the composite skin of Eq. (6.240). A number of theoretical developments have provided means of estimating s_a. A practical protocol for such an implementation was suggested by Odeh (1968). Odeh proposed the following expression to compute the skin due to restricted entry:

$$s_a = \frac{\left(\ln\frac{r_d}{r_w} - 0.75\right)\left(1 - \frac{q}{q_r}\right)}{\frac{q}{q_r}} \tag{6.241}$$

where r_d is the radial extension of the reservoir (ft) and r_w is the wellbore radius (ft).

The ratio q/q_r is the ratio of the rate of flow obtained with the specific open interval to the rate of flow when the total productive interval is open to flow. It is a function of the total thickness of the formation (h_t), thickness of the perforated interval (h_p), distance between the top of the formation and the top of the perforated interval (z_1), and the radii r_d and r_w as well. Odeh (1968) presented several graphs for q/q_r for a variety of combinations of the previously mentioned parameters.

6.11.6 Wellbore Storage Effects

In our ideal PDA formulation, we assume that flow rates (q) are imposed at the *sandface* ($r = r_w$) at $t = 0$ instantaneously. In the real practice, however, well flow rates are actually manipulated at the surface—several feet away from the sandface. This means that, when the well is shut in *at the surface,* after a certain period of production, it is going to take some time until the formation ceases producing into the wellbore. In other words, it takes some time for the sandface to become *"aware"* that the well is shut in (choked) at the wellhead. In this case, although the flow rate at the wellhead is zero, it is not zero at the sandface. This effect is known as *wellbore storage or afterflow*.

The idealized well test model only considers the ideal case, where the formation stops production immediately after the well is shut in at the surface. Accordingly, it is of paramount importance to determine for how long this wellbore storage is going to take place as when the wellbore storage effects are dominant, the collected well test data becomes noisy. It is important for the well test analyst to know how long the well test data should be collected so that noisy part of the test data can be neglected and the relevant data is analyzed (relevant data implies the segment of the data collected after the wellbore storage effects are diminished).

A similar class of problem occurs when the well is put back on production at a specified rate. It will not be possible to have the desired rate of flowing at the sandface immediately after it is recorded at the wellhead. In fact, the flow at reservoir conditions has to build up from 0 *STB/D* to the desired $q_{sandface}$. While the flow rate at the sandface is building up to the specified production rate, some part of the production at the wellhead will be from the compressed fluid that is in storage in the well but not from the reservoir. This disparity between the wellhead and sandface flow rates is called the *wellbore unloading* effect. Figure 6.30 shows schematically the delayed response of the sandface to the wellhead-controlled rate changes for the two cases that are discussed above.

To explore the wellbore storage concept, it will be necessary to write a volumetric balance over the wellbore:

$$q_{wellhead} = q_{sandface} + q_{wellbore} \tag{6.242}$$

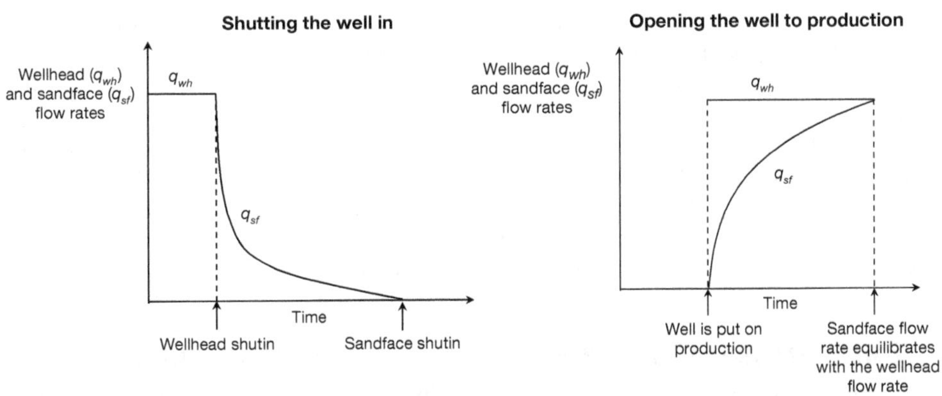

Figure 6.30 Disparity in wellhead and sandface flow rates due to wellbore storage and wellbore unloading effects.

The wellbore depletion can be expressed in terms of the wellbore storage constant (C_s):

$$q_{wellbore} = C_s \frac{d}{dt}(p_i - p_{wf}) \tag{6.243}$$

The wellbore storage constant (C_s) can be defined using the compressibility of the wellbore fluid evaluated at the mean wellbore pressure and temperature:

$$C_s = c_{ws} V_{ws} \tag{6.244}$$

where C_s is the wellbore storage constant (bbl/psi), c_{ws} is the compressibility of the wellbore fluid (psi^{-1}), and V_{ws} is the total wellbore volume, which is the volume of the wellbore tubing and the annulus if there is no packer (bbl).

A more empirical development for the analysis of wellbore storage provides the following expression for dimensionless wellbore storage time:

$$t_{wsD} = 60 \cdot C_{sD} \tag{6.245}$$

where t_{wsD} is the dimensionless time when wellbore storage effects disappear and C_{SD} is the dimensionless wellbore storage constant.

The previous dimensionless variables are redefined as:

$$t_{wsD} = \frac{\lambda k}{\phi \mu c_{ws} r_w^2} t_{ws} \tag{6.246}$$

and

$$C_{sD} = \frac{0.894}{\phi h c_{ws} r_w^2} C_s \tag{6.247}$$

Introducing Eqs. (6.244), (6.246), and (6.247) into Eq. (6.245), we obtain an expression for the wellbore storage time:

$$t_{ws} = \frac{53.64}{\lambda} \frac{\mu c_{ws}}{kh} V_{ws} (\text{bbl}) \tag{6.248}$$

Equation (6.248) provides a practical expression for estimating the time when storage effects disappear. For gas reservoirs, wellbore storage effects are expected to last longer because of their larger μc product value as compared to that of liquids. Gas reservoirs at high pressures dissipate wellbore storage effects much faster because of its smaller compressibility values encountered at high pressures. Equation (6.248) allows visualizing that wellbore storage effects are also function of the permeability of the formation (k). This is an important relationship that is not easily seen at a first glance. Accordingly, tight formations (due to their small permeability characteristics) are expected to have longer wellbore storage effects.

6.12 Basic Pressure Transient Data Analysis (PTA) Techniques

So far in this chapter we have discussed the fundamentals of *PTA* theory. At the beginning of our discussion we stated a highly idealized problem and came up with the necessary formalisms. Then, we have introduced a number of concepts and mathematical tools to ensure that the idealized problem is brought to reality to some degree. In describing the idealized reservoir-well system, we have started with some sweeping assumptions such as infinitely large reservoir, with one single well which has been produced at a constant rate, and we have removed these constraints later by implementing the superposition principle in various ways. We have also discussed how some of the other deviations from the idealized reservoir-well system could be handled. Along these lines, topics such as wellbore-storage/wellbore-unloading, presence of skin, existence of inertial-turbulent flow effects and multiphase flow conditions, and skin due to well completion strategies were discussed. At this point what is left to discuss is specific well testing and related analysis protocols so that a set of meaningful conclusions can be reached, which can later be used in making critical decisions in the process of reservoir management.

In petroleum engineering literature, there are literally hundreds of testing protocols and their analysis procedures. Within the confines of this chapter, we will be discussing two fundamental well test analysis techniques. These are the pressure buildup (PBU) and the drawdown (PDD) tests.

6.12.1 Pressure Buildup (PBU) Analysis

Pressure buildup (PBU) analysis is an important field application of one of the concepts we studied earlier in this chapter, *superposition in time*. The idea is that the well subject to testing has been producing at a constant rate "q" for a certain period of time, and it is suddenly shut in—that is, a new rate, $q = 0$, is imposed. During production at rate "q", sandface pressure declines. When the test "starts" ($t = 0$)—when the well is shut in, the pressure starts to *build up* and we monitor the pressure building up as a function of time. Figure 6.31 shows flow rate and pressure history for a typical buildup test. With this data, and the tools of well testing, our objective is to infer or estimate the properties of the system. This is what we call *inverse solution analysis*: knowing the output or the response of the system to a prescribed input (the perturbation we imposed); we are now ready to estimate the characteristics of the system using the measured response function (pressure vs. time).

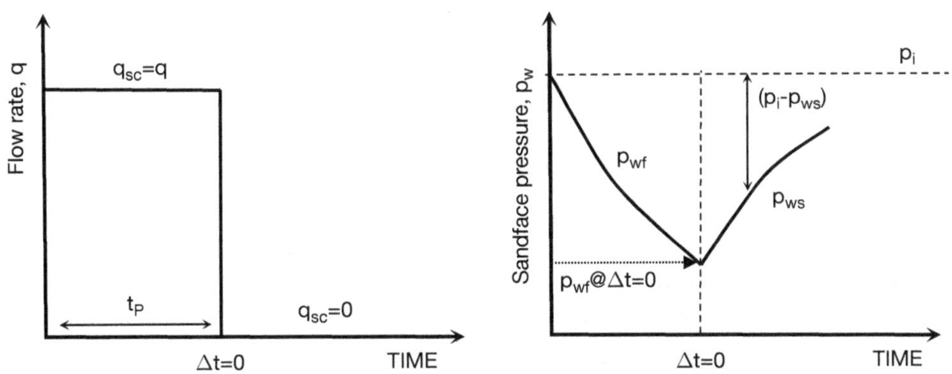

FIGURE 6.31 Flow rate and sandface pressure history of a buildup test.

In this treatment, we will be using the following terminology:

$_s$ = shut-in conditions,

$_f$ = flowing conditions,

p_{wf} = flowing well pressure,

p_{ws} = shut-in well pressure,

t_p = total production time,

Δt = shut-in time.

The data recorded during a buildup test would look as follows:

Δt	p_w
0	$p_{wf}(\Delta t = 0)$
Δt_1	p_{ws1}
Δt_2	p_{ws2}
Δt_3	p_{ws3}
...	...

Using the superposition in time, we now know that if we wanted to calculate the total pressure drop at the sandface for any shut-in time ($\Delta t > 0$), we can write:

$$\Delta p|_{@\Delta t} = (\Delta p_1)'' + (\Delta p_2)'' \tag{6.249}$$

where:

$(\Delta p_1)''$ = Pressure drop that would have been created by the well operating at the constant rate q during the entire time under consideration "$t_p + \Delta t$"

$(\Delta p_2)''$ = Pressure drop created by the flow rate difference $(0 - q)$ acting on the reservoir during the shut-in time "Δt"

It is possible to calculate each of these terms using the infinite-acting solution (yes, this assumes that the pressure transients have not reached any boundaries during the shut-in time "Δt"). Then, we have:

$$\Delta p_{D1} = \frac{(\Delta p_1)''}{p_i q_D} = -\frac{1}{2}E_i\left[-\frac{r_D^2}{4(t_p + \Delta t)_D}\right] \tag{6.250}$$

$$\Delta p_{D2} = \frac{(\Delta p_2)''}{p_i(0 - q)_D} = -\frac{1}{2}E_i\left[-\frac{r_D^2}{4(\Delta t)_D}\right] \tag{6.251}$$

At this point, it is useful to recall that well test data (pressure values) are recorded at the wellbore ($r_D = 1$), and hence the logarithmic approximation can be applied. Then, instead of using the previous equations, the logarithmic simplification is introduced.

$$\Delta p_{D1} = \frac{(\Delta p_1)''}{p_i q_D} = \frac{1}{2}[\ln(t_p + \Delta t)_D + 0.80907] \tag{6.252}$$

$$\Delta p_{D2} = \frac{(\Delta p_2)''}{p_i(0-q)_D} = \frac{1}{2}[\ln(\Delta t)_D + 0.80907] \tag{6.253}$$

Now, one can substitute Eqs. (6.252) and (6.253) into Eq. (6.249) to obtain:

$$\Delta p|_{@\Delta t} = \frac{1}{2}p_i q_D[\ln(t_p + \Delta t)_D + 0.80907] - \frac{1}{2}p_i q_D[\ln(\Delta t)_D + 0.80907] \tag{6.254}$$

Equation (6.254) can be manipulated to yield:

$$\frac{(p_i - p_{ws})}{p_i} = \frac{1}{2}q_D\left[\ln\left(\frac{t_p + \Delta t}{\Delta t}\right)\right] \tag{6.255}$$

Equation (6.255) is the dimensionless *pressure buildup analysis equation*. Notice that the ratio of dimensionless variables $(t_p + \Delta t)_D/(\Delta t)_D$ is the same as that of the ratio of the corresponding time dimensional variables.

If we want to derive the pressure buildup equation as it applies to oil reservoirs (liquid systems), we substitute the appropriate definitions for the dimensionless variables. For *liquid* systems, we had obtained:

$$t_D = \frac{\lambda k}{\phi \mu c_t r_w^2} t \tag{6.256}$$

and,

$$q_D = \frac{\gamma \mu B}{khp_i} q_{sc} \tag{6.257}$$

where $\lambda = 2.637.10^{-4}$ and $\gamma = 141.2$ in practical field units. Substituting into Eq. (6.255) yields,

$$(p_i - p_{ws}) = \frac{70.6Bq_{sc}\mu}{kh}\left[\ln\left(\frac{t_p + \Delta t}{\Delta t}\right)\right] \tag{6.258}$$

For plotting purposes, it is more convenient to express the above equation in common logarithms with base 10, simply because logarithmic graph papers are typically based on log10. Recall the transformation between the natural (Naperian) and the common logarithm with base 10 is:

$$\ln(x) = 2.3026 \times \log(x) \tag{6.259}$$

$$(p_i - p_{ws}) = \frac{162.6Bq_{sc}\mu}{kh}\left[\log\left(\frac{t_p + \Delta t}{\Delta t}\right)\right] \tag{6.260}$$

Equation (6.260) is the *classical buildup equation* (expressed in practical field units), used for the well test analysis of oil reservoirs. This is a simple-looking equation that we will use to extract the information we want. It allows the tracking of the evolution of p_{ws} with time, as it can also be expressed as follows:

$$p_{ws} = p_i - \frac{162.6Bq_{sc}\mu}{kh}\left[\log\left(\frac{t_p + \Delta t}{\Delta t}\right)\right] \tag{6.261}$$

It should be kept in mind that "t_p" represents the total production time at the moment that the well is shut in. Equation (6.261), the pressure buildup equation, simply tells us that we have a *linear relationship* between p_{ws} (the recorded shut in pressure) and the logarithmic ratio $\log[(t_p + \Delta t)/\Delta t]$ (Horner time). Here is where the semi-logarithmic plot paper becomes handy!

Let us recall the type of data we are going to get from the field:

Δt	p_w
0	$p_{wf}(\Delta t = 0)$
Δt_1	p_{ws1}
Δt_2	p_{ws2}
...	...

During the test, the information about "p_{ws}" versus "Δt" is collected. We have just seen that if we want to analyze this data, we would have to construct a new column for the ratio $(t_p + \Delta t)/\Delta t$:

p_w	Δt	$(t_p + \Delta t)/\Delta t$
$p_{wf}(\Delta t = 0)$	0	—
p_{ws1}	Δt_1	—
p_{ws2}	Δt_2	—
...	...	...

Figure 6.32 shows what we expect to see when plot buildup test data [p_{ws} vs. $(t_p + \Delta t)/\Delta t$) on a semi-log paper: *a straight line.*

It is important to note that the values in the abscissa of the plot have been *reversed* in order to preserve the logic of shut-in time increasing in the positive direction of the x-axis. The truth of the matter is that the ratio $(t_p + \Delta t)/\Delta t$ decreases with increasing Δt because the denominator prevails (the numerator is essentially constant because t_p values are usually large, $t_p >> \Delta t$). By reversing the scale, we are preserving the association that points *to the left* of the graph correspond to earliest measurements and points *to the right* are late-time measurements during the test—just as intuition would tell. As with any other straight line technique, we go after its *slope* and *intercept*. Figure 6.33 illustrates this calculation.

The *intercept* will be given by the value of "p_{ws}" when $\log[(t_p + \Delta t)/\Delta t] = 0$; this is when the argument becomes unity. By looking at Eq. (6.222), the intercept should be close to the value of initial reservoir pressure. To calculate the *slope* of the graph, we apply the basic rule.

From Fig. 6.33, we will have

$$\text{Slope: } m = (p_{ws2} - p_{ws1}) \quad [\text{psi/cycle}] \tag{6.262}$$

Since this is a semi-log paper, the easiest way to go is to choose one cycle and look for the pressures at the end and at the beginning of one cycle. The pressure difference dictates, directly, the slope of the line [see Eq. (6.262)]. Because $p_{ws2} > p_{ws1}$, the slope is

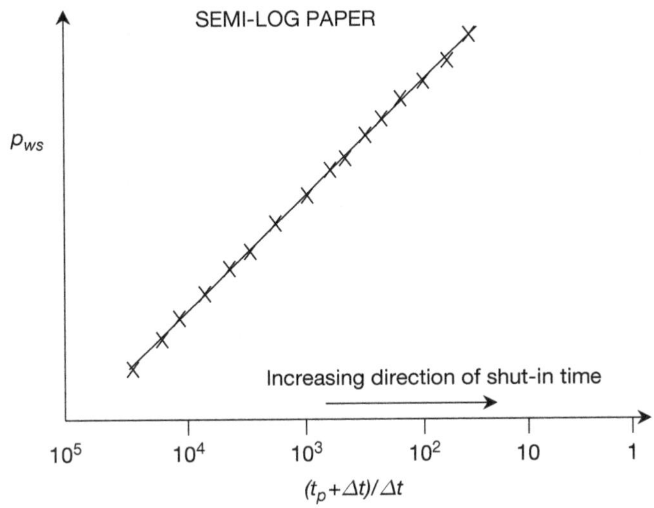

FIGURE 6.32 Ideal behavior of p_{ws} vs. $\log[(t_p + \Delta)/\Delta t]$.

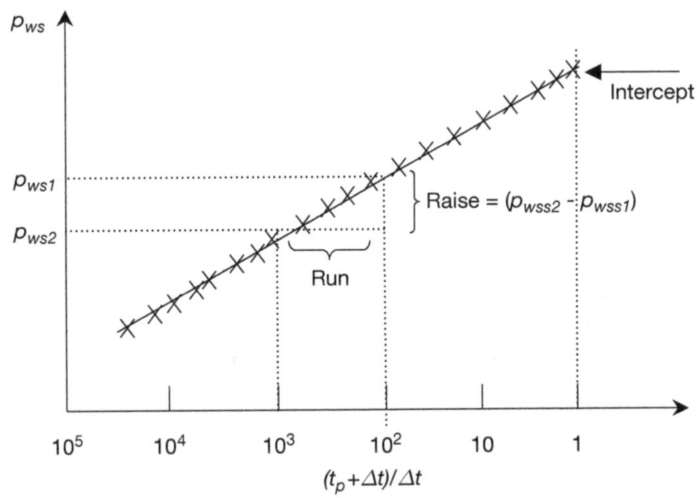

FIGURE 6.33 Slope and intercept of the ideal p_{ws} vs. $\log[(t_p + \Delta t)/\Delta t]$ plot.

always negative (recall that the abscissa was reversed). In fact, inspecting Eq. (6.261) closely, we see that the slope is given by the expression:

$$m = -\frac{162.6Bq_{sc}\mu}{kh} \tag{6.263}$$

Equation (6.263) tells us that we could calculate the permeability of the formation (k) by knowing the slope of the buildup test straight line. In fact, this is the single most important information that we want to get when we run a well test.

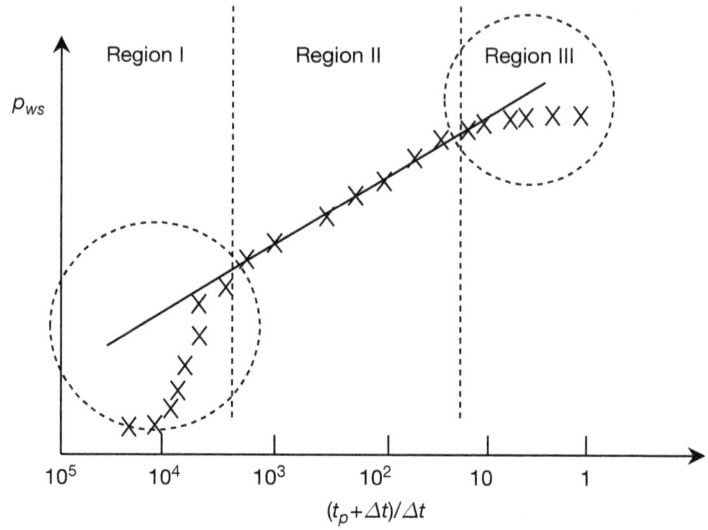

FIGURE 6.34 Pressure buildup analysis in an actual reservoir.

Basically, if we were to conduct a pressure buildup test for an ideal infinitely large system, it should be remembered that, at all times, Eq. (6.261), on which we have based all this analysis, is strictly applicable to an idealized infinite-acting system. However, field data is far from ideal. For an actual system using field data, Fig. 6.32 will look like the plot shown in Fig. 6.34.

As we see in Fig. 6.34, field data contains some noises from what we can identify as *early-time* effects and *late-time effects*. Both effects introduce important deviations from the ideal (infinite-acting) behavior that is represented by the straight line. Early-time effects are associated with wellbore storage and skin problems in the well (both deviations of this type were discussed earlier). Late-time effects are associated with pressure transients reaching the reservoir boundaries. The more "useful" data which do not violate the basic assumptions of the well test theory are found in the middle region. In fact, these aforementioned effects clearly demarcate three regions in the pressure buildup analysis plot:

Region I—Early Time Region (ETR)

Region II—Middle Time Region (MTR)

Region III—Late Time Region (LTR)

Ideally, if all the assumptions of the well test theory would have held, Regions I and III would not exist. Ideally, every single point of the test should fall on the straight line, as observed in the middle-time region. However, actual pressure transient behavior usually shows important deviations from the idealized model. Even worse, if the wellbore storage effects are strongly dominant and the well is situated close to the reservoir boundary, Regions I and III may overlap and the well test analyst would not be able to see the Region-II type of ideal behavior. We may even run a test for a too short

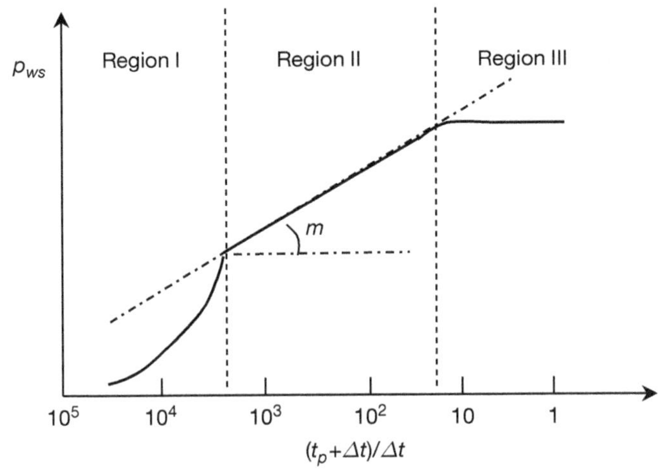

FIGURE 6.35 Horner plot for a buildup test.

period of time where wellbore storage effects would be completely masking the data. For both cases, no useful data would be available to conduct a valuable engineering analysis. The basic assumption (or wish?) when we run a well test is that at some point during the test, the pressure transient would capture some infinite-acting behavior. In consequence, all well tests must be designed for an amount of time that guarantees that wellbore storage effects have ended for our classical well test equations to be readily applicable.

Good judgment calls are required for the correct identification of infinite-acting behavior from field data. Otherwise, the well test analyst could come up with a completely false analysis, if he/she applies the pressure buildup equation in a wrong region.

Buildup Horner Plot

The analysis of pressure buildup tests using the procedure outlined in the previous paragraphs is known as Horner buildup analysis. The semi-logarithmic plot we constructed—p_{ws} in the ordinate and the Horner time ratio in the reversed-direction of abscissa—is called the Horner plot. He was the first to suggest its usefulness for petroleum reservoirs' buildup analysis and point out the estimation of permeability through the calculation of the slope of the plot (Horner, 1951). A typical Horner plot is shown in Fig. 6.35, based on our discussion on pressure transient behavior in actual field tests.

As we have stated earlier in the previous section, we can distinguish three zones in a typical Horner plot that describes a specific effect on the pressure transient:

Region I	Early-Time Region (*ETR*)	Early-time effects (storage, skin effects)
Region II	Middle-Time Region (*MTR*)	Infinite-acting period
Region III	Late-Time Region (*LTR*)	Late-time effects (boundary effects)

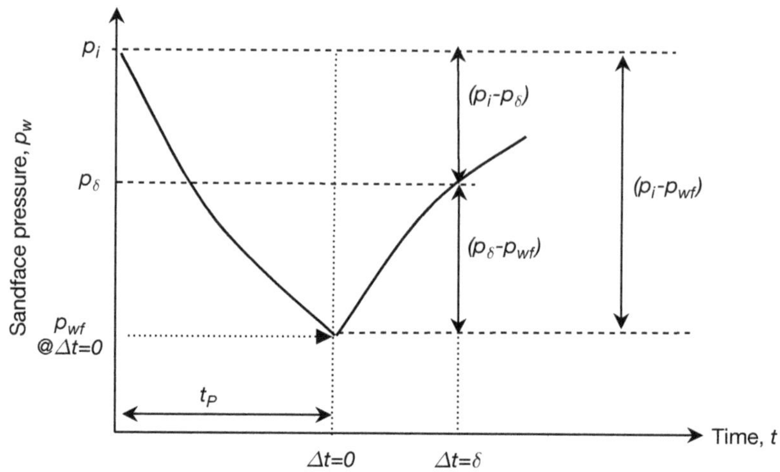

Figure 6.36 Pressure behavior during "$\Delta t = \delta$" in a buildup test.

The most useful data in terms of determination of reservoir conductivity (permeability) can be obtained through the analysis of the middle time region. However, the main difficulty in the Horner analysis is the correct identification of the middle-time region (*MTR*)— and in most of the cases some judgment calls will be required. Often times, we can search even further with the data and use the early and late time data to gather some more information about the reservoir. Typical information we can obtain from the analysis of each region may be as follows:

Region I	→	magnitude of skin, duration of wellbore storage
Region II	→	Permeability
Region III	→	Size of the reservoir, initial reservoir pressure

Skin Factor and PBU Data

In the previous section, we have just said that we may use the early-time data in a Horner plot for the estimation of the skin factor around the wellbore where the buildup test has been conducted. In this section, we will discuss how this can be done. For this, let us recall the typical behavior of the sandface pressure with time in a buildup test, as shown in Fig. 6.36. The Cartesian plot displayed in Fig. 6.36 gives a better view of this behavior and illustrates the pressure drops at a particular time "$\Delta t = \delta$" during the duration of the test ($\Delta t > 0$).

As discussed earlier, actual, measured pressure drop in a well is written in terms of dimensionless variables by incorporating the concept of skin. At the wellbore:

$$(\Delta p_D)_{well} = p_t + s \tag{6.264}$$

Introducing the logarithmic approximation for infinite-acting behavior at the wellbore, we obtain:

$$(\Delta p_D)_{well} = \frac{\Delta p}{p_i q_D} = \frac{1}{2}[\ln t_D + 0.80907] + s \tag{6.265}$$

Therefore, the pressure drop at a wellbore with skin can be calculated through the expression:

$$\Delta p = p_i q_D \left[\frac{1}{2}(\ln t_D + 0.80907) + s\right] \tag{6.266}$$

Recalling the definitions of the dimensionless variables [Eqs. (6.256) and (6.257), with $\lambda = 2.637.10^{-4}$ and $\gamma = 141.2$ for practical field units], the pressure drawdown in the reservoir by the time the buildup test starts—$(p_i - p_{wf})$ in Fig. 6.36—is given by the expression:

$$p_i - p_{wf} = \frac{141.2Bq_{sc}\mu}{kh}\left[\frac{1}{2}\left(\ln\frac{2.637 \times 10^{-4} kt_p}{\phi\mu c_t r_w^2} + 0.80907\right) + s\right] \tag{6.267}$$

On the other hand, the pressure buildup equation at time "$\Delta t = \delta$" gives [Eq. (6.258)]:

$$(p_i - p_\delta) = \frac{141.2Bq_{sc}\mu}{kh}\left[\frac{1}{2}\ln\left(\frac{t_p + \delta}{\delta}\right)\right] \tag{6.268}$$

From Fig. 6.36, it is clearly seen that the extent of the buildup in pressure at time "$\Delta t = \delta$", or $(p_\delta - p_{wf})$, can be calculated as:

$$(p_\delta \ p_{wf}) = (p_i - p_{wf}) - (p_i - p_\delta) \tag{6.269}$$

Introducing Eqs. (6.267) and (6.268) into Eq. (6.269), we obtain:

$$p_\delta - p_{wf} = \frac{70.6Bq_{sc}\mu}{kh}\left[\ln\frac{2.637 \times 10^{-4} kt_p}{\phi\mu c_t r_w^2} + 0.80907 + 2s - \ln\left(\frac{t_p + \delta}{\delta}\right)\right] \tag{6.270}$$

Using common logarithms with base 10, and assuming that "δ" is an early measurement in the test such that:

$$\frac{t_p + \delta}{\delta} \approx \frac{t_p}{\delta} \quad (t_p >> \delta) \tag{6.271}$$

one will end up with,

$$p_\delta - p_{wf} = \frac{162.6Bq_{sc}\mu}{kh}\left[\log 2.637 \times 10^{-4} + \log\frac{k}{\phi\mu c_t r_w^2} + \frac{0.80907}{2.3026} + \frac{2s}{2.3026} + \log\delta\right] \tag{6.272}$$

or,

$$(p_\delta - p_{wf}) = \frac{162.6Bq_{sc}\mu}{kh}\left[\log\frac{k}{\phi\mu c_t r_w^2} + 0.8685s - 3.2275 + \log\delta\right] \tag{6.273}$$

It should be noted at this junction by inspecting Eqs. (6.273) and (6.263), we realize that the coefficient that multiplies the bracket in Eq. (6.273) is simply the slope of the middle-region straight line of the Horner plot. Therefore, after the permeability and slope of the middle-region in the Horner plot have been calculated, Eq. (6.273) could be applied to a measurement "$\Delta t = \delta$, $p_{ws} = p_\delta$" in order to get skin "s"—as long as Eqs. (6.268) and (6.272) hold for such a measurement.

A very convenient choice here is $\delta = 1$ hour, not only because Eq. (6.272) would be certainly true, but also because $\log(1) = 0$, and Eq. (6.273) simplifies to:

$$(p_{1hr} - p_{wf}) = -m\left[\log\frac{k}{\phi\mu c_t r_w^2} + 0.8685s - 3.2275\right] \tag{6.274}$$

In Eq. (6.274), skin is the only unknown once the slope of the middle region straight line is known (and thus, the permeability "k"). Then, finally solving for "s" gives:

$$s = 1.1513\left(\frac{p_{1hr} - p_{wf@\Delta t=0}}{-m} - \log\frac{k}{\phi\mu c_t r_w^2} + 3.2275\right) \tag{6.275}$$

There is one more critically important point that needs to be discussed here. Equation (6.274) strictly applies if infinite-acting behavior is taking place at $\Delta t = \delta = 1$ hr and Eq. (6.268) holds. This is not usually the case, because after such short period of time ($\Delta t = \delta = 1$ hr), the buildup data may be still masked by wellbore storage effects. What we do here is that, after the actual infinite-acting behavior has been identified (start of the middle-time region), it will be necessary to extrapolate the straight line to $\Delta t = 1$ hr to read the value of "p_{1hr}" we are searching for. This process is illustrated in Fig. 6.37.

To reinforce the protocol outlined here we would like to reiterate that "p_{1hr}" must always be read off the straight line of infinite-acting behavior—and usually this requires extrapolation. The only possibility for the value of "p_{1hr}" to be equal to the pressure value recorded during the test (p_{ws} @ $\Delta t = 1$ hr) is that if the middle-time region has started within the first hour of the test.

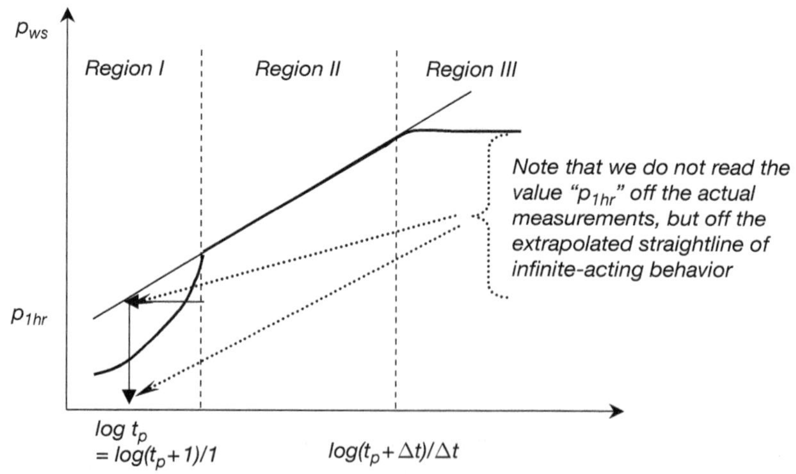

Figure 6.37 Reading of "p_{1hr}" for skin calculations (buildup).

It is important to keep in mind that the value of "skin" that we obtain as described here is actually the *composite* skin. It should be recognized that if a higher pressure drop than what the ideal model would predict is measured, the well test analyst is in a position to attribute it to a composite skin factor. This composite skin includes the combined effects of inertial-turbulent flow, restricted entry, and physical damage/stimulation.

PBU Analysis with Variable Flow Rates Prior to Testing

An important assumption of the buildup equation is that before the buildup test is conducted ($\Delta t = 0$), the reservoir has been producing at a constant rate "q_{sc}"—as shown in the left part of Fig. 6.31. Even though this is highly desirable in terms of the reliability of the buildup equations, this is usually not the case in actual field applications. Throughout its productive life, a well is expected to experience variable flow rates to follow a production schedule for optimized reservoir exploitation.

When a well has been subjected to variable flow rates prior to buildup testing, one possibility is to use the most recent "stabilized" flow rate "q_{sc}" at which the well was producing right before shutdown and manipulate the value of total production time, t_p. With this approach actual volume of fluids that have been produced out of the reservoir prior to testing is divided by the most recent stabilized rate. In other words, it is possible to apply the buildup equations using a *synthetic* material balance production time. Figure 6.38 illustrates this approach.

PBU Diagnostic Plots

As stated earlier, the most challenging task in a pressure buildup analysis is the identification of the middle-time region. To overcome this challenge, a "diagnostic plot"

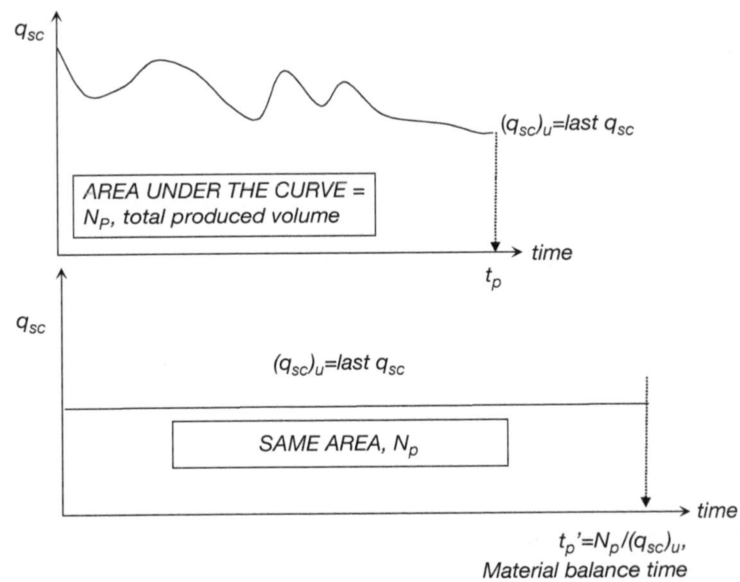

FIGURE 6.38 Synthetic material-balance time for buildup analysis with variable flow rate prior to testing.

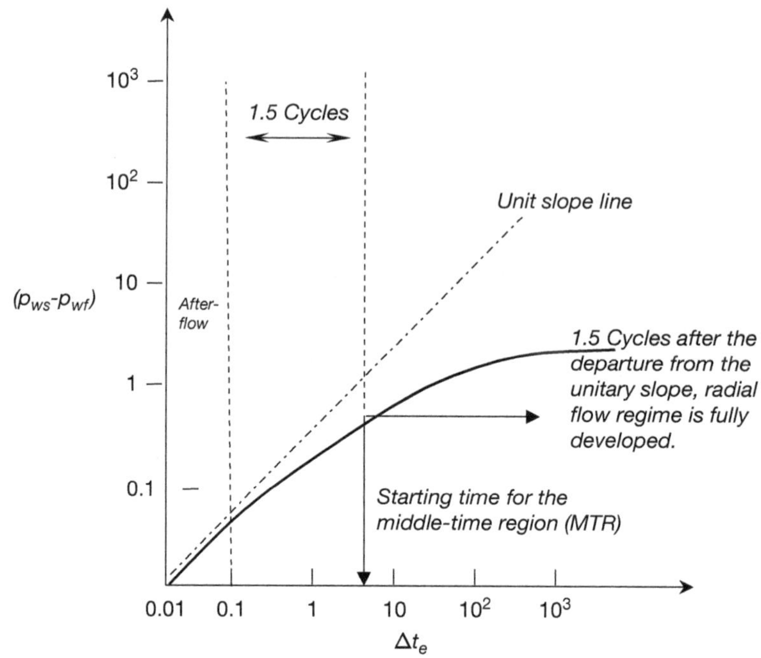

FIGURE 6.39 Diagnostic plot for a buildup test.

is utilized. The well-test analyst should review the diagnostic plot before constructing the Horner plot. Diagnostic plot is a log-log plot of ($p_{ws} - p_{wf}$) versus Δt_e—where p_{wf} is the measured bottomhole pressure just before the well is shut in and Δt_e is the equivalent time defined as:

$$\Delta t_e = \frac{\Delta t}{1 + \Delta t / t_p} \quad (\text{if } \ \Delta t < 0.1 t_p \text{ then simply set } \Delta t_e = \Delta t\,) \tag{6.276}$$

Figure 6.39 shows a typical diagnostic plot for a buildup test. When wellbore storage effects dominate (*afterflow*), the slope of the buildup curve is *equal to unity* in a log-log diagnostic plot. *Afterflow time* is defined as the time required for wellbore storage effects to cease and not to disturb the data. Afterflow is considered to be over when the graph starts to deviate from the unit slope line. The point where the curve starts to depart from the unit slope line, it will be necessary to move 3/2 cycles away to avoid pseudo-radial flow regime and find the time when actual radial flow regime starts. This procedure is instrumental for the identification of the *MTR*. Otherwise, finding the start of the *MTR* is usually challenging. The unit slope represents a 45° angle straight line in a log-log paper, if the scaling of the cycles on both axes are the same.

6.12.2 Pressure Drawdown (PDD) Analysis

One disadvantage of the buildup test we studied in the previous section is that we need to stop production while testing which in practice may not be desirable. A pressure drawdown test allows the engineer to conduct a well test while producing. The

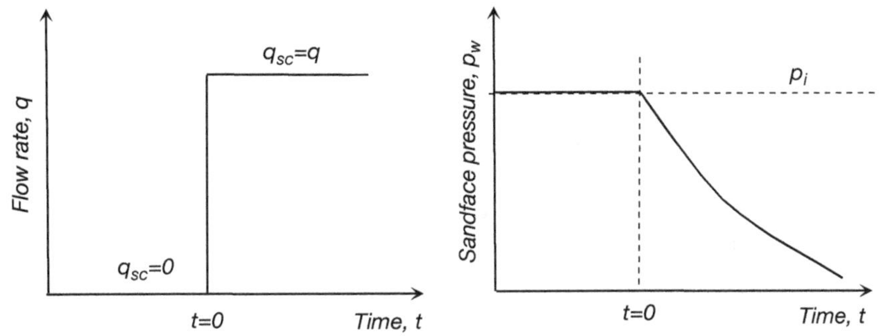

FIGURE 6.40 Flow rate and sandface pressure history of a PDD test.

requirement is that—ideally—an initial uniform pressure distribution throughout the reservoir is needed to be in place. However, in terms of field implementation, it should also be duly recognized that controlling the production from a well at a constant rate during a drawdown test is significantly more challenging than conducting a buildup test when the well is shut in.

The PDD Test Equation

A recently drilled and completed well which has not been brought on line ideally is the best candidate for a drawdown test as the test requires uniform pressure distribution at the beginning of the test. Then, at $t = 0$, the test is started by putting the well on production at a constant rate "q_{sc}" and the declining sandface pressure against time is recorded. Figure 6.40 shows flow rate and pressure histories for a typical drawdown test. Again, our objective is to infer or estimate the properties of the reservoir-well system using the pressure transient data collected during the test.

The data that we would record by conducting this test looks as follows:

t	p_{wf}
0	$p_i(t = 0)$
t_1	p_{wf1}
t_2	p_{wf2}
t_3	p_{wf3}
...	...

Some features of the drawdown test can be highlighted as follows:

- The duration of the flow period may be from a few hours to a number of days, mainly depending on the permeability of the formation.
- Drawdown test provides sound information about welllbore storage.
- Production is not lost during the test.
- It is difficult to maintain a constant flow rate during the test.
- The analysis of the data provides a possibility for estimating reservoir volume and shape of the drainage area.

Brand-new wells and wells that have been shut in for maintenance operations for extended periods of time (e.g., workover operations) are excellent candidates for a drawdown test. This is because a crucial condition in this analysis is that the pressure distribution in the reservoir at the onset of the test should be uniform. Therefore, a pressure drawdown test cannot be conducted effectively for a well that has been producing until recently.

The solution that we have studied for pressure drawdown in a producing well is the exponential-integral function:

$$\Delta p_D = -\frac{1}{2}E_i\left(-\frac{r_D^2}{4t_D}\right) \tag{6.277}$$

Since pressure is recorded at the wellbore ($r_D = 1$), the logarithmic approximation can be used to express Eq. (6.277) in a simpler form such as:

$$\Delta p_D = p_t = \frac{\Delta p}{p_i q_D} = \frac{1}{2}[\ln t_D + 0.80907] \tag{6.278}$$

If we consider the skin of the well, using the superposition principle, the skin factor can be added as an additional pressure drop. Then, we will have:

$$\Delta p = p_i q_D\left[\frac{1}{2}(\ln t_D + 0.80907) + s\right] \tag{6.279}$$

Recalling the definitions of the dimensionless groups for liquid-filled reservoirs with constants $\lambda = 2.637\times10^{-4}$ and $\gamma = 141.2$ which are good for practical field units, the pressure drawdown equation becomes:

$$p_i - p_{wf} = \frac{141.2Bq_{sc}\mu}{kh}\left[\frac{1}{2}\left(\ln\frac{2.637\times10^{-4}kt}{\phi\mu c_t r_w^2} + 0.80907\right) + s\right] \tag{6.280}$$

Switching to common logarithms with base 10, rearranging, and solving for the flowing bottomhole pressure yield:

$$p_{wf} = p_i - \frac{162.6Bq_{sc}\mu}{kh}\left[\log t + \log\frac{k}{\phi\mu c_t r_w^2} - 3.2275 + 0.87s\right] \tag{6.281}$$

Equation (6.281) is the *drawdown test equation* for liquid reservoir systems. From this equation, it is clear that a plot of p_{wf} versus t in a semilog plot should give a straight line with a slope:

$$m = -\frac{162.6Bq_{sc}\mu}{kh} \tag{6.282}$$

Figure 6.41 shows the Horner plot for an idealized drawdown test. Figure 6.41 qualitatively displays the behavior of the pressure transient data as expressed by Eq. (6.281) (as long as the pressure transient data is collected while the assumptions for well test analysis formulation are not violated). As we can see, we do not need to invert the

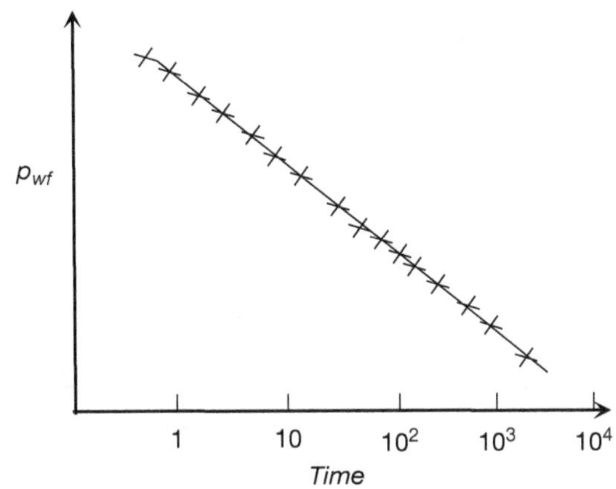

FIGURE 6.41 Idealized behavior of p_{wf} versus log(t).for a drawdown test.

x-axis as we did for the buildup test case. Similar to the buildup formulation the sign of the slope is negative (p_{wf} is decreasing with time).

It is clear that the slope of the straight line of Fig. 6.41 can be determined and using Eq. (6.282) the *permeability* of the formation can be calculated. It should be reiterated that Fig. 6.41 displays the behavior of a data set originating from an idealized drawdown test. In other words, assumptions such as reservoir is infinite-acting, no wellbore storage effects are taking place, there is no skin of any kind are all satisfied. In actual field applications such an idealized test behavior can hardly be encountered. The behavior of an actual drawdown test is described in the next section.

Drawdown Horner Plot

In the previous section, we discussed how an *ideal* Horner plot for a drawdown test looks. As indicated, Horner plot for a drawdown test analysis is a semi-logarithmic plot of p_{wf} in the ordinate and time in the logarithmically scaled abscissa. However, just as we discussed for buildup analysis, we expect to see some deviations of the collected data from ideal behavior. Figure 6.42 shows a Horner plot for actual drawdown data and illustrates the most common deviations that we can expect.

In Fig. 6.42, three regions of a typical Horner plot that highlight some specific effects on the recorded pressure transient data are recognized as described below:

Region I	*Early-Time Region (ETR)*	Early-time effects (storage and skin effects)
Region II	*Middle-Time Region (MTR)*	Infinite-acting period
Region III	*Late-Time Region (LTR)*	Late-time effects (boundary effects)

As discussed for buildup analysis, the most useful data in terms of determination of reservoir conductivity (permeability) can be obtained through the analysis of the middle time region. The main difficulty of Horner plot analysis is the correct identification of the *MTR*. As shown in Fig. 6.42, the analysis of early and late data also provides

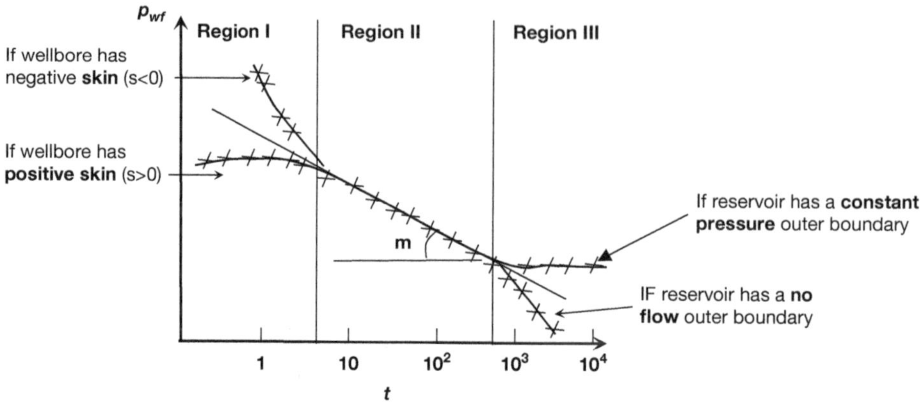

FIGURE 6.42 Horner plot for an actual drawdown test.

useful data for wellbore and reservoir characterization. Typical information we can obtain from the analysis of each region may be as follows:

Region I	→	Skin, duration of wellbore storage
Region II	→	Permeability
Region III	→	Size of the reservoir and type of the outer boundary and shape of the reservoir/well configuration

For the identification of the middle time region, we will again resort to a diagnostic plot. In the analysis of the early time data (*ETR*), we see that if the well is stimulated ($s < 0$), the actual value of p_{wf} is higher than what the ideal model would predict (smaller pressure drop than the one for the ideal case is observed). If the well is damaged ($s > 0$), there will be a higher pressure drop than that predicted by the ideal model—and thus, a smaller value for the value of p_{wf} than the one predicted by the ideal model, which is represented by the straight line in the plot.

When transients encounter a boundary, the *LTR* data starts by departing from the infinite-acting straight line. If the transients reach a closed, no-flow boundary, pressure starts to decline faster (sharper slope of the *LTR* as compared to the *MTR*). If a constant pressure outer boundary is encountered, the external boundary will discharge fluids to the formation and compensate for the fluids withdrawn from it. When the rate of influx from the outer boundary compensates the well production rate, system reaches the true steady-state conditions and the value of pressure stabilizes everywhere in the reservoir. Therefore, in this case, the *LTR* slope diminishes indicating that pressure is not changing with time.

Skin Factor Analysis from PDD Data

A quick inspection of the drawdown equation [Eq. (6.283)] reveals that the value of skin (s) is already embedded into the equation.

$$p_{wf} = p_i - \frac{162.6Bq_{sc}\mu}{kh}\left[\log t + \log\frac{k}{\phi\mu c_t r_w^2} - 3.2275 + 0.87s\right] \qquad (6.283)$$

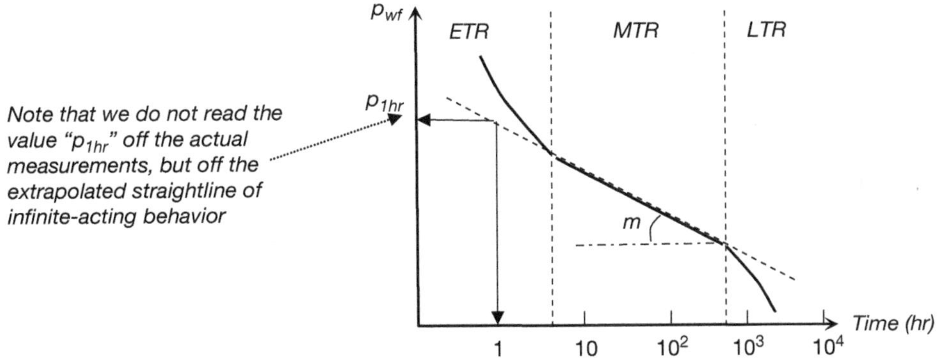

FIGURE 6.43 Reading of "p_{1hr}" for skin calculations (PDD data).

Since log(1) = 0, and to preserve an analogy to skin factor analysis from buildup data, $t = 1$ hr would be again a convenient choice to calculate the value of skin from Eq. (6.283). Then,

$$p_{1hr} = p_i + m\left[\log\frac{k}{\phi\mu c_t r_w^2} - 3.2275 + 0.87s\right] \tag{6.284}$$

Solving Eq. (6.284) for "s" gives:

$$s = 1.1513\left(\frac{p_{1hr} - p_i}{+m} - \log\frac{k}{\phi\mu c_t r_w^2} + 3.2275\right) \tag{6.285}$$

Equation (6.285) is the skin factor equation for drawdown data. This equation is very much similar to that for skin factor analysis using buildup data [Eq. (6.275)]—the difference being that the slope (m) preserves its sign in the drawdown equation. Just as for the case of buildup analysis, "p_{1hr}" must always be read off the straight line of infinite-acting behavior even if this requires an extrapolation of the *MTR* infinite-acting straight line. This situation is illustrated in Fig. 6.43 for the case of analysis of drawdown data from a no-flow outer boundary reservoir with a wellbore that presents stimulation ($s > 0$).

PDD Diagnostic Plots

At this point it is clear that the most difficult task in a pressure drawdown analysis is the identification of the middle time region. To help us out with this identification, we again refer to the use of *diagnostic plot*. The diagnostic plot should be constructed before preparing the Horner plot. For pressure transient data from a drawdown test, a diagnostic plot is prepared as log $(p_i - p_{wf})$ versus log(t) time, as shown in Fig. 6.44.

Figure 6.44 shows a typical diagnostic plot for a drawdown test. During afterflow period, wellbore storage effects dominate and the log-log curve displays a unit slope. At the point where the curve starts to depart from the unit-slope line, it will be necessary to move 3/2 cycles away to determine the time when the radial flow regime starts.

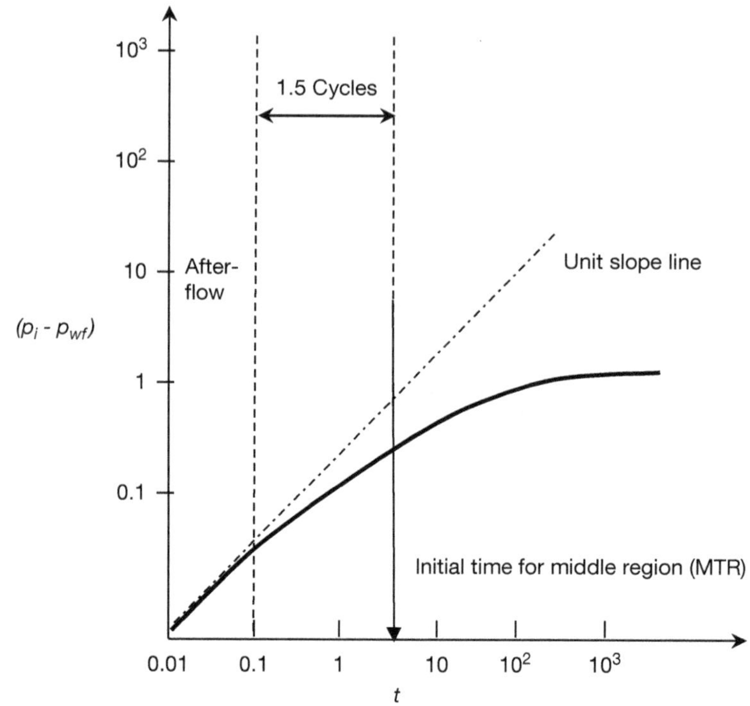

FIGURE 6.44 Diagnostic plot for a drawdown test.

Reservoir Limit Test Analysis

In the case we are dealing with a no-flow outer boundary reservoir, and if the drawdown test is run for a long enough time, we would be able to record transient data extended into pseudo-steady-state flow regime. If one recalls the discussion in the earlier parts of this chapter involving the behavior of a well in a closed, regular-polygon-shaped reservoir, the existence of three flow regimes was recognized. These flow regimes are experienced by the well throughout its production life: *transient flow regime, late transient flow regime,* and *pseudo-steady state or stabilized flow regime.* Figure 6.45 shows these three important flow periods for the case of a drawdown test conducted in a closed reservoir.

It is important to stress that the data set used in Fig. 6.45 is plotted on Cartesian coordinates. The data collected during the *late-transient* and *stabilized flow* periods correspond to the data in the *LTR* section of the Horner plot. The end of the late transient flow period marks the start of the *pseudo-steady state flow,* which represents the *stabilization time* or the time when all the points at the outer boundary have encountered a measurable pressure transient. During this flow period, reservoir pressure will decline at the same rate everywhere in the drainage region and therefore, this flow period can be characterized by a straight line of slope m^* and intercept "$p_{intercept}$", as shown in Fig. 6.45.

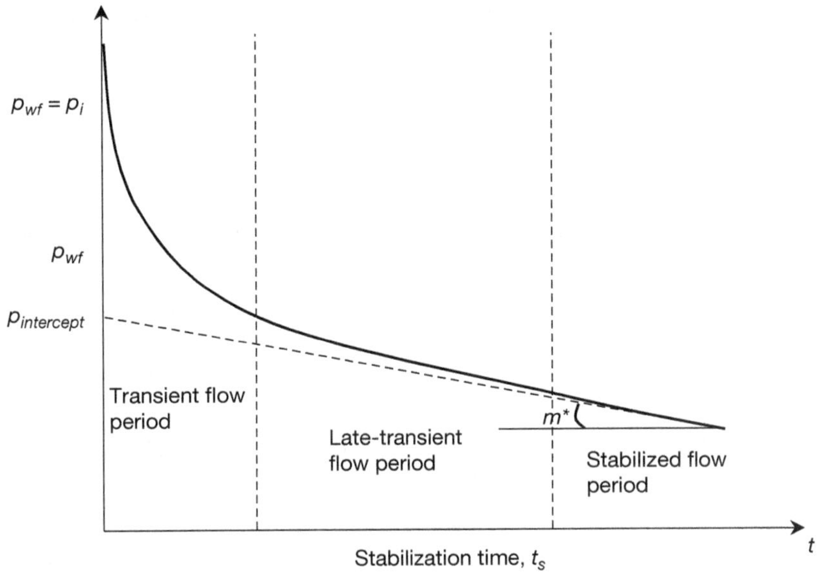

FIGURE 6.45 Cartesian plot of "p_{wf}" versus production time for a drawdown test conducted in a closed reservoir.

In the analysis of reservoir size using drawdown data, it will be necessary to focus on the data that is collected during the *pseudo-steady-state flow period*. Earlier, we were able to derive the equation that models the behavior of "p_{wf}" with time during this flow period [Eq. (6.228)]. Let us have a second look at this equation:

$$\Delta p\Big|_{\text{ideal,at sandface}} = \frac{1}{2} p_i (q_A)_D \left[\ln\left(\frac{4A}{1.781 r_w^2 C_A} \right) + 4\pi t_{DA} \right] \tag{6.286}$$

or,

$$(\Delta p_D)_{ideal,at\ sandface} = \frac{1}{2} \ln\left(\frac{4A}{1.781 r_w^2 C_A} \right) + 2\pi t_{DA} \tag{6.287}$$

Equation (6.287) describes the pressure changes during the *pseudo-steady-state flow period* for a producing well drilled in a reservoir surrounded with no-flow outer boundaries. It assumes that the flowing pressure at the wellbore follows the ideal model behavior; that is, no skin effects are considered. If we wanted to incorporate the concept of skin, it will be necessary to superimpose the dimensionless pressure drop due to skin (s):

$$(\Delta p_D)_{actual} = (\Delta p_D)_{ideal} + s$$

Then, the actual pressure drop during the *pseudo-steady-state flow period* for a producing well drilled in a reservoir with no-flow outer boundary is described by Eq. (6.288):

$$\Delta p_D = \frac{1}{2}\ln\left(\frac{4A}{1.781 r_w^2 C_A}\right) + 2\pi t_{DA} + s \tag{6.288}$$

Rearranging and substituting the definitions for the dimensionless variables, one will get:

$$\frac{p_i - p_{wf}}{p_i \dfrac{\gamma B q_{sc}\mu}{p_i kh}} = 2\pi\frac{\lambda kt}{\phi\mu c_t A} + \frac{1}{2}\ln\left(\frac{A}{r_w^2}\right) + \frac{1}{2}\ln\left(\frac{2.2458}{C_A}\right) + s \tag{6.289}$$

Using practical field units with λ=2.637×10^{-4} and $\gamma = 141.2$, the final form of the pseudo-steady-state equation becomes:

$$p_{wf} = p_i - \frac{141.2 B q_{sc}\mu}{kh}\left[2\pi\frac{2.637\times 10^{-4} kt}{\phi\mu c_t A} + \frac{1}{2}\ln\left(\frac{A}{r_w^2}\right) + \frac{1}{2}\ln\left(\frac{2.2458}{C_A}\right) + s\right] \tag{6.290}$$

or,

$$p_{wf} = p_i - \frac{0.23395 B q_{sc}}{c_t A h\phi}t - \frac{70.6 B q_{sc}\mu}{kh}\left(\ln\left(\frac{A}{r_w^2}\right) + \ln\left(\frac{2.2458}{C_A}\right) + 2s\right) \tag{6.291}$$

It is clear from Eq. (6.291) that the graph of p_{wf} versus time on a Cartesian plot yields a straight line during the *pseudo-steady-state flow regime* period. The slope of that line is "m^*", which from Eq. (6.291) is given by:

$$m^* = -\frac{0.23395 q_{sc} B}{c_t A h\phi} \tag{6.292}$$

We can see that the product "$Ah\phi$" in the denominator represents the entire pore volume (V_p) of the reservoir under consideration. Because we are dealing with a single-phase liquid reservoir, this quantity represents the total amount of oil in place in the system (N)—as we can safely assume that the pore space must be completely saturated with oil. Hence, the analysis of the late time data allows us to determine the size of the reservoir, as shown below:

$$N(ft^3) = V_P(ft^3) = -\frac{0.23395 q_{sc} B}{c_t m^*} \tag{6.293}$$

We could express the initial oil in place in barrels by applying the conversion 1 bbl = 5.615 ft^3. The value of "N" and "V_p" are always positive quantities because the slope "m^*" is always negative. Another possibility is to solve for the lateral extent of the reservoir (A) in Eq. (6.292) instead of pore volume.

Additionally, we could use the value of the intercept of the *PPS* (pseudo-steady state) straight line to get more information. By inspecting Eq. (6.291), we see that the value of "$p_{intercept}$"—the intercept or value of "p_{wf}" in the pseudo-steady-state straight line at $t = 0$—is equal to:

$$p_{intercept} = p_i - \frac{70.6Bq_{sc}\mu}{kh}\left[\ln\left(\frac{A}{r_w^2}\right) + \ln\left(\frac{2.2458}{C_A}\right) + 2s\right] \quad (6.294)$$

Recalling the definition of "P_{1hr}" and "m" from Horner plot analysis,

$$p_{1hr} = p_i + m\left[\log\frac{k}{\phi\mu c_t r_w^2} - 3.2275 + 0.87s\right] \quad (6.295)$$

or,

$$p_{1hr} = p_i - \frac{70.6Bq_{sc}\mu}{kh}\left[\ln\frac{k}{\phi\mu c_t r_w^2} - 7.4216 + 2s\right] \quad (6.296)$$

After subtracting Eq. (6.294) from Eq. (6.296) one obtains:

$$p_{1hr} - p_{intercept} = \frac{m}{2.3026}\left[\ln\frac{k}{\phi\mu c_t r_w^2} - 7.4216 + 2s - \ln\left(\frac{A}{r_w^2}\right) - \ln\left(\frac{2.2458}{C_A}\right) - 2s\right] \quad (6.297)$$

or,

$$p_{1hr} - p_{intercept} = \frac{m}{2.3026}\ln\left(e^{-7.4216}\frac{k}{\mu c_t A\phi}\frac{C_A}{2.2458}\right) \quad (6.298)$$

At this point, we see that the ratio of slopes m/m* is equal to:

$$\frac{m^*}{m} = \frac{0.23395}{162.6}\frac{k}{\mu c_t A\phi} \quad (6.299)$$

Introducing Eq. (6.299) into Eq. (6.298), and solving for "C_A" we obtain:

$$C_A = 5.456\frac{m}{m^*}\exp\left[2.3026\left(\frac{p_{1hr} - p_{initial}}{m}\right)\right] \quad (6.300)$$

Equation (6.300) allows us calculating the shape factor (C_A) of the reservoir using the information from the late-time *PSS* straight line ($p_{intercept}$, m^*) and from the Horner plot *MTR* straight line (p_{1hr}, m). Once the value of C_A for the reservoir is determined, then, Table 6.5, which provides a summary of the shape factors for different reservoir shapes, can be used to find the best match between the calculated C_A and the shape with a listed C_A value that gives the closest match. It should be carefully noted that the accuracy of the analysis of the data collected during the pseudo-steady-state regime critically hinges upon forcefully on the correct identification of the *PSS* straight line.

6.13 Conformal Mapping Techniques

There are a number of analytical techniques available to solve transport equations of porous media. For the case of solution of the steady-state version of the single-phase liquid flow equations, we would be rather interested in seeking analytical solutions to the Laplace equation:

$$\frac{\partial^2 p}{\partial x^2} + \frac{\partial^2 p}{\partial y^2} = 0 \tag{6.301}$$

Conformal mapping is a well-known technique used to solve the Laplace equation. In Chap. 2, we discussed a coordinate transformation or mapping which established a correspondence between points in the rectangular coordinates and radial-cylindrical coordinates (x–y and r–θ planes). The set of equations $u = u(x,y)$ and $v = v(x,y)$ defines in general a transformation between points in the u–v and x–y planes. As shown in Fig. 6.46, we consider the mapping of point (x_0, y_0) of the x–y plane into the point (u_0, v_0) of the u–v plane. If the transformation is such that the angle "θ" at (x_0, y_0) between c_1 and c_2 is equal to the angle "θ'" at (u_0, v_0) between c_1' and c_2' both in magnitude and sense, the transformation is said to be conformal (one can think the phrase "conformal" as angle preserving). Accordingly, conformal mapping can be a useful mathematical tool for the solution of the steady-state flow problems in porous media. In this case of conformal mapping, it will be necessary to consider the properties of the complex potential.

$$z(z) = u(x,y) + iv(x,y) \tag{6.302}$$

Equation (6.302) generates two families of curves $u(x, y) =$ *constant* and $v(x, y) =$ *constant* and furthermore these two sets of curves are orthogonal (see Fig. 6.47). If the curvilinear quadrangle in the z-plane also has the same shape form in the w-plane, the transformation is known as conformal.

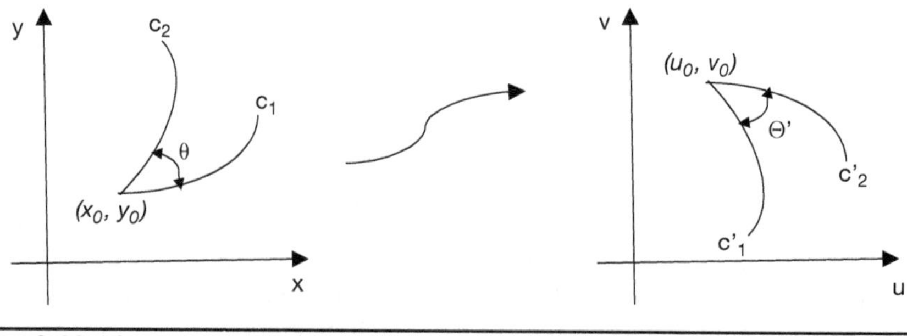

Figure 6.46 Conformal mapping of point (x_0, y_0) of the x-y plane to the point (u_0, v_0) of the u–v plane.

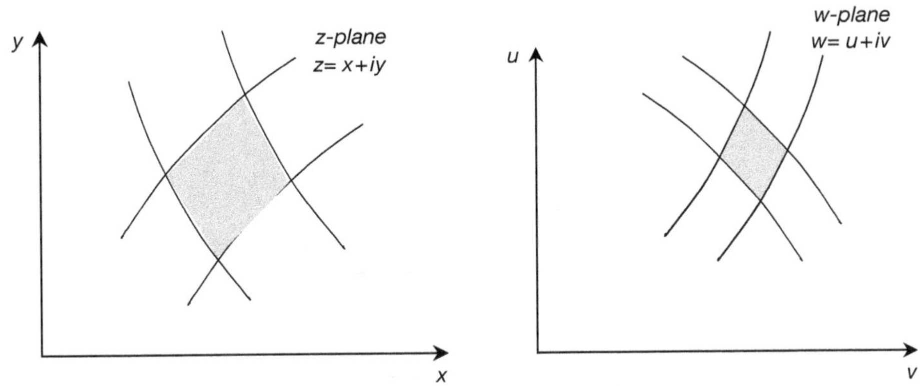

FIGURE 6.47 Conformal mapping of a curvilinear quadrangle in z-plane into the w-plane.

Example 6.1 Let us consider the problem of plane radial flow $\frac{\partial^2 p}{\partial x^2} + \frac{\partial^2 p}{\partial y^2} = 0$ within the domain $r_w^2 \leq \left(x^2 + y^2\right) \leq r_e^2$, with boundary conditions (BC):

$$\text{BC (1)}\, p = p_w \text{ on } x^2 + y^2 = r_w^2 \tag{6.303}$$

$$\text{BC (2)}\, p = p_e \text{ on } x^2 + y^2 = r_e^2 \tag{6.304}$$

A graphical representation of the problem on the x–y plane is shown in Fig. 6.48.

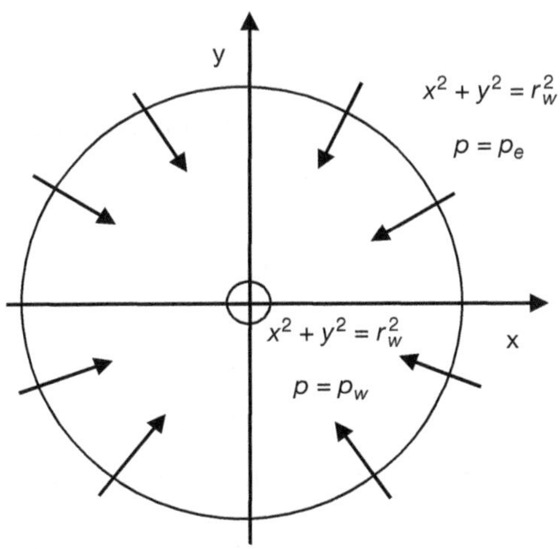

FIGURE 6.48 Radial flow problem on the z-plane.

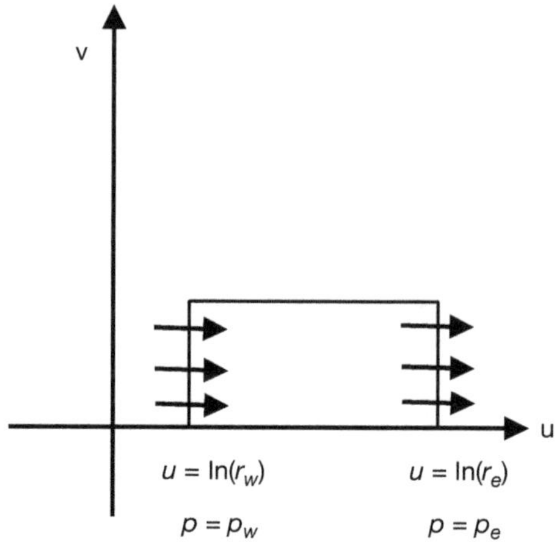

FIGURE 6.49 Radial flow problem on the *w*-plane.

Now, we would like to perform a conformal mapping of $w = f(z)$ which would map the flow domain into a simpler geometric form for which the solution is already known. Such a mapping in this case can be achieved by $w = \ln(z) = \ln(re^{i\theta})$. In other words,

$$u + iv = \ln(r) + i\theta \tag{6.305}$$

Then, $u = \ln(r)$ and $v = \theta$. Knowing that $r_w \leq r \leq r_e$ and $0 \leq \theta \leq 2\pi$.

The mapping on the u–v plane will be as shown in Fig. 6.49. Now, the solution of the flow problem on the w-plane is straightforward since it is a linear flow problem.

Therefore,

$$p = A + Bu \tag{6.306}$$

Now using the boundary conditions as specified,

$$p_w = A + B\ln(r_w) \tag{6.307}$$

$$p_e = A + B\ln(r_e) \tag{6.308}$$

From the above two equations A and B can be solved simultaneously to find that

$$A = p_w - \frac{p_e - p_w}{\ln\left(\frac{r_e}{r_w}\right)}\ln(r_w) \tag{6.309}$$

$$B = \frac{p_e - p_w}{\ln\left(\frac{r_e}{r_w}\right)} \tag{6.310}$$

Substituting Eqs. (6.309) and (6.310) into Eq. (6.306) yields:

$$p = p_w + \frac{p_e - p_w}{\ln\left(\frac{r_e}{r_w}\right)}[\ln(r) - \ln(r_w)] \tag{6.311}$$

Solution Eq. (6.311) in the x–y plane can be expressed by simply setting:

$$r = \left(x^2 + y^2\right)^{\frac{1}{2}} \tag{6.312}$$

Then,

$$p = p_w + \frac{p_e - p_w}{\ln\left(\dfrac{r_e}{r_w}\right)}\left[\ln\left(x^2 + y^2\right)^{\frac{1}{2}} - \ln(r_w)\right] \tag{6.313}$$

or

$$p = p_w + \frac{p_e - p_w}{\ln\left(\dfrac{r_e}{r_w}\right)}\left[\ln\frac{\left(x^2 + y^2\right)^{\frac{1}{2}}}{r_w}\right] \tag{6.314}$$

Other General Conformal Mapping Transformations

Table 6.6 gives a summary of general transformations that are used in conformal mapping. See the following examples of transformations described.

Transformation name	Transformation	Remarks
Translation	$w = z - i\beta$	Shapes on the z-plane are translated in the direction of the vector β
Rotation	$w = e^{i\theta_0} z$	Shapes in the z-plane are rotated through an angle θ_0 ($\theta_0 > 0$ rotation is counter-clockwise $\theta_0 < 0$ rotation is clockwise)
Stretching (linear amplification)	$w = az$	Shapes in the z-plane stretched or contracted $0 < a < 1$ contraction $a > 1$ stretching
Inversion and reflection	$w = \dfrac{1}{z}$	Creates negative image of the shape
Successive transformation	$w = \alpha z + \beta$	Translation + rotation + stretching
Bilinear or fractional transformation	$w = \dfrac{\alpha z + \beta}{\gamma' z + \delta}$	Translation + rotation + stretching + inversion
Schwarz–Christoffel transformation		Converts the interior of a polygon in the z-plane into the upper half of the w-plane

Table 6.6 Widely Used Conformal Transformations
Note: α and β are complex constants and a, θ_0 are real constants.

Example 6.2 Inversion/reflection transformation $w = \frac{1}{z}$

One seeks to solve for the pressure distribution inside the porous media domain, as shown in Fig. 6.50 (upper section). The inversion transformation is considered to be promising. In Fig. 6.50, at the origin two circles, one with a radius $r = \frac{1}{2}$ and the other one with a radius $r = \infty$ are observed. Since both pass through the origin with $w = \frac{1}{z}$ transformation, both circles get sent into two straight lines on the w-plane. Since only two points needed to describe a straight line, it will be necessary to pick two points from $x = 0$ and two from the $\left|z = \frac{1}{2}\right| = \frac{1}{2}$ boundaries. From the $x = 0$ boundary if one chooses $z_1 = iy_1$ and $z_1 = -iy$ then through the $w = \frac{1}{z}$ transformation, $w_1 = \frac{1}{z_1}$ so i maps to $-i$.

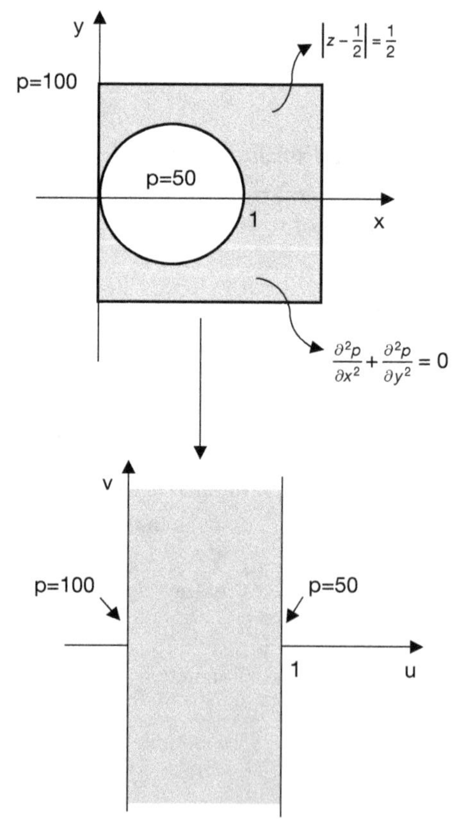

FIGURE 6.50 Example on inversion transformation.

$w_2 = \frac{1}{z_2}$ then $-i$ maps to i. Similarly, on the $\left|z - \frac{1}{2}\right| = \frac{1}{2}$ boundary (the circle with a radius of $\frac{1}{2}$):

$$w_1 = \frac{1}{z_1} \text{ so } x = 1 \text{ maps to } u_1 = 1 \text{ and,}$$

$$x = \frac{1}{2} + i\frac{1}{2} \text{ maps to } u + iv = \frac{1}{\frac{1}{2} + i\frac{1}{2}} = \frac{2}{1+i} = 1 - i$$

which implies that $u_2 = 1$ and $v_2 = -1$.

As shown in Fig. 6.50 (lower section), the mapped region on the x–y plane maps to a semi-infinite strip. Then, the problem reduces to: $\frac{\partial^2 p}{\partial u^2} = 0$ which gives a solution $p = c_1 u + c_2$. Using the prescribed boundary conditions, $p(0) = 100$ and $p(1) = 50$ one can find that:

$c_1 = -50$ and $c_2 = 100$. Finally on the u–v plane, the equation that describes the distribution of p is $p = 100 - 50u$. Now, it will be necessary to bring the solution back to the x–y plane.

Since,

$$w = u + iv \text{ and } w = \frac{1}{z} \tag{6.315}$$

$$u + iv = \frac{1}{x + iy} \text{ or} \tag{6.316}$$

$$u + iv = \frac{1}{x + iy}\frac{x - iy}{x - iy} \tag{6.317}$$

$$u + iv = \frac{x}{x^2 + y^2} - i\frac{y}{x^2 + y^2} \tag{6.318}$$

which implies that

$$u = \frac{x}{x^2 + y^2} \text{ and } v = -\frac{y}{x^2 + y^2} \tag{6.319}$$

And finally,

$$p(x,y) = 100 - 50\frac{x}{x^2 + y^2} \tag{6.320}$$

Example 6.3 The bilinear transformation $w = \dfrac{\alpha z + \beta}{\gamma z + \delta}$

This example illustrates that the transformation $w = \dfrac{\alpha z + \beta}{\gamma z + \delta}$ (also known as Möbius transformation). In this transformation there are four complex parameters that need to be determined. This transformation, similar to $w = \ln(z)$ transformation, maps circles in the x–y plane to circles in the u–v plane in which circles become concentric (see Fig. 6.51).

If one lets $\Gamma = \gamma z + \delta$ and $\theta = \dfrac{1}{\Gamma}$ then original transformation can be expressed as:

$$w = \frac{\alpha z + \beta}{\gamma z + \delta} = \frac{\beta\gamma - \alpha\delta}{\gamma}\left(\frac{1}{\gamma z + \delta}\right) + \frac{\alpha}{\gamma} \tag{6.321}$$

In the above representation of the proposed transformation we see three different transformations applied in sequence.

i. First, linear transformation $\Gamma = \gamma z + \delta$ which maps circles into circles on the u–v plane first scaling by γ and then shifting by δ
ii. Second, inversion transformation through $\theta = \dfrac{1}{\Gamma}$ again mapping circles to circles
iii. Third, a second linear transformation

$$w = \frac{\beta\gamma - \alpha\delta}{\gamma}\theta + \left(\frac{\alpha}{\gamma}\right) \tag{6.322}$$

once again sends circles into circles.

Since each individual transformation sends circles to circles, the entire transformation has to do the same. To specify a circle it will be necessary to assign three points on the circle. The other circle can easily get sent to a circle with center at $w = 0$ by simply setting $|z| = 1$. At this stage, the goal is to send the three selected points on the inner (smaller) circle $\left|z - \dfrac{1}{4}\right| = \dfrac{1}{4}$ on the u–v plane to another circle which will be concentric with the outer circle in the u–v plane. From a table of transformation of domains, one can find that $w = \dfrac{z - a}{az - 1}$ with $a = 2 + \sqrt{3}$ maps the smaller circle of the original problem of Fig. 6.51 to another circle which is concentric with the original unit circle.

Since problem on the u–v plane is one-dimensional flow (radial-cylindrical coordinates, r-direction only):

$$p = A + B\ln(r) \tag{6.323}$$

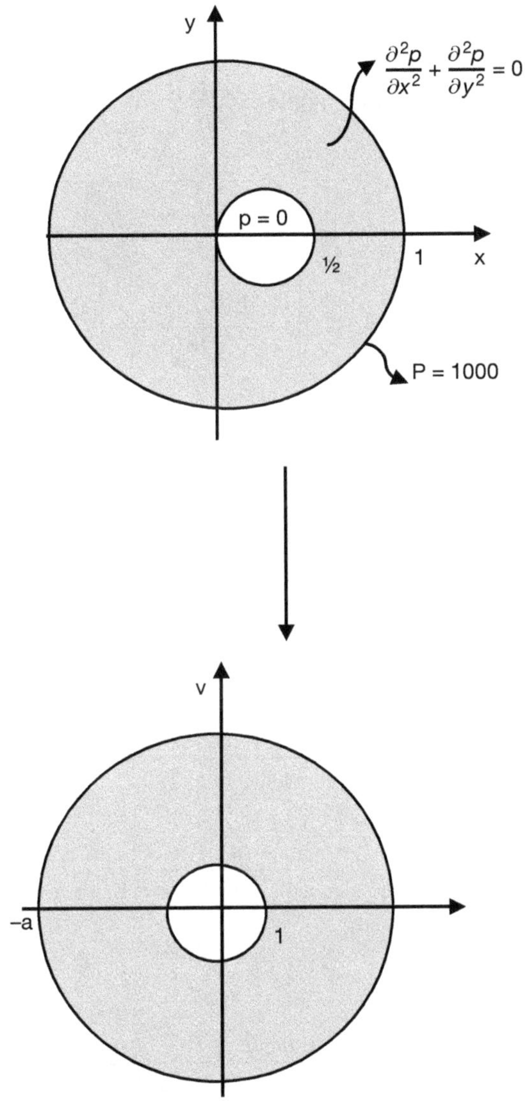

Figure 6.51 Bilinear transformation application.

Now assume $r = 1, p = 1000$ and at $r = a, p = 0$. Using the above boundary conditions, it can be found that $A = 1000$ and $B = -\frac{1000}{\ln(r_a)}$. Then,

$$p = 1000 - \frac{1000}{\ln(r_a)}\ln(r) \tag{6.324}$$

or

$$p = 1000\left(1 - \frac{\ln(r)}{r_a}\right) \tag{6.325}$$

which is still on the u–v plane.

$$p(u,v) = 1000\left(1 - \frac{\ln\left(u^2 + v^2\right)^{\frac{1}{2}}}{\ln(a)}\right) \tag{6.326}$$

or,

$$p(u,v) = 1000\left(1 - \frac{\ln\left(u^2 + v^2\right)}{2\ln(a)}\right) \tag{6.327}$$

Now, one needs to bring the solution to the original x–y plane. Since,

$$u + iv = \frac{x + iy - a}{a(x + iy - 1)} \tag{6.328}$$

$$u + iv = \frac{(x - a) + iy}{(ax - 1) + iay} \tag{6.329}$$

or,

$$u + iv = \frac{(ax - 1)(x - 1) + iy(ax - 1) - iay(x - a) - ay^2}{(ax - 1)^2 + \left(a^2y^2\right)} \tag{6.330}$$

Finally, solution on the x–y plane will be by substituting

$$u = \frac{(ax - 1)(x - a) - ay^2}{(ax - 1) + \left(a^2y^2\right)} \text{ and } v = \frac{\left(a^2 - 1\right)y}{(ax - 1)^2 + \left(a^2y^2\right)} \tag{6.331}$$

into the solution expressed in the u–v plane, with $a = 2 + \sqrt{3}$.

Example 6.4 The Schwartz-Christoffel transformation

The Schwartz–Christoffel transformation can be used successfully when the flow domain in the z-plane has the form of a single connected polygon. The boundaries provided by the polygon divides the z-plane into two domains, which are interior to the boundary and exterior to the boundary. As shown in Fig. 6.52, boundary of the polygon may extend to infinity. To map the polygon boundary of the polygon shown in Fig. 6.52 on the w-plane, it will be necessary to translate each vertex of the polygon to a point on the u-axis. However, at the same time it will be necessary to convert angle of the polygon at each vertex into a straight angle:

$$\frac{dz}{dw} = A(w-u_1)^{\frac{\alpha_1}{\pi}-1}(w-u_2)^{\left(\frac{\alpha_2}{\pi}\right)-1}\cdots(w-u_n)^{\left(\frac{\alpha_n}{\pi}\right)-1} \text{ or}$$

$$z = A\int(w-u_1)^{\frac{\alpha_1}{\pi}-1}(w-u_2)^{\left(\frac{\alpha_2}{\pi}\right)-1}\cdots(w-u_n)^{\left(\frac{\alpha_n}{\pi}\right)-1} + B \quad (6.332)$$

In the above equation, A and B are two complex constants. Once the polygon in the original x–y plane is mapped to the upper-half of the x–y plane with its vertices on the real axis of the w-plane, it will be possible to map it into a second u'–v' plane with a shape, for which the solution is readily available.

We now find the distribution of the potential function inside of the unit circle $|z| = x^2 + y^2 = 1$ with the boundary condition shown in Fig. 6.53. It is noted that the transformation $w = \frac{1-z}{1+z}i = \frac{1-x-iy}{1+x+iy}i$ will send the unit circle of the z-plane to a rectangular shape, as shown in Fig. 6.53.

In the w-plane,

$$\Phi(u,v) = A\theta + B \text{ where } \theta = \tan^{-1}\left(\frac{v}{u}\right) \quad (6.333)$$

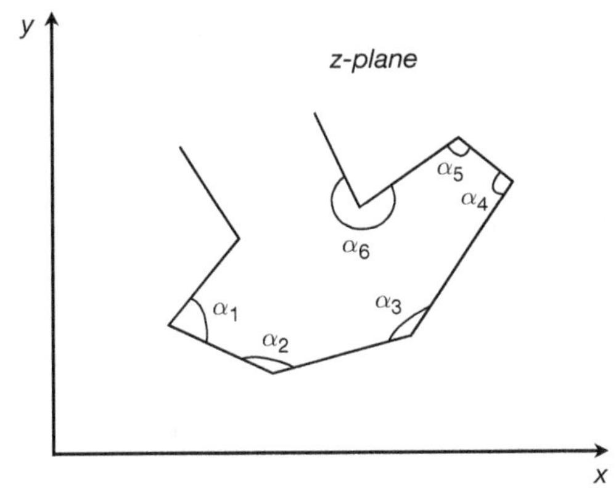

FIGURE 6.52 Polygon in the z-plane.

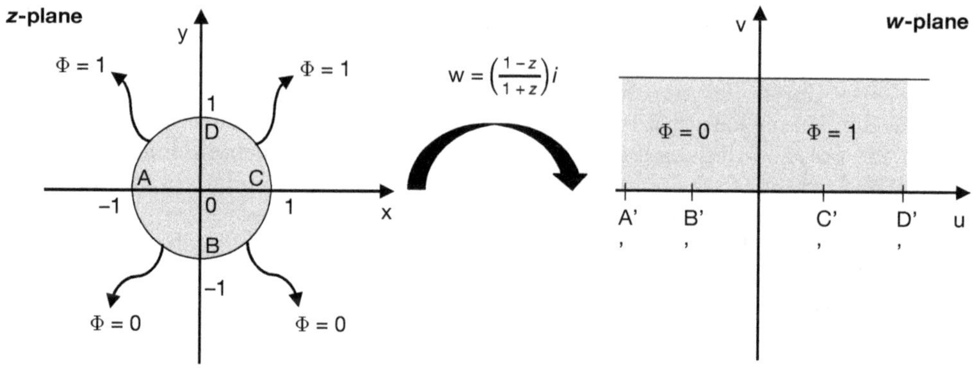

FIGURE 6.53 Example on Schwartz–Christoffel transformation.

So when

$$\theta = 0, \Phi = 1 \rightarrow B = 1 \tag{6.334}$$

and when

$$\theta = \pi, \ \Phi = 0 \rightarrow A = \frac{1}{\pi} \tag{6.335}$$

Then, $\Phi = -\frac{1}{\pi}\theta + 1$ or $\Phi = -\frac{1}{\pi}\tan^{-1}\left(\frac{v}{u}\right) + 1$ is the solution in the w-plane. Since $w = \frac{(1-z)}{1+z}i$,

$$u + iv = i\frac{(1-x) - iy}{(1+x) + iy} = \frac{\left(1-x^2\right) - (1-x)yi - (1+x)yi - y^2}{1 + x^2 + 2x - y^2}i \tag{6.336}$$

or,

$$u + iv = \frac{i\left(1 - x^2 - y^2\right) + (1-x)y + (1+x)y}{\left(1+x\right)^2 + y^2} \tag{6.337}$$

or,

$$u + iv = \frac{2y + \left(1 - x^2 - y^2\right)i}{\left(1+x\right)^2 + y^2} \tag{6.338}$$

or,

$$u + iv = \frac{2y}{(1+x)^2 + y^2} + i\frac{1 - x^2 - y^2}{(1+x)^2 + y^2} \tag{6.339}$$

resulting in,

$$u = \frac{2y}{(1+x)^2 + y^2} \text{ and } v = \frac{1 - x^2 - y^2}{(1+x)^2 + y^2} \tag{6.340}$$

Finally, the solution in the x–y plane takes the form:

$$\Phi(x, y) = 1 - \frac{1}{\pi}\tan^{-1}\frac{1 - x^2 - y^2}{2y} \tag{6.341}$$

References

Agarwal, R.G., "Real Gas Pseudo-Time: A New Function for Pressure Buildup of MHF Gas Wells," SPE Paper 8279 presented at the *54th Annual SPE Fall Technical Conference and Exhibition*, Las Vegas, NV.

Al-Hussainy, R., H.J. Ramey, and P.B. Crawford, "The Flow of Real Gases Through Porous Media," *SPE J.* 18(5), May 1996.

Aziz, K., and D.L. Flock, "Unsteady State Gas Flow: Use of Drawdown Data in the Prediction of Gas Well Behavior," *J. Can. Pet. Tech.*, 2(1):9–15, 1963.

Earlougher, R.C., H.J. Ramey, F.G. Miller, and T.D. Mueller, "Pressure Distributions in Rectangular Reservoirs," *J. Pet. Tech.* 20:199–208, 1968.

Earlougher, R.C., *Advances in Well Test Analysis*, SPE Monograph Series Vol. 5, Society of Petroleum Engineers, Dallas, TX, 1977.

Energy Resources Conservation Board (ERCB), *Theory and Practice of the Testing of Gas Wells*, 3rd ed., Report ERCB 75-34, Calgary, Alberta, Canada, 1975.

Horner, D.R., "Pressure Buildup in Wells," *Proceedings of the Third World Petroleum Congress*, The Hague, Sec. II, pp. 503–521, 1951.

Klinkenberg, L.J., *The Permeability of Porous Media to Liquids and Gases*, Drilling and Production Practice, American Petroleum Institute, pp. 200–213, New York, NY, 1941.

Martin, J. C., "Simplified Equations of Flow in Gas Drive Reservoirs and the Theoretical Foundation of Multiphase Pressure Buildup Analysis," SPE Paper 1235-G, *Petrol Trans.* AIME, 216:321–323, 1959.

Matthews, C.S., F. Brons, and P. Hazebroek, "A Method for Determination of Average Pressure in a Bounded Reservoir," *Trans AIME.* 209:192–189, 1954.

Medhat, K., *Transient Well Testing*, SPE Monograph Vol. 23, Henry L. Doherty Series, Society of Petroleum Engineers, Richardson, TX, 2009.

Nisle, R.G., "How to Use the Exponential Integral," *Pet. Eng.*, 171–173, August 1956.

Odeh, A.S., "Steady-State Flow Capacity of Wells with Limited Entry to Flow," *SPE J.*, 43–51, March 1968.

Perrine, R. L., "Analysis of Pressure Buildup Curves," *Drilling and Production Practices*, API, 482–509, 1956.

Raghavan, R., *Well Test Analysis*, Prentice Hall Petroleum Engineering Series, Prentice Hall, Englewood Cliffs, NJ, 1993.

Raghavan, R., "Testing Under Multiphase-Flow Conditions" (Chapter 15), In: Medhat, K. (ed.), *Transient Well Testing*, SPE Monograph Vol. 23, Henry L. Doherty Series, Richardson, TX, 2009.

Spivey, J.P., and J. Lee, *Applied Well Test Interpretation*, SPE Texbook Series Vol. 13, Society of Petroleum Engineers, Richardson, TX, 2013.

CHAPTER 7

Numerical Reservoir Engineering Solutions

In the previous chapter, we discussed how solutions for governing equations for flow in porous media could be obtained by analytical means for certain number of simplified, yet important, cases of interest. For any applications where such simplifications are no longer valid, solution by *analytical* means becomes impossible and *a numerical* solution of the resulting nonlinear partial differential equations is commonly the only way that a solution can be obtained.

Contrary to analytical solutions, where a solution becomes available at all points in the domain, a numerical solution produces answers only at some preselected *discrete points* within the system, as illustrated in Fig. 7.1. In the numerical approach, the original problem is mapped into this discretized domain. Both space and time are discretized:

- *The spatial domain is broken up into a number of cells, grids, or blocks by implementing over the domain of interest some type of grid or mesh.*
- *The time domain is also discretized into a number of time steps, during which the problem is solved to obtain new values of the unknowns.*

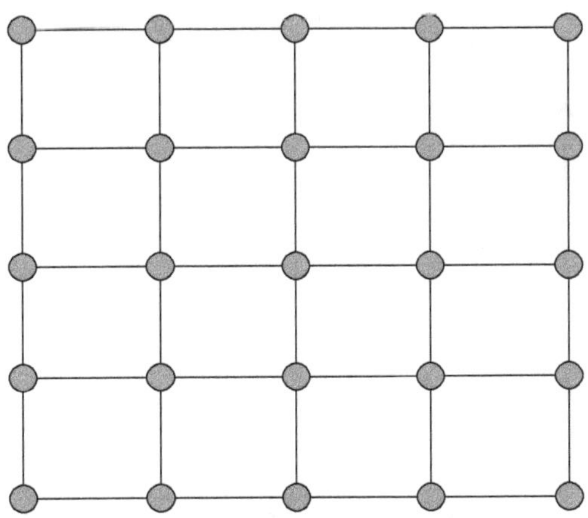

FIGURE 7.1 Discretized spatial domain for a numerical solution.

Partial differential equations are to be transformed into discrete or algebraic forms on the basis of different types of formulations. Such discretization methods can be based on:

- A *Taylor Series* formulation, resulting in the ***Finite-Difference Method (FDM).***
- An *Integral* formulation over a control volume, leading to the ***Finite-Volume Method (FVM).***
- A *variational* formulation that uses basis functions, also an integral scheme, which is the basis of the ***Finite-Element Method (FEM).***

Regardless of the formulation used, the numerical transformation results in an assembly of algebraic equations that needs to be resolved in order to model the entire flow problem.

7.1 Taylor Series and Numerical Schemes

A Taylor Series Expansion is an important mathematical concept routinely used in numerical formulations. The FDM heavily relies on Taylor Series to transform all derivatives in partial differential equations into their discrete algebraic analogs. FVM and FEM may also invoke Taylor Series when dealing with boundary conditions or other nonlinear terms requiring linearization.

A **Taylor Series** provides a way of rewriting a continuous mathematical function as an infinite series of the form:

$$p(x) = p(x_o) + \sum_{n=1}^{\infty} \frac{(x - x_o)^n}{n!} \frac{d^n p}{\partial x^n} \tag{7.1}$$

written about point $x = x_o$. For example, let us consider the pressure response profile, $p(x)$, shown in Fig. 7.2.

Taylor Series Expansions could be written to estimate $p(x + \Delta x)$ or $p(x - \Delta x)$ using $p(x)$ as a starting point. Depending on whether we are taking a forward step $(+\Delta x)$ or backward step $(-\Delta x)$ away from the point of interest "x", we could write forward or backward formulations away from the point $(x, p(x))$. For example, expansion of Eq. (7.1) yields the following expression for $p(x + \Delta x)$ (*"forward step"*):

$$p(x + \Delta x) = p(x) + \Delta x \frac{\partial p(x)}{\partial x} + \frac{(\Delta x)^2}{2!} \frac{\partial^2 p(x)}{\partial x^2} + \frac{(\Delta x)^3}{3!} \frac{\partial^3 p(x)}{\partial x^3} + \ldots \tag{7.2}$$

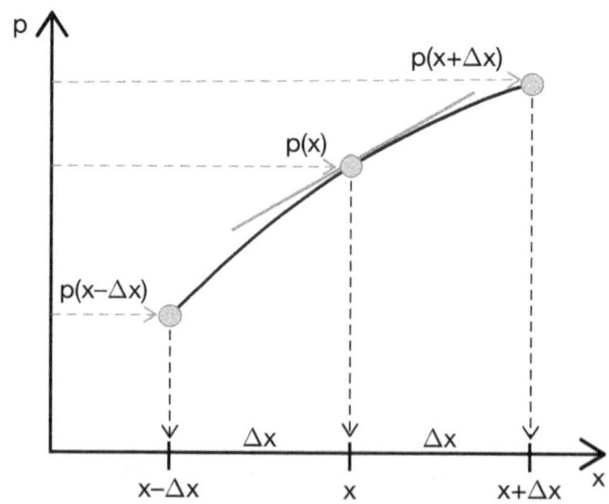

FIGURE 7.2 Pressure surface and locations of $p(x)$, $p(x + \Delta x)$ and $p(x - \Delta x)$.

Alternatively, expansion of Eq. (7.1) yields the following expression for $p(x - \Delta x)$ (*"backward step"*):

$$p(x - \Delta x) = p(x) + (-\Delta x)\frac{\partial p(x)}{\partial x} + \frac{(-\Delta x)^2}{2!}\frac{\partial^2 p(x)}{\partial x^2} + \frac{(-\Delta x)^3}{3!}\frac{\partial^3 p(x)}{\partial x^3} + \ldots \tag{7.3}$$

We note that we could "solve" for $\frac{\partial p(x)}{\partial x}$ using either Eq. (7.2) or (7.3). If we used the forward step in Eq. (7.2), we would obtain:

$$\frac{\partial p}{\partial x} = \frac{p(x + \Delta x) - p(x)}{\Delta x} + O(\Delta x) \tag{7.4}$$

which is the *forward-difference* approximation to $\frac{\partial p(x)}{\partial x}$ using Taylor Series. By comparison against the original Taylor expansion, it is noted that unspecified extra terms in this expression are of order $O(\Delta x)$. For the *backward-difference* approximation to $\frac{\partial p(x)}{\partial x}$, we use the backward step in Eq. (7.3) to obtain:

$$\frac{\partial p}{\partial x} = \frac{p(x) - p(x - \Delta x)}{\Delta x} + O(\Delta x) \tag{7.5}$$

The formulation of a third approximation to $\frac{\partial p(x)}{\partial x}$ is also possible when we combine information from these forward and backward steps. If we *subtract* backward from forward steps, we obtain:

$$p(x + \Delta x) - p(x - \Delta x) = [p(x) - p(x)] + 2(\Delta x)\frac{\partial p(x)}{\partial x}$$

$$+\left[\frac{(\Delta x)^2}{2!}\frac{\partial^2 p(x)}{\partial x^2} - \frac{(-\Delta x)^2}{2!}\frac{\partial^2 p(x)}{\partial x^2}\right] + 2\frac{(\Delta x)^3}{3!}\frac{\partial^3 p(x)}{\partial x^3} + \ldots \tag{7.6}$$

or,

$$\frac{\partial p(x)}{\partial x} = \frac{p(x + \Delta x) - p(x - \Delta x)}{2\Delta x} + O(\Delta x^2) \tag{7.7}$$

which is the *central-difference* approximation to $\frac{\partial p(x)}{\partial x}$ using Taylor Series. It is noted that the errors associated with these approximations are different: $O(\Delta x)$ for forward and backward approximations but $O(\Delta x^2)$ for central approximations. The error associated with the truncation of Taylor Series is called *truncation* error.

Figure 7.3 illustrates the graphical interpretation of the three (forward, backward, and central) difference approximations to the tangent $\frac{\partial p(x)}{\partial x}$ at point x. It is important to note that as Δx steps get vanishingly small ($\Delta x \rightarrow 0$), the three chord approximations will tend to approach each other to ultimately converge to the actual value of the tangent at that point. For all other finite Δx values, each approximation would yield relatively close but different estimations for $\frac{\partial p(x)}{\partial x}$.

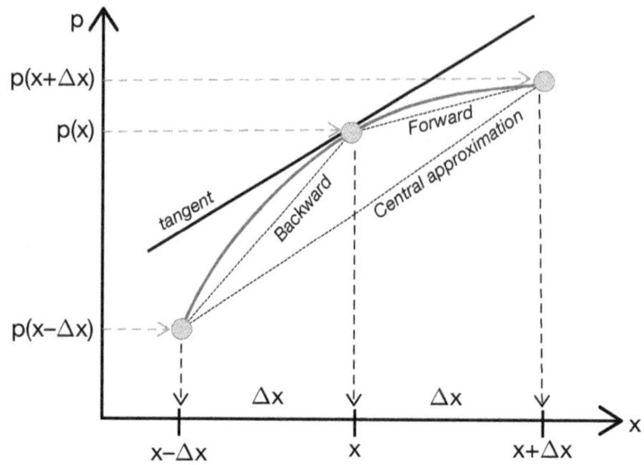

Figure 7.3 Forward, backward, and central difference approximations to $\frac{\partial p(x)}{\partial x}$.

Approximations for ***second order derivatives*** are also possible when we consider the *addition* of forward and backward Taylor Expansions. For the case of uniform Δx values, we obtain:

$$p(x+\Delta x)+p(x-\Delta x)=2p(x)+\left[(\Delta x)\frac{\partial p(x)}{\partial x}+(-\Delta x)\frac{\partial p(x)}{\partial x}\right]$$
$$+2\left[\frac{(\Delta x)^2}{2!}\frac{\partial^2 p(x)}{\partial x^2}\right]+\left[\frac{(\Delta x)^3}{3!}\frac{\partial^3 p(x)}{\partial x^3}+\frac{(-\Delta x)^3}{3!}\frac{\partial^3 p(x)}{\partial x^3}\right]+O(\Delta x^4) \tag{7.8}$$

Thus,

$$\frac{\partial^2 p(x)}{\partial x^2}=\frac{p(x+\Delta x)-2p(x)+p(x-\Delta x)}{(\Delta x)^2}+O(\Delta x^2) \tag{7.9}$$

which is the *central-difference* approximation to $\frac{\partial p(x)}{\partial x}$ using Taylor Series when spacing Δx is uniform. It is a second-order accurate ($O(\Delta x^2)$) approximation.

7.2 Finite-Volume Method Representations of Flow Equations

Numerical formulations must be implemented over some discrete network of nodes. Dependent variables (unknowns) are calculated at such discrete locations once all governing partial differential equations are transformed into their discrete or algebraic analogs.

The **Finite-Volume Method** (*FVM*) is the numerical method of choice used in this chapter for illustration purposes. Ertekin et al. (2001) discussed the subject of Finite-Difference Method (FDM) at length in their work. In the *FVM*, the spatial domain is discretized into a number of finite control volumes. The classical formulation uses a cell- or block-centered formulation, and each element is a control volume (*CV*) representing a discrete region of the domain. The *FVM* then integrates the applicable governing equation over each control volume in order to formulate a discretized equation written at each of the nodal points. The procedure involves the evaluation of volumetric sources and interblock fluxes that enter and leave the control volume. Excellent reviews about the deployment of *FVM* applications, that can be used to further the information presented in this chapter, can be found in the works of Versteeg and Malalasekera (2007), Zikanov (2010), and Mazumder (2016). Mazumder (2016), in particular, covers the important subject of unstructured grid implementation, which is one of the most significant advantages of *FVM* over *FDM*. We limit our *FVM* derivations to structured grids to simplify our introductory presentation.

The most straightforward application of *FVM* is the one made on structural orthogonal grids, such as the three-dimensional Cartesian system of brick-shaped elements presented in Fig. 7.4. The integral procedure in *FVM*, however, is not restricted to such structured numerical grids. In fact, one of the key advantages of *FVM* over classical finite-difference methods is its easy extension to unstructured and nonorthogonal grids. *FDMs* are easiest and most straightforward to apply when formulated for grids using orthogonal coordinate systems (Cartesian, cylindrical or spherical), such as the Cartesian presented in the figure, unless curvilinear transforms are applied to extend it to nonorthogonal systems.

For the three-dimensional Cartesian system in Fig. 7.4., each *CV* located at position (i,j,k), as shown in Fig. 7.5, is surrounded by six neighboring nodes and control volumes: an (*E*)ast neighbor at position $(i+1,j,k)$; a (*W*)est neighbor at position $(i-1,j,k)$; a (*N*)orth neighbor at position $(i,j+1,k)$; a (*S*)outh neighbor at position $(i,j-1,k)$; an (*A*)bove neighbor at position $(i,j,k+1)$; and a (*B*)elow neighbor at position $(i,j,k-1)$. Each node has its own control volume associated with it. Note that boundaries (interfaces) of each *CV* are located midway between neighboring nodes. The (*E*)ast boundary is found at $x = x_e$ or $(i+1/2,j,k)$ location, the (*W*)est boundary at $x = x_w$ or $(i-1/2,j,k)$ location, the (*N*)orth boundary at $y = y_n$ or $(i,j+1/2,k)$ location, the (*S*)outh boundary at $y = y_s$ or $(i,j-1/2,k)$ location, the (*A*)bove boundary at $z = z_a$ or $(i,j,k+1/2)$ location, and the (*B*)elow boundary at $z = z_b$ or $(i,j,k-1/2)$ location. The (C)entral node is located at position (i,j,k) with coordinates (x_C, y_C, z_C).

7.3 FVM Representations of Single-Phase Flow Equations

7.3.1 Single-Phase Incompressible Flow

In Chap. 5, it was shown that the governing equation for *incompressible fluid flow* in a porous media is written as follows (written here for *incompressible* water flow):

$$\nabla \cdot \left(\frac{\bar{\bar{k}}}{\mu_w B_w} \nabla \Phi \right) + \frac{q_{wsc}}{V_b} = 0 \tag{7.10}$$

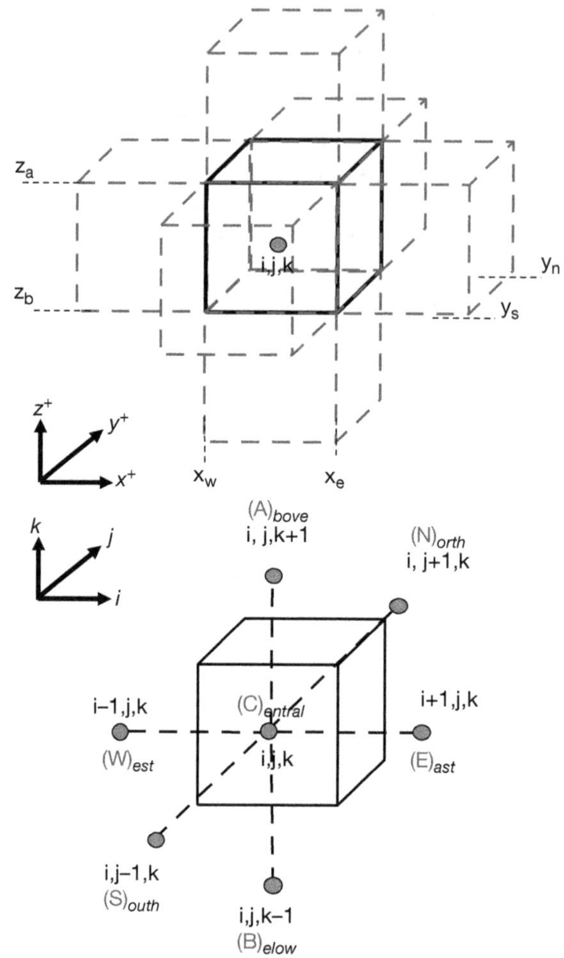

FIGURE 7.4 Cell-centered discretized CV in 3D Cartesian coordinates.

We seek to derive the *finite-volume* representation of Eq. (7.10) for the *3D* Cartesian representation shown in Fig. 7.4. The governing equation of interest is:

$$\frac{\partial}{\partial x}\left(\frac{k_x}{\mu_w B_w}\frac{\partial \Phi}{\partial x}\right)+\frac{\partial}{\partial y}\left(\frac{k_y}{\mu_w B_w}\frac{\partial \Phi}{\partial y}\right)+\frac{\partial}{\partial z}\left(\frac{k_z}{\mu_w B_w}\frac{\partial \Phi}{\partial z}\right)+\frac{q_{wsc}}{V_b}=0 \tag{7.11}$$

They key idea behind *FVM* is the integration of the governing equation [Eq. (7.11) for this case] over the entire control volume depicted in Fig. 7.5. This gives:

$$\int_{V_b}\frac{\partial}{\partial x}\left(\frac{k_x}{\mu_w B_w}\frac{\partial \Phi}{\partial x}\right)dV+\int_{V_b}\frac{\partial}{\partial y}\left(\frac{k_y}{\mu_w B_w}\frac{\partial \Phi}{\partial y}\right)dV+\int_{V_b}\frac{\partial}{\partial z}\left(\frac{k_z}{\mu_w B_w}\frac{\partial \Phi}{\partial z}\right)dV+\int_{V_b}\frac{q_{wsc}}{V_b}dV=0 \tag{7.12}$$

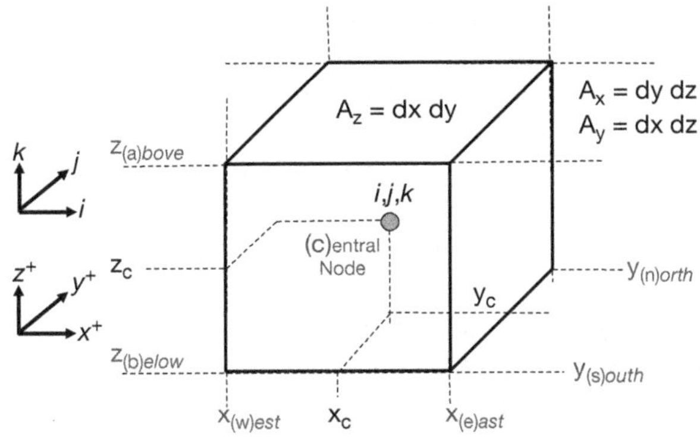

FIGURE 7.5 A control volume (CV) in a cell-centered *3D* Cartesian domain.

In *FVM*, these volume integrals are transformed into surface integrals for each surface in the control volume using Gauss divergence theorem (discussed in Chap. 2), which is applicable to any arbitrary closed volume. Mazumder (2016) demonstrates how the divergence theorem can be applied to this transformation for both structured and unstructured grids. For the particular Cartesian *CV* of interest (Fig. 7.5), however, this integral transformation is quite straightforward given that volumes and surface areas are readily related as follows in such structured geometry:

$$dV = dx\,dy\,dz = A_x\,dx = A_y\,dy = A_z\,dz \tag{7.13}$$

where $A_x = dydz$ is uniform within $x_w < x < x_e$, $A_y = dxdz$ is uniform within $y_s < y < y_n$, and $A_z = dxdy$ is uniform within $z_b < z < z_a$. Thus, into Eq. (7.12):

$$\int_{x_w}^{x_e} \frac{\partial}{\partial x}\left(\frac{k_x}{\mu_w B_w}\frac{\partial \Phi}{\partial x}A_x\right)dx + \int_{y_s}^{y_n} \frac{\partial}{\partial y}\left(\frac{k_y}{\mu_w B_w}\frac{\partial \Phi}{\partial y}A_y\right)dy + \int_{z_b}^{z_a} \frac{\partial}{\partial z}\left(\frac{k_z}{\mu_w B_w}\frac{\partial \Phi}{\partial z}A_z\right)dz + \int_{V_b} \frac{q_{wsc}}{V_b}dV = 0 \tag{7.14}$$

Performing the integrations,

$$\left[\left(\frac{k_x A_x}{\mu_w B_w}\frac{\partial \Phi}{\partial x}\right)_{x_e} - \left(\frac{k_x A_x}{\mu_w B_w}\frac{\partial \Phi}{\partial x}\right)_{x_w}\right] + \left[\left(\frac{k_y A_y}{\mu_w B_w}\frac{\partial \Phi}{\partial y}\right)_{y_n} - \left(\frac{k_y A_y}{\mu_w B_w}\frac{\partial \Phi}{\partial y}\right)_{y_s}\right]$$
$$+ \left[\left(\frac{k_z A_z}{\mu_w B_w}\frac{\partial \Phi}{\partial z}\right)_{z_a} - \left(\frac{k_z A_z}{\mu_w B_w}\frac{\partial \Phi}{\partial z}\right)_{z_b}\right] + (q_{wsc})_{i,j,k} = 0 \tag{7.15}$$

In this equation, the source term q_{wsc} is evaluated assuming it is uniformly distributed (constant) across the *CV*. Thus, by using the definition of a volume-averaged property $\left(\bar{X} = \frac{1}{V_b}\int_{V_b} X dV\right)$, we were able to write:

$$\frac{1}{V_b}\int_{V_b} q_{wsc}\,dV = (q_{wsc})_{i,j,k} \tag{7.16}$$

where $(q_{wsc})_{i,j,k}$ is the volume-averaged value of the source term applicable to the entire *CV* at the (C)entral node location (i,j,k). It should be noted that the rest of the terms represent the fluxes going through each of the boundaries (*E, W, N, S, A, B*) into the *CV*. The net effect of all fluxes and source terms must add to zero, as per the steady-state requirements of this incompressible flow equation without an accumulation term.

One additional step required by the *FVM* is the estimation of all gradients of the integrated equation $\left(\dfrac{\partial\Phi}{\partial x}, \dfrac{\partial\Phi}{\partial y}, \dfrac{\partial\Phi}{\partial z}\right)$ as a function of nodal information. The typical choice is to use Taylor Series for such first-derivative approximations at each of the interfaces using central differencing [Eq. (7.7)]. When such approximation is embraced at each of the *CV* boundaries, boundary gradients become:

$$\left(\frac{\partial\Phi}{\partial x}\right)_{x_e} = \frac{\Phi_{i+1,j,k} - \Phi_{i,j,k}}{x_{i+1,j,k} - x_{i,j,k}} = \frac{\Phi_{i+1,j,k} - \Phi_{i,j,k}}{(\Delta x)_{x_e}} \tag{7.17}$$

$$\left(\frac{\partial\Phi}{\partial x}\right)_{x_w} = \frac{\Phi_{i,j,k} - \Phi_{i-1,j,k}}{x_{i,j,k} - x_{i-1,j,k}} = \frac{\Phi_{i,j,k} - \Phi_{i-1,j,k}}{(\Delta x)_{x_w}} \tag{7.18}$$

$$\left(\frac{\partial\Phi}{\partial y}\right)_{y_n} = \frac{\Phi_{i,j+1,k} - \Phi_{i,j,k}}{y_{i,j+1,k} - y_{i,j,k}} = \frac{\Phi_{i,j+1,k} - \Phi_{i,j,k}}{(\Delta y)_{y_n}} \tag{7.19}$$

$$\left(\frac{\partial\Phi}{\partial y}\right)_{y_s} = \frac{\Phi_{i,j,k} - \Phi_{i,j-1,k}}{y_{i,j,k} - y_{i,j-1,k}} = \frac{\Phi_{i,j,k} - \Phi_{i,j-1,k}}{(\Delta y)_{y_s}} \tag{7.20}$$

$$\left(\frac{\partial\Phi}{\partial z}\right)_{z_a} = \frac{\Phi_{i,j,k+1} - \Phi_{i,j,k}}{z_{i,j,k+1} - z_{i,j,k}} = \frac{\Phi_{i,j,k+1} - \Phi_{i,j,k}}{(\Delta z)_{z_a}} \tag{7.21}$$

$$\left(\frac{\partial\Phi}{\partial z}\right)_{z_b} = \frac{\Phi_{i,j,k} - \Phi_{i,j,k-1}}{z_{i,j,k} - z_{i,j,k-1}} = \frac{\Phi_{i,j,k} - \Phi_{i,j,k-1}}{(\Delta z)_{z_b}} \tag{7.22}$$

Substituting Eqs. (7.17) through (7.22) into Eq. (7.15):

$$\begin{aligned}
&\left(\frac{k_x A_x}{\mu_w B_w \Delta x}\right)_{x_e}\left(\Phi_{i+1,j,k} - \Phi_{i,j,k}\right) - \left(\frac{k_x A_x}{\mu_w B_w \Delta x}\right)_{x_w}\left(\Phi_{i,j,k} - \Phi_{i-1,j,k}\right)\\
&+\left(\frac{k_y A_y}{\mu_w B_w \Delta y}\right)_{y_n}\left(\Phi_{i,j+1,k} - \Phi_{i,j,k}\right) - \left(\frac{k_y A_y}{\mu_w B_w \Delta y}\right)_{y_s}\left(\Phi_{i,j,k} - \Phi_{i,j-1,k}\right)\\
&+\left(\frac{k_z A_z}{\mu_w B_w \Delta z}\right)_{z_a}\left(\Phi_{i,j,k+1} - \Phi_{i,j,k}\right) - \left(\frac{k_z A_z}{\mu_w B_w \Delta z}\right)_{z_b}\left(\Phi_{i,j,k} - \Phi_{i,j,k-1}\right)\\
&+(q_{wsc})_{i,j,k} = 0
\end{aligned} \tag{7.23}$$

where coefficients in front of the *flow potentials* represent the following water (incompressible-fluid) transmissibility coefficients:

$$T_{W,x_e} = \left.\frac{k_x A_x}{\mu_w B_w \Delta x}\right|_{x_e} = E_{i,j,k}, \text{ where } x_e \text{ is the } i+\frac{1}{2}, j, k \text{ location} \tag{7.24}$$

$$T_{W,x_w} = \left.\frac{k_x A_x}{\mu_w B_w \Delta x}\right|_{x_w} = W_{i,j,k}, \text{ where } x_w \text{ is the } i-\frac{1}{2}, j, k \text{ location} \tag{7.25}$$

$$T_{W,y_n} = \left.\frac{k_y A_y}{\mu_w B_w \Delta y}\right|_{y_n} = N_{i,j,k}, \text{ where } y_n \text{ is the } i, j+\frac{1}{2}, k \text{ location} \tag{7.26}$$

$$T_{W,y_s} = \left.\frac{k_y A_y}{\mu_w B_w \Delta y}\right|_{y_s} = S_{i,j,k}, \text{ where } y_s \text{ is the } i, j-\frac{1}{2}, k \text{ location} \tag{7.27}$$

$$T_{W,z_a} = \left.\frac{k_z A_z}{\mu_w B_w \Delta z}\right|_{z_a} = A_{i,j,k}, \text{ where } z_a \text{ is the } i, j, k+\frac{1}{2} \text{ location} \tag{7.28}$$

$$T_{W,z_b} = \left.\frac{k_z A_z}{\mu_w B_w \Delta z}\right|_{z_b} = B_{i,j,k}, \text{ where } z_b \text{ is the } i, j, k-\frac{1}{2} \text{ location} \tag{7.29}$$

By grouping terms, Eq. (7.23) can be alternatively written in shorthand notation as:

$$\begin{aligned} &B_{i,j,k}\Phi_{i,j,k-1} + S_{i,j,k}\Phi_{i,j-1,k} + W_{i,j,k}\Phi_{i-1,j,k} + C_{i,j,k}\Phi_{i,j,k} + E_{i,j,k}\Phi_{i+1,j,k} \\ &+ N_{i,j,k}\Phi_{i,j+1,k} + A_{i,j,k}\Phi_{i,j,k+1} = Q_{i,j,k} \end{aligned} \tag{7.30}$$

where:

$$C_{i,j,k} = -\left(B_{i,j,k} + S_{i,j,k} + W_{i,j,k} + E_{i,j,k} + N_{i,j,k} + A_{i,j,k}\right) \tag{7.31}$$

$$Q_{i,j,k} = -(q_{wsc})_{i,j,k} \tag{7.32}$$

It should be noted that boundary gradients could be written directly in terms of *pressure and depth gradients* by deploying the definition of the flow potential gradient (see Chap. 3). For instance, potential gradient approximations in Eqs. (7.17) to (7.22) would be written instead as:

$$\left(\frac{\partial \Phi}{\partial x}\right)_{x_e} = \left(\frac{\partial p}{\partial x}\right)_{x_e} - \gamma_{w,x_e}\left(\frac{\partial G}{\partial x}\right)_{x_e} = \frac{p_{i+1,j,k} - p_{i,j,k}}{(\Delta x)_{x_e}} - \gamma_{w,x_e}\frac{G_{i+1,j,k} - G_{i,j,k}}{(\Delta x)_{x_e}} \tag{7.33}$$

$$\left(\frac{\partial \Phi}{\partial x}\right)_{x_w} = \left(\frac{\partial p}{\partial x}\right)_{x_w} - \gamma_{w,x_w}\left(\frac{\partial G}{\partial x}\right)_{x_w} = \frac{p_{i,j,k} - p_{i-1,j,k}}{(\Delta x)_{x_w}} - \gamma_{w,x_w}\frac{G_{i+1,j,k} - G_{i,j,k}}{(\Delta x)_{x_w}} \tag{7.34}$$

$$\left(\frac{\partial \Phi}{\partial y}\right)_{y_n} = \left(\frac{\partial p}{\partial y}\right)_{y_n} - \gamma_{w,y_n}\left(\frac{\partial G}{\partial y}\right)_{y_n} = \frac{p_{i,j+1,k} - p_{i,j,k}}{(\Delta y)_{y_n}} - \gamma_{w,y_n}\frac{G_{i,j+1,k} - G_{i,j,k}}{(\Delta y)_{y_n}} \tag{7.35}$$

$$\left(\frac{\partial \Phi}{\partial y}\right)_{y_s} = \left(\frac{\partial p}{\partial y}\right)_{y_s} - \gamma_{w,y_s}\left(\frac{\partial G}{\partial y}\right)_{y_s} = \frac{p_{i,j,k} - p_{i,j-1,k}}{(\Delta y)_{y_s}} - \gamma_{w,y_s}\frac{G_{i,j,k} - G_{i,j-1,k}}{(\Delta y)_{y_s}} \tag{7.36}$$

$$\left(\frac{\partial \Phi}{\partial z}\right)_{z_a} = \left(\frac{\partial p}{\partial z}\right)_{z_a} - \gamma_{w,z_a}\left(\frac{\partial G}{\partial z}\right)_{z_a} = \frac{p_{i,j,k+1} - p_{i,j,k}}{(\Delta z)_{z_a}} - \gamma_{w,z_a}\frac{G_{i,j,k+1} - G_{i,j,k}}{(\Delta z)_{z_a}} \tag{7.37}$$

$$\left(\frac{\partial \Phi}{\partial z}\right)_{z_b} = \left(\frac{\partial p}{\partial z}\right)_{z_b} - \gamma_{w,z_b}\left(\frac{\partial G}{\partial z}\right)_{z_b} = \frac{p_{i,j,k} - p_{i,j,k-1}}{(\Delta z)_{z_b}} - \gamma_{w,z_b}\frac{G_{i,j,k} - G_{i,j,k-1}}{(\Delta z)_{z_b}} \tag{7.38}$$

where γ_w is fluid gravity (in psi/ft), calculated in terms of fluid density as $\gamma_w = \frac{1}{144}\frac{g}{g_c}\rho_w$ (see Chap. 3). As a result, in terms of pressure and depth gradients, Eq. (7.23) would become:

$$\begin{aligned}
&\left(\frac{k_x A_x}{\mu_w B_w \Delta x}\right)_{x_e}\left(p_{i+1,j,k} - p_{i,j,k}\right) - \left(\frac{k_x A_x}{\mu_w B_w \Delta x}\right)_{x_w}\left(p_{i,j,k} - p_{i-1,j,k}\right) \\
&+ \left(\frac{k_y A_y}{\mu_w B_w \Delta y}\right)_{y_n}\left(p_{i,j+1,k} - p_{i,j,k}\right) - \left(\frac{k_y A_y}{\mu_w B_w \Delta y}\right)_{y_s}\left(p_{i,j,k} - p_{i,j-1,k}\right) \\
&+ \left(\frac{k_z A_z}{\mu_w B_w \Delta z}\right)_{z_a}\left(p_{i,j,k+1} - p_{i,j,k}\right) - \left(\frac{k_z A_z}{\mu_w B_w \Delta z}\right)_{z_b}\left(p_{i,j,k} - p_{i,j,k-1}\right) \\
&- \left[\begin{array}{l}
\left.\frac{k_x A_x \gamma_w}{\mu_w B_w \Delta x}\right|_{x_e}\left(G_{i+1,j,k} - G_{i,j,k}\right) - \left.\frac{k_x A_x \gamma_w}{\mu_w B_w \Delta x}\right|_{x_w}\left(G_{i,j,k} - G_{i-1,j,k}\right) \\
+ \left.\frac{k_y A_y \gamma_w}{\mu_w B_w \Delta y}\right|_{y_n}\left(G_{i,j+1,k} - G_{i,j,k}\right) - \left.\frac{k_y A_y \gamma_w}{\mu_w B_w \Delta y}\right|_{y_s}\left(G_{i,j,k} - G_{i,j-1,k}\right) \\
+ \left.\frac{k_z A_z \gamma_w}{\mu_w B_w \Delta z}\right|_{z_a}\left(G_{i,j,k+1} - G_{i,j,k}\right) - \left.\frac{k_z A_z \gamma_w}{\mu_w B_w \Delta z}\right|_{z_b}\left(G_{i,j,k} - G_{i,j,k-1}\right)
\end{array}\right] \\
&+ (q_{lsc})_{i,j,k} = 0
\end{aligned} \tag{7.39}$$

Hence, the *FVM* approximation for the seven-point numerical stencil of Fig. 7.4 is now written in terms of nodal pressures as:

$$\begin{aligned}
B_{i,j,k}p_{i,j,k-1} + S_{i,j,k}p_{i,j-1,k} + W_{i,j,k}p_{i-1,j,k} + C_{i,j,k}p_{i,j,k} + E_{i,j,k}p_{i+1,j,k} \\
+ N_{i,j,k}p_{i,j+1,k} + A_{i,j,k}p_{i,j,k+1} = Q_{i,j,k}
\end{aligned} \tag{7.40}$$

using the definitions:

$$C_{i,j,k} = -\left(B_{i,j,k} + S_{i,j,k} + W_{i,j,k} + E_{i,j,k} + N_{i,j,k} + A_{i,j,k}\right) \tag{7.41}$$

$$Q_{i,j,k} = -(q_{lsc})_{i,j,k} + \left[E_{i,j,k}\gamma_{w,x_e}\left(G_{i+1,j,k} - G_{i,j,k}\right) - W_{i,j,k}\gamma_{w,x_w}\left(G_{i,j,k} - G_{i-1,j,k}\right) + N_{i,j,k}\gamma_{w,y_n}\left(G_{i,j+1,k} - G_{i,j,k}\right) - S_{i,j,k}\gamma_{w,y_s}\left(G_{i,j,k} - G_{i,j-1,k}\right) + A_{i,j,k}\gamma_{w,z_a}\left(G_{i,j,k+1} - G_{i,j,k}\right) - B_{i,j,k}\gamma_{w,z_b}\left(G_{i,j,k} - G_{i,j,k-1}\right)\right] \quad (7.42)$$

Extensive discussions of well treatment choices for the $(q_{lsc})_{i,j,k}$ source term (i.e., wellbore modeling) are found in Ertekin et al. (2001). Treatment of transmissibility values for no-flow and constant-pressure outer boundary conditions is also discussed extensively in that work. In Eq. (7.42), for the purposes of this presentation, it is assumed that the well source term is constant and known.

Transmissibility Calculations

Before one can proceed with a numerical solution, all terms and coefficients in the flow equation must be estimated. One of such coefficients are all the transmissibility coefficients enlisted in Eqs. (7.24) to (7.29). For incompressible fluid simulation, the variables $\mu_w B_w$ in these expressions are constant and pressure-independent. However, the remaining variables (k, A, and Δx, Δy, Δz) may vary through the domain depending on the type of grid used for discretization. We thus seek to calculate the appropriate value of the "constant part" (grid-related or pressure-independent) of these *boundary* transmissibility, which is defined at the boundary between any two adjacent control volumes or gridblocks:

$$T^* = \left(\frac{k_x A_x}{\Delta x}\right) \quad (7.43)$$

While domain discretization will yield values of k_x, A_x, and Δx for every gridblock in the system, the values of such quantities have not been defined at the boundaries of adjacent gridblocks. For the case of the east boundary transmissibility, e.g., we may seek to estimate at $x = x_e$:

$$T_e^* = \left(\frac{k_x A_x}{\Delta x}\right)_{x_e} \quad (7.44)$$

as a function of the values of transmissibilities for the adjacent control volumes (i,j,k) and $(i + 1,j,k)$:

$$T_{i,j,k}^* = \left(\frac{k_x A_x}{\Delta x}\right)_{i,j,k} \quad (7.45)$$

$$T_{i+1,j,k}^* = \left(\frac{k_x A_x}{\Delta x}\right)_{i+1,j,k} \quad (7.46)$$

To do so, we invoke Darcy's law, which prescribes that flow of a generic incompressible liquid (l) from the central node $\Phi_{i,j,k}$ to the east boundary location (Φ_e) is given by:

$$q_l = \left(\frac{k_x A_x}{\mu_l B_l}\right)_{i,j,k} \frac{\left(\Phi_e - \Phi_{i,j,k}\right)}{(\Delta x/2)_{i,j,k}} \quad (7.47)$$

while the flow from the east boundary (Φ_e) to the east neighboring node $\Phi_{i+1,j,k}$ is:

$$q_l = \left(\frac{k_x A_x}{\mu_l B_l}\right)_{i+1,j,k} \frac{\left(\Phi_{i+1,j,k} - \Phi_e\right)}{(\Delta x/2)_{i+1,j,k}} \tag{7.48}$$

Since:

$$\left(\Phi_{i+1,j,k} - \Phi_{i,j,k}\right) = \left(\Phi_{i+1,j,k} - \Phi_{x_e}\right) + \left(\Phi_{x_e} - \Phi_{i,j,k}\right) \tag{7.49}$$

and where the *FVM* flow through the east boundary is given by:

$$q_l = \left(\frac{k_x A_x}{\Delta x}\right)_{x_e} \left(\frac{1}{\mu_l B_l}\right)_{x_e} \left(\Phi_{i+1,j,k} - \Phi_{i,j,k}\right) = T^*_{x_e}\left(\frac{1}{\mu_l B_l}\right)_{x_e} \left(\Phi_{i+1,j,k} - \Phi_{i,j,k}\right) \tag{7.50}$$

corresponding to fluid flow from the central node $\Phi_{i,j,k}$ to the east neighbor node $\Phi_{i+1,j,k}$ locations. Substituting Eqs. (7.47), (7.48), and (7.50) into (7.49):

$$\left(\frac{q_l \mu_l B_l}{T^*_e}\right) = \left(\frac{q_l \mu_l B_l}{2T^*_{i+1,j,k}}\right) + \left(\frac{q_l \mu_l B_l}{2T^*_{i,j,k}}\right) \tag{7.51}$$

or, for a constant $\mu_l B_l$ fluid we have:

$$\frac{1}{T^*_e} = \frac{1}{2}\left(\frac{1}{T^*_{i,j,k}} + \frac{1}{T^*_{i+1,j,k}}\right) \tag{7.52}$$

which is also written as:

$$T^*_e = \frac{T^*_{i,j,k} + T^*_{i+1,j,k}}{2T^*_{i,j,k} T^*_{i+1,j,k}} \tag{7.53}$$

These expressions indicate that the constant part of any *boundary* transmissibility should be calculated using the ***harmonic average*** of such values calculated at adjacent control volumes.

7.3.2 Single-Phase Slightly Compressible Flow

Chapter 5 demonstrated that the governing equation for *slightly compressible fluid flow* in a porous media can be written as:

$$\nabla \cdot \left[\frac{\bar{\bar{k}}}{\mu_o B_o} \nabla \Phi\right] + \frac{q_{osc}}{V_b} = \frac{\partial}{\partial t}(\phi / B_o) = \frac{\phi c_o}{B_o^{rf}} \frac{\partial p}{\partial t} \tag{7.54}$$

With the last equality in Eq. (7.54) being valid if the rock is incompressible (constant porosity). We seek to derive the *finite-volume* representation of Eq. (7.53) for the control volume shown in Fig. 7.5. The *3D* Cartesian expansion of Eq. (7.54) is:

$$\frac{\partial}{\partial x}\left(\frac{k_x}{\mu_o B_o} \frac{\partial \Phi}{\partial x}\right) + \frac{\partial}{\partial y}\left(\frac{k_y}{\mu_o B_o} \frac{\partial \Phi}{\partial y}\right) + \frac{\partial}{\partial z}\left(\frac{k_z}{\mu_o B_o} \frac{\partial \Phi}{\partial z}\right) + \frac{q_{osc}}{V_b} = \frac{\phi c_o}{\alpha_c B_o^{rf}} \frac{\partial p}{\partial t} \tag{7.55}$$

Before we proceed with the *FVM* integration of this governing equation over the domain, we must realize the existence of a new domain dimension (time) through the presence of the accumulation term in the right-hand side of this equation—a term which was completely absent in our previous FVM development for incompressible systems. Therefore, the finite-volume integration must now consider an additional integration in time over the entire discretized time step of interest (Δt). This leads to two integrals for each terms of the flow equations, as shown:

$$\int_{t}^{t+\Delta t}\left(\int_{V_b}\frac{\partial}{\partial x}\left(\frac{k_x}{\mu_o B_o}\frac{\partial \Phi}{\partial x}\right)dV\right)dt+\int_{t}^{t+\Delta t}\left(\int_{V_b}\frac{\partial}{\partial y}\left(\frac{k_y}{\mu_o B_o}\frac{\partial \Phi}{\partial y}\right)dV\right)dt+\int_{t}^{t+\Delta t}\left(\int_{V_b}\frac{\partial}{\partial z}\left(\frac{k_z}{\mu_o B_o}\frac{\partial \Phi}{\partial z}\right)dV\right)dt$$
$$+\int_{t}^{t+\Delta t}\left(\int_{V_b}\frac{q_{osc}}{V_b}dV\right)dt=\frac{\phi c_o}{\alpha_c B_o^{rf}}\int_{t}^{t+\Delta t}\int_{V_b}\frac{\partial p}{\partial t}dVdt \tag{7.56}$$

Spatial integrations over the control volume would follow the *FVM* procedure described in the incompressible fluid section, which for the *3D* Cartesian representation of the control volume in Fig. 7.5 straightforwardly yields:

$$\int_{t}^{t+\Delta t}\left(\int_{x_w}^{x_e}\frac{\partial}{\partial x}\left(\frac{k_x}{\mu_o B_o}\frac{\partial \Phi}{\partial x}A_x\right)dx\right)dt+\int_{t}^{t+\Delta t}\left(\int_{y_s}^{y_n}\frac{\partial}{\partial y}\left(\frac{k_y}{\mu_o B_o}\frac{\partial \Phi}{\partial y}A_y\right)dy\right)dt$$
$$+\int_{t}^{t+\Delta t}\left(\int_{z_b}^{z_a}\frac{\partial}{\partial z}\left(\frac{k_z}{\mu_o B_o}\frac{\partial \Phi}{\partial z}A_z\right)dz\right)dt+\int_{t}^{t+\Delta t}(q_{osc})_{i,j,k}dt=\frac{\phi c_o}{\alpha_c B_o^{rf}}V_b\int_{t}^{t+\Delta t}\frac{\partial p_{i,j,k}}{\partial t}dt \tag{7.57}$$

It is noted that for both sources and accumulation terms, the definition of volume-averaged property has been applied $\left(\overline{X}=\frac{1}{V_b}\int_{V_b}XdV\right)$. For the particular case of the accumulation term, we recognize:

$$\int_{V_b}\frac{\partial p}{\partial t}dV=V_b\frac{\partial p_{i,j,k}}{\partial t} \tag{7.58}$$

given that $p_{i,j,k}$ in a *FVM* formulation represents the pressure assumed to exist over the *entire* control volume at location (i,j,k). By performing the spatial integrations in Eq. (7.57), and that of the right-hand side, we further obtain:

$$\int_{t}^{t+\Delta t}\left[\left(\frac{k_xA_x}{\mu_o B_o}\frac{\partial \Phi}{\partial x}\right)_{x_e}-\left(\frac{k_xA_x}{\mu_o B_o}\frac{\partial \Phi}{\partial x}\right)_{x_w}\right]dt+\int_{t}^{t+\Delta t}\left[\left(\frac{k_yA_y}{\mu_o B_o}\frac{\partial \Phi}{\partial y}\right)_{y_n}-\left(\frac{k_yA_y}{\mu_o B_o}\frac{\partial \Phi}{\partial y}\right)_{y_s}\right]dt$$
$$+\int_{t}^{t+\Delta t}\left[\left(\frac{k_zA_z}{\mu_o B_o}\frac{\partial \Phi}{\partial z}\right)_{z_a}-\left(\frac{k_zA_z}{\mu_o B_o}\frac{\partial \Phi}{\partial z}\right)_{z_b}\right]dt+\int_{t}^{t+\Delta t}(q_{osc})_{i,j,k}dt=\frac{\phi c_o}{\alpha_c B_o^{rf}}V_b\int_{t}^{t+\Delta t}dp_{i,j,k} \tag{7.59}$$

In Eq. (7.59), we are now left with time integrals in its left-hand-side only. The right-hand-side time integral is simple to resolve, i.e.:

$$\int_{t}^{t+\Delta t} dp_{i,j,k} = \left(p_{i,j,k}^{n+1} - p_{i,j,k}^{n}\right) \tag{7.60}$$

where $p_{i,j,k}^{n}$ is the pressure of the entire *CV* at location (i,j,k) at the *starting* time of integration, i.e., its value at the end of the old time step n. $p_{i,j,k}^{n+1}$ is said pressure (i,j,k) at the *end* time of integration, i.e., its value at the new time level $n+1$. Please note that conditions at the previous or old time step are represented with a time evolution superscript "n"—and thus value of a variable X at the old-time level is X^n, while values of a time-dependent variable X at the new time level uses the time level evolution superscript "$n+1$" (X^{n+1}).

All remaining integrals in Eq. (7.59) require the formulation of a time-integration approach, which must be based on our assumption of how variables evolve with time within the discretized time step Δt. In any numerical discretization and within a time step Δt, the only time-dependent values available for this integration for an arbitrary variable X are either X^n (X-value at old time level n, $t = t$) or X^{n+1} (X-value at new time level $n+1$, corresponding to $t = t + \Delta t$). As a result, when evaluating time integrals numerically, we have three choices. We could express them in terms of: (1) X^n, (2) X^{n+1}, or (3) a combination of X^n and X^{n+1}.

When time steps Δt are restricted to remain very small, no significant evolution for variable X may be expected. Hence, its value could be said to remain close to whatever value it had at initial time t, i.e., at the beginning of the time step, or old time level "n", X^n. For such situation, time integrals such as the one below could be dealt with very simply:

$$\int_{t}^{t+\Delta t} X dt \approx X^{n}\left[(t+\Delta t) - t\right] = X^{n}(\Delta t) \tag{7.61}$$

If significantly restricting time steps Δt is not desirable, we would need to incorporate information about X^{n+1} in order to better capture time-evolution of X. We could either use X^{n+1} as the best representation of X-evolution with the time interval and write:

$$\int_{t}^{t+\Delta t} X dt \approx X^{n+1}(\Delta t) \tag{7.62}$$

or, we could use a combination of information from X^n and X^{n+1}.

The typically proposed *FVM time-integral treatment* is then generalized as:

$$\int_{t}^{t+\Delta t} X dt \approx X^{m}(\Delta t) \tag{7.63}$$

where

$$X^{m} = X^{n+\theta} = \theta X^{n+1} + (1-\theta) X^{n} \tag{7.64}$$

where $m = n + \theta$ and θ is a time-weighting factor. When $\theta = 0$, $X^m = X^n$ information from the old time-level n is used in the time-integration. When $\theta = 1$, $X^m = X^{n+1}$ information at the new time level $n + 1$ is used for this integration. If $\theta = 1/2$, $X^m = X^{n+1/2} = \frac{1}{2}\left(X^n + X^{n+1}\right)$ and equally weighted information from X^n and X^{n+1} is being used. Therefore, our time-integration choices are:

- $\theta = 0$, $m = n$: leading to a *fully explicit scheme* when used for all variables;
- $\theta = 1/2$, $m = n + 1/2$: leading to an *implicit scheme,* also known as the Crank–Nicolson scheme;
- $\theta = 1$, $m = n + 1$, leading to *fully implicit scheme,* when used for all variables.

The application of the time-integral treatment in Eq. (7.62) transforms Eq. (7.59) into:

$$\left[\left(\frac{k_x A_x}{\mu_o B_o}\frac{\partial\Phi}{\partial x}\right)\Bigg|_{x_e}^{m} - \left(\frac{k_x A_x}{\mu_o B_o}\frac{\partial\Phi}{\partial x}\right)\Bigg|_{x_w}^{m} + \left(\frac{k_y A_y}{\mu_o B_o}\frac{\partial\Phi}{\partial y}\right)\Bigg|_{y_n}^{m} - \left(\frac{k_y A_y}{\mu_o B_o}\frac{\partial\Phi}{\partial y}\right)\Bigg|_{y_s}^{m}\right.$$

$$\left. + \left(\frac{k_z A_z}{\mu_o B_o}\frac{\partial\Phi}{\partial z}\right)\Bigg|_{z_a}^{m} - \left(\frac{k_z A_z}{\mu_o B_o}\frac{\partial\Phi}{\partial z}\right)\Bigg|_{z_b}^{m}\right]\Delta t + \left(q_{osc}\right)_{i,j,k}^{m}\Delta t = \frac{\phi c_o}{\alpha_c B_o^{rf}} V_b\left(p_{i,j,k}^{n+1} - p_{i,j,k}^{n}\right) \qquad (7.65)$$

Deployment of central-difference approximations for all flow potential gradients introduced in Eqs. (7.33) to (7.38) leads to:

$$\frac{k_x A_x}{\mu_o B_o \Delta x}\Bigg|_{x_e}^{m}\left(p_{i+1,j,k}^{m} - p_{i,j,k}^{m}\right) - \frac{k_x A_x}{\mu_o B_o \Delta x}\Bigg|_{x_w}^{m}\left(p_{i,j,k}^{m} - p_{i-1,j,k}^{m}\right)$$

$$+ \frac{k_y A_y}{\mu_o B_o \Delta y}\Bigg|_{y_n}^{m}\left(p_{i,j+1,k}^{m} - p_{i,j,k}^{m}\right) - \frac{k_y A_y}{\mu_o B_o \Delta y}\Bigg|_{y_s}^{m}\left(p_{i,j,k}^{m} - p_{i,j-1,k}^{m}\right)$$

$$+ \frac{k_z A_z}{\mu_o B_o \Delta z}\Bigg|_{z_a}^{m}\left(p_{i,j,k+1}^{m} - p_{i,j,k}^{m}\right) - \frac{k_z A_z}{\mu_o B_o \Delta z}\Bigg|_{z_b}^{m}\left(p_{i,j,k}^{m} - p_{i,j,k-1}^{m}\right)$$

$$- \left[\begin{array}{l} \dfrac{k_x A_x \gamma_o}{\mu_o B_o \Delta x}\Bigg|_{x_e}^{m}\left(G_{i+1,j,k} - G_{i,j,k}\right) - \dfrac{k_x A_x \gamma_o}{\mu_o B_o \Delta x}\Bigg|_{x_w}^{m}\left(G_{i,j,k} - G_{i-1,j,k}\right) \\ + \dfrac{k_y A_y \gamma_o}{\mu_o B_o \Delta y}\Bigg|_{y_n}^{m}\left(G_{i,j+1,k} - G_{i,j,k}\right) - \dfrac{k_y A_y \gamma_o}{\mu_o B_o \Delta y}\Bigg|_{y_s}^{m}\left(G_{i,j,k} - G_{i,j-1,k}\right) \\ + \dfrac{k_z A_z \gamma_o}{\mu_o B_o \Delta z}\Bigg|_{z_a}^{m}\left(G_{i,j,k+1} - G_{i,j,k}\right) - \dfrac{k_z A_z \gamma_o}{\mu_o B_o \Delta z}\Bigg|_{z_b}^{m}\left(G_{i,j,k} - G_{i,j,k-1}\right) \end{array}\right]$$

$$+ \left(q_{osc}\right)_{i,j,k}^{m} = \frac{V_b \phi c_o}{\alpha_c B_o^{rf} \Delta t}\left(p_{i,j,k}^{n+1} - p_{i,j,k}^{n}\right) \qquad (7.66)$$

Note that depth gradients (G) are not evaluated at any time step level as they remain constant during the computations (this statement assumes that there is no significant compaction at any point in the reservoir so any deformation of the cells do not take place). In other words, these are inputs and not unknown variables, and thus they are not subject to numerical updates. The *final discretized FVM form* of Eq. (7.66) would depend on our selection of θ to resolve *each* of the time superscripts "$m = n + \theta$" (Eq. 7.64) for *each* of: (1) transmissibility coefficients, and (2) unknowns of this equation.

For slightly compressible fluids, a commonly held assumption is that pressure changes and density and viscosity changes are not strongly coupled. As a result, μ_o, B_o, and fluid gravity (γ_o) are considered to remain very weak functions of pressure for liquids, and the following oil (slightly compressible fluid) transmissibility calculations:

$$\left.\frac{k_x A_x}{\mu_o B_o \Delta x}\right|_{x_e}^{m} = T_{O,x_e}^{m} = E_{i,j,k}^{m}, \quad \textit{where } x_e \textit{ is the } i+\frac{1}{2}, j,k \textit{ location,} \tag{7.67}$$

$$\left.\frac{k_x A_x}{\mu_o B_o \Delta x}\right|_{x_w}^{m} = T_{O,x_w}^{m} = W_{i,j,k}^{m}, \quad \textit{where } x_w \textit{ is the } i-\frac{1}{2}, j,k \textit{ location,} \tag{7.68}$$

$$\left.\frac{k_y A_y}{\mu_o B_o \Delta y}\right|_{y_n}^{m} = T_{O,y_n}^{m} = N_{i,j,k}^{m}, \quad \textit{where } y_n \textit{ is the } i,j+\frac{1}{2}, k \textit{ location,} \tag{7.69}$$

$$\left.\frac{k_y A_y}{\mu_o B_o \Delta y}\right|_{y_s}^{m} = T_{O,y_s}^{m} = S_{i,j,k}^{m}, \quad \textit{where } y_s \textit{ is the } i,j-\frac{1}{2}, k \textit{ location,} \tag{7.70}$$

$$\left.\frac{k_z A_z}{\mu_o B_o \Delta z}\right|_{z_a}^{m} = T_{O,z_a}^{m} = A_{i,j,k}^{m}, \quad \textit{where } z_a \textit{ is the } i,j,k+\frac{1}{2} \textit{ location,} \tag{7.71}$$

$$\left.\frac{k_z A_z}{\mu_o B_o \Delta z}\right|_{z_b}^{m} = T_{O,z_b}^{m} = B_{i,j,k}^{m}, \quad \textit{where } z_b \textit{ is the } i,j,k-\frac{1}{2} \textit{ location,} \tag{7.72}$$

can be greatly simplified. These transmissibilities, when calculated for liquids and within a time step, are commonly assumed to approximately remain at their values at the beginning of the time step (time step "n"). This approach lets transmissibility calculations remain one-time level behind ("n", $\theta = 0$); but pressure changes could still be calculated at the time level "$n + 1$" ($\theta = 1$). The following *final discretized FVM form* is formulated in such a scenario:

$$\left.\frac{k_x A_x}{\mu_o B_o \Delta x}\right|_{x_e}^{n} \left(p_{i+1,j,k}^{n+1} - p_{i,j,k}^{n+1}\right) - \left.\frac{k_x A_x}{\mu_o B_o \Delta x}\right|_{x_w}^{n} \left(p_{i,j,k}^{n+1} - p_{i-1,j,k}^{n+1}\right)$$

$$+ \left.\frac{k_y A_y}{\mu_o B_o \Delta y}\right|_{y_n}^{n} \left(p_{i,j+1,k}^{n+1} - p_{i,j,k}^{n+1}\right) - \left.\frac{k_y A_y}{\mu_o B_o \Delta y}\right|_{y_s}^{n} \left(p_{i,j,k}^{n+1} - p_{i,j-1,k}^{n+1}\right)$$

$$
+\left.\frac{k_zA_z}{\mu_o B_o\Delta z}\right|_{z_a}^{n}\left(p_{i,j,k+1}^{n+1}-p_{i,j,k}^{n+1}\right)-\left.\frac{k_zA_z}{\mu_o B_o\Delta z}\right|_{z_b}^{n}\left(p_{i,j,k}^{n+1}-p_{i,j,k-1}^{n+1}\right)
$$

$$
-\left[\begin{array}{l}
\left.\frac{k_xA_x\gamma_o}{\mu_oB_o\Delta x}\right|_{x_e}^{n}\left(G_{i+1,j,k}-G_{i,j,k}\right)-\left.\frac{k_xA_x\gamma_o}{\mu_oB_o\Delta x}\right|_{x_w}^{n}\left(G_{i,j,k}-G_{i-1,j,k}\right)\\
+\left.\frac{k_yA_y\gamma_o}{\mu_oB_o\Delta y}\right|_{y_n}^{n}\left(G_{i,j+1,k}-G_{i,j,k}\right)-\left.\frac{k_yA_y\gamma_o}{\mu_oB_o\Delta y}\right|_{y_s}^{n}\left(G_{i,j,k}-G_{i,j-1,k}\right)\\
+\left.\frac{k_zA_z\gamma_o}{\mu_oB_o\Delta z}\right|_{z_a}^{n}\left(G_{i,j,k+1}-G_{i,j,k}\right)-\left.\frac{k_zA_z\gamma_o}{\mu_oB_o\Delta z}\right|_{z_b}^{n}\left(G_{i,j,k}-G_{i,j,k-1}\right)
\end{array}\right]
$$

$$
+(q_{osc})_{i,j,k}^{n}=\frac{V_b\phi c_o}{\alpha_c B_o^{rf}\Delta t}\left(p_{i,j,k}^{n+1}-p_{i,j,k}^{n}\right) \tag{7.73}
$$

where we have ended up with an ***implicit*** scheme for the calculation of nodal pressures. This approach requires solving for all block pressures simultaneously by deploying the following *FVM* approximation for the seven-point numerical stencil of Fig. 7.4:

$$
\begin{aligned}
&B_{i,j,k}^{n}p_{i,j,k-1}^{n+1}+S_{i,j,k}^{n}p_{i,j-1,k}^{n+1}+W_{i,j,k}^{n}p_{i-1,j,k}^{n+1}+C_{i,j,k}^{n}p_{i,j,k}^{n+1}+E_{i,j,k}^{n}p_{i+1,j,k}^{n+1}\\
&\quad+N_{i,j,k}^{n}p_{i,j+1,k}^{n+1}+A_{i,j,k}^{n}p_{i,j,k+1}^{n+1}=Q_{i,j,k}^{n}
\end{aligned} \tag{7.74}
$$

where Eqs. (7.67) to (7.72) have been used along with the following additional definitions :

$$
C_{i,j,k}^{n}=-\left(B_{i,j,k}^{n}+S_{i,j,k}^{n}+W_{i,j,k}^{n}+E_{i,j,k}^{n}+N_{i,j,k}^{n}+A_{i,j,k}^{n}+\left(\frac{V_b\phi c_o}{\alpha_c B_o^{rf}\Delta t}\right)\right) \tag{7.75}
$$

$$
\begin{aligned}
Q_{i,j,k}^{n}&=-(q_{osc})_{i,j,k}^{n}-\left(\frac{V_b\phi c_o}{\alpha_c B_o^{rf}\Delta t}\right)p_{i,j,k}^{n}\\
&\quad+\Big[E_{i,j,k}^{n}\gamma_{o,x_e}^{n}\left(G_{i+1,j,k}-G_{i,j,k}\right)-W_{i,j,k}^{n}\gamma_{o,x_w}^{n}\left(G_{i,j,k}-G_{i-1,j,k}\right)\\
&\qquad+N_{i,j,k}^{n}\gamma_{o,y_n}^{n}\left(G_{i,j+1,k}-G_{i,j,k}\right)-S_{i,j,k}^{n}\gamma_{o,y_s}^{n}\left(G_{i,j,k}-G_{i,j-1,k}\right)\\
&\qquad+A_{i,j,k}^{n}\gamma_{o,z_a}^{n}\left(G_{i,j,k+1}-G_{i,j,k}\right)-B_{i,j,k}^{n}\gamma_{o,z_b}^{n}\left(G_{i,j,k}-G_{i,j,k-1}\right)\Big]
\end{aligned} \tag{7.76}
$$

Please note that all liquid transmissibility and property calculations are being performed at time level "n" in Eqs. (7.74) to (7.76). Again, detailed discussions of well treatment choices for the $(q_{osc})_{i,j,k}^{n}$ source term can be found in Ertekin et al. (2001), along with treatment choices for transmissibility for no-flow and constant-pressure outer boundary conditions. In Eq. (7.76) and elsewhere, well source terms are being assumed constant and known unless otherwise noted.

The formulation of an ***explicit*** scheme for the calculation of nodal pressures is also possible for slightly compressible fluids—albeit at the expense of severely restricting allowable time steps (Δt) in some numerical computations. The *explicit* scheme is derived by formulating a *final discretized FVM form* where *both* liquid transmissibilities *and* nodal pressures in Eq. (7.66) are calculated at time level $m = n$ ($\theta = 0$), yielding:

$$\left.\frac{k_x A_x}{\mu_o B_o \Delta x}\right|_{x_e}^{n}\left(p_{i+1,j,k}^{n}-p_{i,j,k}^{n}\right)-\left.\frac{k_x A_x}{\mu_o B_o \Delta x}\right|_{x_w}^{n}\left(p_{i,j,k}^{n}-p_{i-1,j,k}^{n}\right)$$

$$+\left.\frac{k_y A_y}{\mu_o B_o \Delta y}\right|_{y_n}^{n}\left(p_{i,j+1,k}^{n}-p_{i,j,k}^{n}\right)-\left.\frac{k_y A_y}{\mu_o B_o \Delta y}\right|_{y_s}^{n}\left(p_{i,j,k}^{n}-p_{i,j-1,k}^{n}\right)$$

$$+\left.\frac{k_z A_z}{\mu_o B_o \Delta z}\right|_{z_a}^{n}\left(p_{i,j,k+1}^{n}-p_{i,j,k}^{n}\right)-\left.\frac{k_z A_z}{\mu_o B_o \Delta z}\right|_{z_b}^{n}\left(p_{i,j,k}^{n}-p_{i,j,k-1}^{n}\right)$$

$$-\left[\begin{array}{l}\left.\frac{k_x A_x \gamma_o}{\mu_o B_o \Delta x}\right|_{x_e}^{n}\left(G_{i+1,j,k}-G_{i,j,k}\right)-\left.\frac{k_x A_x \gamma_o}{\mu_o B_o \Delta x}\right|_{x_w}^{n}\left(G_{i,j,k}-G_{i-1,j,k}\right)\\+\left.\frac{k_y A_y \gamma_o}{\mu_o B_o \Delta y}\right|_{y_n}^{n}\left(G_{i,j+1,k}-G_{i,j,k}\right)-\left.\frac{k_y A_y \gamma_o}{\mu_o B_o \Delta y}\right|_{y_s}^{n}\left(G_{i,j,k}-G_{i,j-1,k}\right)\\+\left.\frac{k_z A_z \gamma_o}{\mu_o B_o \Delta z}\right|_{z_a}^{n}\left(G_{i,j,k+1}-G_{i,j,k}\right)-\left.\frac{k_z A_z \gamma_o}{\mu_o B_o \Delta z}\right|_{z_b}^{n}\left(G_{i,j,k}-G_{i,j,k-1}\right)\end{array}\right]$$

$$+\left(q_{osc}\right)_{i,j,k}^{n}=\frac{V_b \phi c_o}{\alpha_c B_o^{rf}\Delta t}\left(p_{i,j,k}^{n+1}-p_{i,j,k}^{n}\right) \tag{7.77}$$

which leads to the *explicit* calculation of all nodal pressures by solving for $p_{i,j,k}^{n+1}$ directly (i.e., explicitly) one cell at a time as shown below:

$$\begin{aligned}p_{i,j,k}^{n+1}=p_{i,j,k}^{n}&+\left(\frac{\alpha_c B_o^{rf}\Delta t}{V_b \phi c_o}\right)\\&\times\left[B_{i,j,k}^{n}p_{i,j,k-1}^{n}+S_{i,j,k}^{n}p_{i,j-1,k}^{n}+W_{i,j,k}^{n}p_{i-1,j,k}^{n}+C_{i,j,k}^{n}p_{i,j,k}^{n}\right.\\&\left.+E_{i,j,k}^{n}p_{i+1,j,k}^{n}+N_{i,j,k}^{n}p_{i,j+1,k}^{n}+A_{i,j,k}^{n}p_{i,j,k+1}^{n}-Q_{i,j,k}^{n}\right]\end{aligned} \tag{7.78}$$

with the following definitions:

$$C_{i,j,k}^{n}=-\left(B_{i,j,k}^{n}+S_{i,j,k}^{n}+W_{i,j,k}^{n}+E_{i,j,k}^{n}+N_{i,j,k}^{n}+A_{i,j,k}^{n}\right) \tag{7.79}$$

$$\begin{aligned}Q_{i,j,k}^{n}=\left(q_{osc}\right)_{i,j,k}^{n}&+\left[E_{i,j,k}^{n}\gamma_{oe}^{n}\left(G_{i+1,j,k}-G_{i,j,k}\right)-W_{i,j,k}^{n}\gamma_{ow}^{n}\left(G_{i,j,k}-G_{i-1,j,k}\right)\right.\\&+N_{i,j,k}^{n}\gamma_{on}^{n}\left(G_{i,j+1,k}-G_{i,j,k}\right)-S_{i,j,k}^{n}\gamma_{os}^{n}\left(G_{i,j,k}-G_{i,j-1,k}\right)\\&\left.+A_{i,j,k}^{n}\gamma_{oa}^{n}\left(G_{i,j,k+1}-G_{i,j,k}\right)-B_{i,j,k}^{n}\gamma_{ob}^{n}\left(G_{i,j,k}-G_{i,j,k-1}\right)\right]\end{aligned} \tag{7.80}$$

Note that no simultaneous solution of equations (i.e., no matrix manipulations) is needed in this explicit scheme. Although this is convenient, it comes at a potentially significant cost. Explicit schemes in fluid flow analysis are conditionally stable; that is, its numerical stability is strongly dependent on time step (Δt) selection, which can significantly restrict allowable Δt (see Ertekin et al., 2001). As a result, fully explicit formulations are not common in reservoir simulation—particularly when associated conditional stability can be severe. In general, some degree of implicitness is desirable in order to achieve greater stability and more reliable numerical performance.

Slightly Compressible Fluid Transmissibility Calculations

Weak nonlinear coefficients in transmissibility calculations are typically computed by arithmetic or volumetric averages. Values of fluid properties weakly dependent with pressure (μ_o, B_o, γ_o) at boundary (interblock) locations are estimated at the block-volume *averaged* pressures at the interface between two blocks. For example, for the east boundary:

$$p_{x_e}^n = \frac{\left(V_{i+1,j,k}p_{i+1,k}^n + V_{i,j,k}p_{i,j,k}^n\right)}{V_{i,j,k} + V_{i+1,j,k}} \tag{7.81}$$

which implies the arithmetic averaging of pressure values between two blocks when neighboring blocks are of equal volume (e.g., $V_{i,j,k} = V_{i+1,j,k}$).

7.3.3 Single-Phase Compressible Flow

Chapter 5 demonstrated that the governing equation for *compressible (dry gas) fluid flow* in a porous media could be written as:

$$\nabla \cdot \left[\frac{\bar{\bar{k}}}{\mu_g B_g}\nabla p\right] + \frac{q_{gsc}}{V_b} = \frac{1}{\alpha_c}\frac{\partial}{\partial t}\left(\phi / B_g\right) \tag{7.82}$$

The 3D Cartesian expansion of Eq. (7.82) is:

$$\frac{\partial}{\partial x}\left(\frac{k_x}{\mu_g B_g}\frac{\partial p}{\partial x}\right) + \frac{\partial}{\partial y}\left(\frac{k_y}{\mu_g B_g}\frac{\partial p}{\partial y}\right) + \frac{\partial}{\partial z}\left(\frac{k_z}{\mu_g B_g}\frac{\partial p}{\partial z}\right) + \frac{q_{gsc}}{V_b} = \frac{1}{\alpha_c}\frac{\partial}{\partial t}\left(\frac{\phi}{B_g}\right) \tag{7.82}$$

To derive the *finite-volume* representation of Eq. (7.82) for the control volume shown in Fig. 7.5, we proceed with the *FVM* integration of this governing equation over the spatial and time domains within the time step of interest (Δt). This is:

$$\begin{aligned} &\int_t^{t+\Delta t}\left(\int_{V_b}\frac{\partial}{\partial x}\left(\frac{k_x}{\mu_g B_g}\frac{\partial p}{\partial x}\right)dV\right)dt + \int_t^{t+\Delta t}\left(\int_{V_b}\frac{\partial}{\partial y}\left(\frac{k_y}{\mu_g B_g}\frac{\partial p}{\partial y}\right)dV\right)dt \\ &+ \int_t^{t+\Delta t}\left(\int_{V_b}\frac{\partial}{\partial z}\left(\frac{k_z}{\mu_g B_g}\frac{\partial p}{\partial z}\right)dV\right)dt + \frac{1}{V_b}\int_t^{t+\Delta t}\left(\int_{V_b} q_{gsc}\,dV\right)dt = \frac{1}{\alpha_c}\int_t^{t+\Delta t}\int_{V_b}\frac{\partial}{\partial t}\left(\frac{\phi}{B_g}\right)dV dt \end{aligned} \tag{7.83}$$

By performing the spatial integrations in Eq. (7.83), introducing the time-integration treatment of Eq. (7.63), and deploying the central-difference approximations for pressure gradients introduced in Eqs. (7.33) to (7.38) (neglecting depth gradients), we readily obtain:

$$\begin{aligned}
&\left.\frac{k_x A_x}{\mu_g B_g \Delta x}\right|_{x_e}^{m}\left(p_{i+1,j,k}^{m}-p_{i,j,k}^{m}\right)-\left.\frac{k_x A_x}{\mu_g B_g \Delta x}\right|_{x_w}^{m}\left(p_{i,j,k}^{m}-p_{i-1,j,k}^{m}\right)\\
&+\left.\frac{k_y A_y}{\mu_g B_g \Delta y}\right|_{y_n}^{m}\left(p_{i,j+1,k}^{m}-p_{i,j,k}^{m}\right)-\left.\frac{k_y A_y}{\mu_g B_g \Delta y}\right|_{y_s}^{m}\left(p_{i,j,k}^{m}-p_{i,j-1,k}^{m}\right)\\
&+\left.\frac{k_z A_z}{\mu_g B_g \Delta z}\right|_{z_u}^{m}\left(p_{i,j,k+1}^{m}-p_{i,j,k}^{m}\right)-\left.\frac{k_z A_z}{\mu_g B_g \Delta z}\right|_{z_b}^{m}\left(p_{i,j,k}^{m}-p_{i,j,k-1}^{m}\right)\\
&+(q_{gsc})_{i,j,k}^{m}=\frac{V_b}{\alpha_c \Delta t}\left[\left(\frac{\phi}{B_g}\right)_{i,j,k}^{n+1}-\left(\frac{\phi}{B_g}\right)_{i,j,k}^{n}\right]
\end{aligned} \tag{7.84}$$

where the definition of a volume-averaged property $\left(\bar{X}=\frac{1}{V_b}\int_{V_b} X dV\right)$ was used to resolve both source and accumulations term integrals:

$$\frac{1}{V_b}\int_{t}^{t+\Delta t}\left(\int_{V_b} q_{gsc} dV\right)dt=\int_{t}^{t+\Delta t}(q_{lsc})_{i,j,k}dt=(q_{gsc})_{i,j,k}^{m}\Delta t \tag{7.85}$$

$$\int_{t}^{t+\Delta t}\int_{V_b}\frac{\partial}{\partial t}\left(\frac{\phi}{B_g}\right)dVdt=V_b\int_{t}^{t+\Delta t} d\left(\frac{\phi}{B_g}\right)_{i,j,k}=\left(\frac{\phi}{B_g}\right)_{i,j,k}^{n+1}-\left(\frac{\phi}{B_g}\right)_{i,j,k}^{n} \tag{7.86}$$

in Eq. (7.84), $m = n + \theta$ where θ is the time-weighting factor used for the time-treatment of each term in the flow equation.

In direct contrast to the slightly compressible case discussed in the preceding section, compressible fluid formulations require a higher degree of implicitness in their numerical treatment. This is because fluid transmissibilities in compressible cases yield a highly nonlinear formulation due to their much higher sensibility to pressure as compared to liquids. Therefore, the de facto choice for compressible fluid formulation is the use of $m = n + 1$ for all terms including the transmissibility coefficients, to yield:

$$\begin{aligned}
&T_{Gx_e}^{n+1}\left(p_{i+1,j,k}^{n+1}-p_{i,j,k}^{n+1}\right)-T_{Gx_w}^{n+1}\left(p_{i,j,k}^{n+1}-p_{i-1,j,k}^{n+1}\right)\\
&+T_{Gy_n}^{n+1}\left(p_{i,j+1,k}^{n+1}-p_{i,j,k}^{n+1}\right)-T_{Gy_s}^{n+1}\left(p_{i,j,k}^{n+1}-p_{i,j-1,k}^{n+1}\right)\\
&+T_{Gz_u}^{n+1}\left(p_{i,j,k+1}^{n+1}-p_{i,j,k}^{n+1}\right)-T_{Gz_b}^{n+1}\left(p_{i,j,k}^{n+1}-p_{i,j,k-1}^{n+1}\right)\\
&+(q_{gsc})_{i,j,k}^{n+1}=\frac{V_b}{\alpha_c \Delta t}\left[\left(\frac{\phi}{B_g}\right)_{i,j,k}^{n+1}-\left(\frac{\phi}{B_g}\right)_{i,j,k}^{n}\right]
\end{aligned} \tag{7.87}$$

where gas (compressible-fluid) transmissibilities are being defined as:

$$E_{i,j,k}^{n+1} = T_{G,x_e}^{n+1} = \left.\frac{k_x A_x}{\mu_g B_g \Delta x}\right|_{x_e}^{n+1} \tag{7.88}$$

$$W_{i,j,k}^{n+1} = T_{G,x_w}^{n+1} = \left.\frac{k_x A_x}{\mu_g B_g \Delta x}\right|_{x_w}^{n+1} \tag{7.89}$$

$$N_{i,j,k}^{n+1} = T_{G,y_n}^{n+1} = \left.\frac{k_y A_y}{\mu_g B_g \Delta y}\right|_{y_n}^{n+1} \tag{7.90}$$

$$S_{i,j,k}^{n+1} = T_{G,y_s}^{n+1} = \left.\frac{k_y A_y}{\mu_g B_g \Delta y}\right|_{y_s}^{n+1} \tag{7.91}$$

$$A_{i,j,k}^{n+1} = T_{G,z_a}^{n+1} = \left.\frac{k_z A_z}{\mu_g B_g \Delta z}\right|_{z_a}^{n+1} \tag{7.92}$$

$$B_{i,j,k}^{n+1} = T_{G,z_b}^{n+1} = \left.\frac{k_z A_z}{\mu_g B_g \Delta z}\right|_{z_b}^{n+1} \tag{7.93}$$

The difficulty faced in solving Eq. (7.87) as proposed is that all information at time level "$n + 1$" is unknown at the start of the numerical procedure for any given time step. All pressures are unknown at $n + 1$ and thus are unavailable to calculate pressure-dependent transmissibilities at time level $n + 1$. Solving this would require some degree of (within-time-step) "decoupling" of transmissibilities and pressures; and while both would remain anchored at time level "$n + 1$", they would not be calculated simultaneously. Thus, pressure-dependent transmissibility terms are typically evaluated *one-iteration behind* the pressure solution. This entails the deployment of inner (within-time-step) iterations which are denoted as "k" iteration levels. This "decoupling" yields the following FVM equation:

$$\begin{aligned}
&\overset{(k)}{T_{G,x_e}^{n+1}}\left(\overset{(k+1)}{p_{i+1,j,k}^{n+1}} - \overset{(k+1)}{p_{i,j,k}^{n+1}}\right) - \overset{(k)}{T_{G,x_w}^{n+1}}\left(\overset{(k+1)}{p_{i,j,k}^{n+1}} - \overset{(k+1)}{p_{i-1,j,k}^{n+1}}\right) \\
&+ \overset{(k)}{T_{G,y_n}^{n+1}}\left(\overset{(k+1)}{p_{i,j+1,k}^{n+1}} - \overset{(k+1)}{p_{i,j,k}^{n+1}}\right) - \overset{(k)}{T_{G,y_s}^{n+1}}\left(\overset{(k+1)}{p_{i,j,k}^{n+1}} - \overset{(k+1)}{p_{i,j-1,k}^{n+1}}\right) \\
&+ \overset{(k)}{T_{G,z_a}^{n+1}}\left(\overset{(k+1)}{p_{i,j,k+1}^{n+1}} - \overset{(k+1)}{p_{i,j,k}^{n+1}}\right) - \overset{(k)}{T_{G,z_b}^{n+1}}\left(\overset{(k+1)}{p_{i,j,k}^{n+1}} - \overset{(k+1)}{p_{i,j,k-1}^{n+1}}\right) \\
&+ \overset{(k)}{(q_{gsc})_{i,j,k}^{n+1}} = \frac{V_b}{\alpha_c \Delta t}\left[\overset{(k+1)}{\left(\frac{\phi}{B_g}\right)_{i,j,k}^{n+1}} - \left(\frac{\phi}{B_g}\right)_{i,j,k}^{n}\right]
\end{aligned} \tag{7.94}$$

Please note that in this equation, transmissibility terms and other pressure-dependent entries are evaluated at "$n+1, (k)$" while pressures are evaluated at "$n+1, (k+1)$", i.e., transmissibilities are evaluated one-iteration behind the pressure solution.

Also, the right-hand-side accumulation term in Eq. (7.94) as currently written is not directly expressed in terms of *nodal pressures*. Before Eq. (7.94) could be used for the simultaneous solution of all nodal pressures, its right-hand side term must need to be manipulated in order to rewrite it in terms of *pressure*, because pressure is the primary unknown being solved for. Using the chain rule we obtain:

$$\frac{\partial}{\partial t}(\phi/B_g) = \frac{\partial}{\partial p}(\phi/B_g) \cdot \frac{\partial p}{\partial t} \tag{7.95}$$

This allows the right-hand side of Eq. (7.94) to be rewritten in terms of backward Taylor approximations of the chain derivatives that appear in Eq. (7.95), as follows:

$$\frac{V_b}{\Delta t}\left[\left(\frac{\phi}{B_g}\right)_{i,j,k}^{\overset{(k+1)}{n+1}} - \left(\frac{\phi}{B_g}\right)_{i,j,k}^{n}\right] \cong \frac{V_b}{\Delta t}\left\{\frac{\left(\frac{\phi}{B_g}\right)_{i,j,k}^{\overset{(k)}{n+1}} - \left(\frac{\phi}{B_g}\right)_{i,j,k}^{n}}{\overset{(k)}{p_{i,j,k}^{n+1}} - p_{i,j,k}^{n}}\right\}\left[\overset{(k+1)}{p_{i,j,k}^{n+1}} - p_{i,j,k}^{n}\right] \tag{7.96}$$

where the backward approximation for $\frac{\partial}{\partial p}(\phi/B_g)$ is purposely evaluated at time level "$n+1$" while lagging one k-iteration behind the pressure solution. Alternatively, the concept of total compressibility could be deployed to approximate this accumulation term. We have shown before that $\frac{d\phi}{dp} = c_\phi \phi$ following definition of pore compressibility (see Chap. 3); but also $\frac{d}{dp}(1/B_g) = c_g/B_g$ (by applying the definition of isothermal compressibility to the dry gas formation volume factor). By expanding $\frac{\partial}{\partial p}(\phi/B_g)$ in Eq. (7.95) and introducing these compressibility equivalencies, Eq. (7.96) could be also written as:

$$\frac{V_b}{\Delta t}\left[\left(\frac{\phi}{B_g}\right)_{i,j,k}^{\overset{(k+1)}{n+1}} - \left(\frac{\phi}{B_g}\right)_{i,j,k}^{n}\right] \cong \frac{V_b}{\Delta t}\left(\frac{\phi c_t}{B_g}\right)_{i,j,k}^{\overset{(k)}{n+1}}\left[\overset{(k+1)}{p_{i,j,k}^{n+1}} - p_{i,j,k}^{n}\right] \tag{7.96}$$

where $c_t = c_g + c_\phi$ is the total compressibility.

Introducing these accumulation term expressions into Eq. (7.94), the FVM approximation for the single-phase compressible flow problem using the seven-point numerical stencil of Fig. 7.4 becomes:

$$\begin{aligned}&\overset{(k)}{B_{i,j,k}^{n+1}}\overset{(k+1)}{p_{i,j,k-1}^{n+1}} + \overset{(k)}{S_{i,j,k}^{n+1}}\overset{(k+1)}{p_{i,j-1,k}^{n+1}} + \overset{(k)}{W_{i,j,k}^{n+1}}\overset{(k+1)}{p_{i-1,j,k}^{n+1}} + \overset{(k)}{C_{i,j,k}^{n+1}}\overset{(k+1)}{p_{i,j,k}^{n+1}} + \overset{(k)}{E_{i,j,k}^{n+1}}\overset{(k+1)}{p_{i+1,j,k}^{n+1}}\\ &+ \overset{(k)}{N_{i,j,k}^{n+1}}\overset{(k+1)}{p_{i,j+1,k}^{n+1}} + \overset{(k)}{A_{i,j,k}^{n+1}}\overset{(k+1)}{p_{i,j,k+1}^{n+1}} = \overset{(k)}{Q_{i,j,k}^{n+1}}\end{aligned} \tag{7.97}$$

where definitions in Eqs. (7.88) to (7.93) have been deployed along with the definitions:

$$\overset{(k)}{C^{n+1}_{i,j,k}} = -\left(\overset{(k)}{B^{n+1}_{i,j,k}} + \overset{(k)}{S^{n+1}_{i,j,k}} + \overset{(k)}{W^{n+1}_{i,j,k}} + \overset{(k)}{E^{n+1}_{i,j,k}} + \overset{(k)}{N^{n+1}_{i,j,k}} + \overset{(k)}{A^{n+1}_{i,j,k}} + \overset{(k)}{\xi^{n+1}_{i,j,k}}\right) \tag{7.98}$$

$$\overset{(k)}{Q^{n+1}_{i,j,k}} = -\overset{(k)}{(q_{gsc})^{n+1}_{i,j,k}} - \overset{(k)}{\xi^{n+1}_{i,j,k}}\, p^{n}_{i,j,k} \tag{7.99}$$

where

$$\overset{(k)}{\xi^{n+1}_{i,j,k}} = \left(\frac{V_b}{\alpha_c \Delta t}\right)^{n+1}_{i,j,k} \left\{ \frac{\overset{(k)}{\left(\dfrac{\phi}{B_g}\right)^{n+1}_{i,j,k}} - \left(\dfrac{\phi}{B_g}\right)^{n}_{i,j,k}}{\overset{(k)}{p^{n+1}_{i,j,k}} - p^{n}_{i,j,k}} \right\} = \left(\frac{V_b}{\alpha_c \Delta t}\right)^{n+1}_{i,j,k} \overset{(k)}{\left(\frac{\phi c_t}{B_g}\right)^{n+1}_{i,j,k}} \tag{7.100}$$

Eq. (7.97) provides the implicit formulation required for solving for all block pressures simultaneously. At the start of every time step calculations ($k=0$), transmissibilities would equal to those calculated at the end of the previous time step. Once new pressures are calculated, transmissibilities are updated. Iterations continue until no further pressure improvements are realized within the k-inner loop. Source term $\left(\overset{(k)}{(q_{gsc})^{n+1}_{i,j,k}}\right)$ and outer-boundary numerical treatment information can be found in Ertekin et al. (2001). In Eq. (7.99) and elsewhere, well source terms are being assumed constant and known unless otherwise noted.

Compressible Fluid Transmissibility Calculations

Nonlinear coefficients in gas transmissibility calculations also computed by arithmetic or volumetric averages, but at time level $n+1$. As μ_g, B_g are rather strong functions of pressure, they must be estimated at average pressure for the *current $n+1$ time level, one iteration level behind (k)* between the two blocks. For example, for the east boundary:

$$\overset{(k)}{p^{n+1}_{x_e}} = \frac{\left(V_{i+1,j,k}\overset{(k)}{p^{n+1}_{i+1,k}} + V_{i,j,k}\,\overset{(k)}{p^{n+1}_{i,j,k}}\right)}{V_{i,j,k} + V_{i+1,j,k}} \tag{7.101}$$

which equals the arithmetic pressure between two blocks when neighboring blocks are of equal volume (e.g., $V_{i,j,k} = V_{i+1,j,k}$).

7.4 Newton–Raphson and Single-Phase Numerical Solutions

The Newton–Raphson formulation is a powerful iterative method widely used to solve systems of simultaneous nonlinear equations. It is quite versatile and well known to provide *quadratic* convergence toward a solution. However, this only happens when the root or actual solution is close by. In other words, "initial guesses" (first trial approximations) deployed to start the iterative process should not be found not too far away from the actual solution. Newton–Raphson procedure "travels" through the solution surface using gradient information at such "trial" points on the solution surface in order to locate better approximations found closer to the actual solution.

The objective function "f" to be solved does not have to be dependent on a single-variable or unknown. More generally, it can be any vector function combining N-functional relationships of the type:

$$\vec{f} = (f_1, f_2, f_3, f_4, \ldots, f_N) \tag{7.102}$$

which are functions of a vector of unknowns $\vec{x}$ (N-unknowns):

$$\vec{x} = (x_1, x_2, x_3, x_4, \ldots, x_N) \tag{7.103}$$

where $\vec{f} = 0$ when evaluated at the value of $\vec{x}$ that satisfies all N-equation functions (solution or root of the system of equation). For such system, its gradient vector is expressed as:

$$\nabla\vec{f}(\vec{x}) = \frac{\partial \vec{f}}{\partial \vec{x}} = \frac{\partial(f_1, f_2, f_3, f_4, \ldots, f_N)}{\partial(x_1, x_2, x_3, x_4, \ldots, x_N)} \tag{7.104}$$

whose components represent the **Jacobian of the system,** which is written in terms of a matrix containing all first-order partial derivatives of vector function $\vec{f}$ with respect to each component of the vector of unknowns $\vec{x}$:

$$\nabla\vec{f}(\vec{x}) = \begin{bmatrix} \frac{\partial f_1}{\partial x_1} & \frac{\partial f_1}{\partial x_2} & \frac{\partial f_1}{\partial x_3} & \frac{\partial f_1}{\partial x_4} & \cdots & \frac{\partial f_1}{\partial x_N} \\ \frac{\partial f_2}{\partial x_1} & \frac{\partial f_2}{\partial x_2} & \frac{\partial f_2}{\partial x_3} & \frac{\partial f_2}{\partial x_4} & \cdots & \frac{\partial f_2}{\partial x_N} \\ \frac{\partial f_3}{\partial x_1} & \frac{\partial f_3}{\partial x_2} & \frac{\partial f_3}{\partial x_3} & \frac{\partial f_3}{\partial x_4} & \cdots & \frac{\partial f_3}{\partial x_N} \\ \frac{\partial f_4}{\partial x_1} & \frac{\partial f_4}{\partial x_2} & \frac{\partial f_4}{\partial x_3} & \frac{\partial f_4}{\partial x_4} & \cdots & \frac{\partial f_4}{\partial x_N} \\ \cdots & \cdots & \cdots & \cdots & \cdots & \cdots \\ \frac{\partial f_N}{\partial x_1} & \frac{\partial f_N}{\partial x_2} & \frac{\partial f_N}{\partial x_3} & \frac{\partial f_N}{\partial x_4} & \cdots & \frac{\partial f_N}{\partial x_N} \end{bmatrix} \tag{7.105}$$

Using the first few terms of a Taylor Series written around a current guess, the **multivariate Newton–Raphson formulation** can be written in terms of the function vector $\vec{f}$, gradient vector $\nabla\vec{f}(\vec{x})$, and a vector of "improvements" $\Delta\vec{x}^{new}$ as follows:

$$\nabla\vec{f}(\vec{x}^{old})\Delta\vec{x}^{new} = -\vec{f}(\vec{x}^{old}) \tag{7.106}$$

where $\vec{x}^{old}$ is the initialization value (or trial) which is currently available for all unknowns, $\nabla\vec{f}(\vec{x}^{old})$ is the value of the Jacobian matrix evaluates using such current values, and $\vec{f}(\vec{x}^{old})$ are the *"residuals"* or value of the each of the components of the function vector when evaluated at those current guesses. As a result, a $\Delta\vec{x}^{new}$ vector of *"improvements"* (updates to be applied to current trial guesses in order to find successively better approximations to the actual solution) can be calculated and applied to advance the iterative solution:

$$\vec{x}^{new} = \vec{x}^{old} + \Delta\vec{x}^{new} \tag{7.107}$$

To deploy the **Newton–Raphson iterative procedure**, the following sequence of steps is typically employed:

- *Set all equations in the Newton–Raphson form: $\vec{f} = 0$ (**residual** form).*
- *Provide a first approximation (guess) to the solution ($\vec{x}^{old}$).*
- *Evaluate the right-hand side of Eq. (7.106) [$-\vec{f}(\vec{x}^{old})$].*
- *Calculate the entries of the Jacobian $\nabla\vec{f}$ [coefficient matrix of partial derivatives, $\nabla\vec{f}(\vec{x}^{old})$].*
- *Solve for $\Delta\vec{x}^{new}$ in Eq. (7.106).*
- *Update the solution using Eq. (7.107).*
- *Recalculate Jacobian and residuals until convergence is attained [all components of $\Delta\vec{x}^{new}$ and/or $\vec{f}(\vec{x}^{old})$ vectors are found below an acceptable tolerance].*

Newton–Raphson Formulation: 3D Compressible Flow

The *implicit* finite-difference approximation of the 3D Cartesian compressible flow problem may be written as:

$$\begin{aligned}
&E_{i,j,k}^{n+1}\left(p_{i+1,j,k}^{n+1} - p_{i,j,k}^{n+1}\right) + W_{i,j,k}^{n+1}\left(p_{i-1,j,k}^{n+1} - p_{i,j,k}^{n+1}\right)\\
&+ N_{i,j,k}^{n+1}\left(p_{i,j+1,k}^{n+1} - p_{i,j,k}^{n+1}\right) + S_{i,j,k}^{n+1}\left(p_{i,j-1,k}^{n+1} - p_{i,j,k}^{n+1}\right)\\
&+ A_{i,j,k}^{n+1}\left(p_{i,j,k+1}^{n+1} - p_{i,j,k}^{n+1}\right) + B_{i,j,k}^{n+1}\left(p_{i,j,k-1}^{n+1} - p_{i,j,k}^{n+1}\right)\\
&+ q_{gsc\ i,j,k}^{n+1} = \left(\frac{V_b}{\alpha_c \Delta t}\right)_{i,j,k}^{n+1}\left[(\phi/B_g)_{i,j,k}^{n+1} - (\phi/B_g)_{i,j,k}^{n}\right]
\end{aligned} \tag{7.108}$$

Therefore, the "residual" function that we would like to make equal to zero at each gridblock in this system would be:

$$\overset{(k)}{R_{i,j,k}^{n+1}}\left(\overset{(k)}{p_{i+1,j,k}^{n+1}}, \overset{(k)}{p_{i-1,j,k}^{n+1}}, \overset{(k)}{p_{i,j+1,k}^{n+1}}, \overset{(k)}{p_{i,j-1,k}^{n+1}}, \overset{(k)}{p_{i,j,k+1}^{n+1}}, \overset{(k)}{p_{i,j,k-1}^{n+1}} \overset{(k)}{p_{i,j,k}^{n+1}}\right)$$

$$= \overset{(k)}{E_{i,j,k}^{n+1}}\left(\overset{(k+1)}{p_{i+1,j,k}^{n+1}} - \overset{(k+1)}{p_{i,j,k}^{n+1}}\right) + \overset{(k)}{W_{i,j,k}^{n+1}}\left(\overset{(k+1)}{p_{i-1,j,k}^{n+1}} - \overset{(k+1)}{p_{i,j,k}^{n+1}}\right)$$

$$+ \overset{(k)}{N_{i,j,k}^{n+1}}\left(\overset{(k+1)}{p_{i,j+1,k}^{n+1}} - \overset{(k+1)}{p_{i,j,k}^{n+1}}\right) + \overset{(k)}{S_{i,j,k}^{n+1}}\left(\overset{(k+1)}{p_{i,j-1,k}^{n+1}} - \overset{(k+1)}{p_{i,j,k}^{n+1}}\right)$$

$$+ \overset{(k)}{A_{i,j,k}^{n+1}}\left(\overset{(k+1)}{p_{i,j,k+1}^{n+1}} - \overset{(k+1)}{p_{i,j,k}^{n+1}}\right) + \overset{(k)}{B_{i,j,k}^{n+1}}\left(\overset{(k+1)}{p_{i,j,k-1}^{n+1}} - \overset{(k+1)}{p_{i,j,k}^{n+1}}\right)$$

$$+ \overset{(k)}{q_{gsc\ i,j,k}^{n+1}} - \left(\frac{V_b}{\alpha_c \Delta t}\right)_{i,j,k}^{n+1}\left[\overset{(k+1)}{(\phi/B_g)_{i,j,k}^{n+1}} - (\phi/B_g)_{i,j,k}^{n}\right] = 0 \tag{7.109}$$

where the "k" subscripts refer to current/previous iteration step "guess" to the actual "$n+1$" time level values. In a compressible fluid system of N gridblocks, there would be "N" unknowns: all gridblock pressures at time level $n+1$ and "N" equations: one residual equation [Eq. (7.109)] at each of the gridblocks in the system. Altogether, a system of "N" equations in "N" unknowns. The Newton–Raphson solution system would be written as:

$$\begin{bmatrix} \frac{\partial R_1}{\partial p_1} & \frac{\partial R_1}{\partial p_2} & \frac{\partial R_1}{\partial p_3} & \frac{\partial R_1}{\partial p_4} & \cdots & \frac{\partial R_1}{\partial p_N} \\ \frac{\partial R_2}{\partial p_1} & \frac{\partial R_2}{\partial p_2} & \frac{\partial R_2}{\partial p_3} & \frac{\partial R_2}{\partial p_4} & \cdots & \frac{\partial R_2}{\partial p_N} \\ \frac{\partial R_3}{\partial p_1} & \frac{\partial R_3}{\partial p_2} & \frac{\partial R_3}{\partial p_3} & \frac{\partial R_3}{\partial p_4} & \cdots & \frac{\partial R_3}{\partial p_N} \\ \frac{\partial R_4}{\partial p_1} & \frac{\partial R_4}{\partial p_2} & \frac{\partial R_4}{\partial p_3} & \frac{\partial R_4}{\partial p_4} & \cdots & \frac{\partial R_4}{\partial p_N} \\ \cdots & \cdots & \cdots & \cdots & \cdots & \cdots \\ \frac{\partial R_N}{\partial p_1} & \frac{\partial R_N}{\partial p_2} & \frac{\partial R_N}{\partial p_3} & \frac{\partial R_N}{\partial p_4} & \cdots & \frac{\partial R_N}{\partial p_N} \end{bmatrix}^{\overset{(k)}{n+1}} \begin{bmatrix} \Delta p_1 \\ \Delta p_2 \\ \Delta p_3 \\ \Delta p_4 \\ \cdots \\ \cdots \\ \Delta p_N \end{bmatrix}^{\overset{(k+1)}{n+1}} = - \begin{bmatrix} R_1 \\ R_2 \\ R_3 \\ R_4 \\ \cdots \\ \cdots \\ R_N \end{bmatrix}^{\overset{(k)}{n+1}} \tag{7.110}$$

where "R_1" stands for residual function of gridblock 1 and p_1 is the pressure in gridblock one ($i = 1$), for example. Example 7.1 examines the study of a simple *1D* gas flow system as a way of illustration of the implementation of the Newton–Raphson methodology.

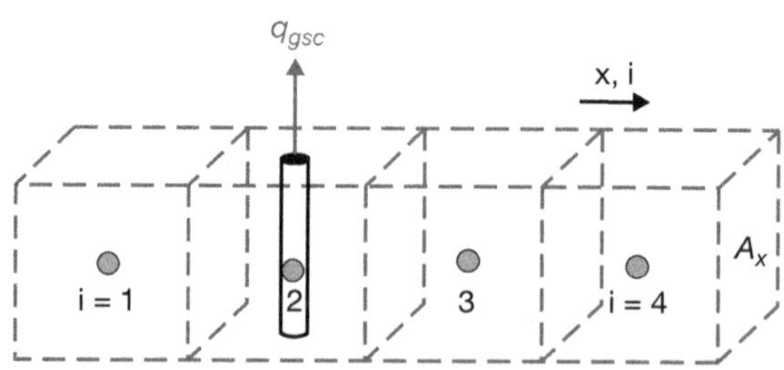

Figure 7.6 1D gas flow in a reservoir system.

Example 7.1: Flow in a *1D* gas reservoir: *Let us consider the case of a four-gridblock, 1D Cartesian flow depicted in Fig. 7.6. All external boundaries are no-flow and a well is producing at gridblock i = 2 at a constant bottomhole pressure. All reservoir, fluid, and well properties are known. Develop the Newton–Raphson solution for this compressible (gas) fluid scenario.*

The *implicit* finite-volume approximation of this *1D* Cartesian compressible flow problem may be written as:

$$E_i^{n+1}\left(p_{i+1}^{n+1} - p_i^{n+1}\right) + W_i^{n+1}\left(p_{i-1}^{n+1} - p_i^{n+1}\right) + q_{gsc\ i}^{n+1}$$

$$= \left(\frac{V_b}{\alpha_c \Delta t}\right)_i^{n+1} \left[(\phi/B_g)_i^{n+1} - (\phi/B_g)_i^n\right] \tag{7.111}$$

where $q_{gsc\ i}^{n+1} = -\Omega_i^{n+1}\left(p_i^{n+1} - p_{wf}^{n+1}\right)$ given the well specification (see Ertekin et al., 2001, for well model treatment and derivations).

To solve this problem using the Newton–Raphson formulation, the residual equation for this problem results from zeroing out Eq. (7.111), for each Newton–Raphson k-iteration:

$$\overset{(k)}{R_i^{n+1}} = \overset{(k)}{E_i^{n+1}}\left(\overset{(k+1)}{p_{i+1}^{n+1}} - \overset{(k+1)}{p_i^{n+1}}\right) + \overset{(k)}{W_i^{n+1}}\left(\overset{(k+1)}{p_{i-1}^{n+1}} - \overset{(k+1)}{p_i^{n+1}}\right)$$

$$+ \overset{(k+1)}{q_{gsc\ i}^{n+1}} - \left(\frac{V_b}{\alpha_c \Delta t}\right)_i^{n+1} \left[\overset{(k+1)}{(\phi/B_g)_i^{n+1}} - (\phi/B_g)_i^n\right] = 0 \tag{7.112}$$

which is the residual governing equation that must be satisfied by all gridblocks in Fig. 7.6. Since all external boundaries are no flow, $W_1 = E_4 = 0$ (zero transmissibility at the external boundaries). Therefore, four residual equations can be written for this system:

$$\overset{(k)}{R_1^{n+1}} = \overset{k}{E_1^{n+1}}\left(\overset{(k+1)}{p_2^{n+1}} - \overset{(k+1)}{p_1^{n+1}}\right)$$

$$- \left(\frac{V_b}{\alpha_c \Delta t}\right)_1^{n+1} \left[\overset{(k+1)}{(\phi/B_g)_1^{n+1}} - (\phi/B_g)_1^n\right] = 0 \tag{7.113}$$

$$\overset{(k)}{R_2^{n+1}} = \overset{(k)}{E_2^{n+1}}\left(\overset{(k+1)}{p_3^{n+1}} - \overset{(k+1)}{p_2^{n+1}}\right) + \overset{(k)}{W_2^{n+1}}\left(\overset{(k+1)}{p_1^{n+1}} - \overset{(k+1)}{p_2^{n+1}}\right)$$

$$- \overset{(k)}{\Omega_2^{n+1}}\left(\overset{(k+1)}{p_2^{n+1}} - p_{wf}^{n+1}\right) - \left(\frac{V_b}{\alpha_c \Delta t}\right)_2^{n+1}\left[\overset{(k+1)}{(\phi/B_g)_2^{n+1}} - (\phi/B_g)_2^n\right] = 0 \tag{7.114}$$

$$\overset{(k)}{R_3^{n+1}} = \overset{(k)}{E_3^{n+1}}\left(\overset{(k+1)}{p_4^{n+1}} - \overset{(k+1)}{p_3^{n+1}}\right) + \overset{(k)}{W_3^{n+1}}\left(\overset{(k+1)}{p_2^{n+1}} - \overset{(k+1)}{p_3^{n+1}}\right)$$

$$- \left(\frac{V_b}{\alpha_c \Delta t}\right)_3^{n+1}\left[\overset{(k+1)}{(\phi/B_g)_3^{n+1}} - (\phi/B_g)_3^n\right] = 0 \tag{7.115}$$

$$\overset{(k)}{R_4^{n+1}} = \overset{k}{W_4^{n+1}}\left(\overset{(k+1)}{p_3^{n+1}} - \overset{(k+1)}{p_4^{n+1}}\right)$$

$$- \left(\frac{V_b}{\alpha_c \Delta t}\right)_4^{n+1}\left[\overset{(k+1)}{(\phi/B_g)_4^{n+1}} - (\phi/B_g)_4^n\right] = 0 \tag{7.116}$$

that can be used to solve for four unknowns: $\overset{(k+1)}{p_1^{n+1}}$, $\overset{(k+1)}{p_2^{n+1}}$, $\overset{(k+1)}{p_3^{n+1}}$, and $\overset{(k+1)}{p_4^{n+1}}$, all gridblock pressures.

The Newton–Raphson solution system would then be written as:

$$\begin{bmatrix} \frac{\partial R_1}{\partial p_1} & \frac{\partial R_1}{\partial p_2} & 0 & 0 \\ \frac{\partial R_2}{\partial p_1} & \frac{\partial R_2}{\partial p_2} & \frac{\partial R_2}{\partial p_3} & 0 \\ 0 & \frac{\partial R_3}{\partial p_2} & \frac{\partial R_3}{\partial p_3} & \frac{\partial R_3}{\partial p_4} \\ 0 & 0 & \frac{\partial R_4}{\partial p_3} & \frac{\partial R_4}{\partial p_4} \end{bmatrix}^{\overset{(k)}{n+1}} \begin{bmatrix} \Delta p_1 \\ \Delta p_2 \\ \Delta p_3 \\ \Delta p_4 \end{bmatrix}^{\overset{(k+1)}{n+1}} = - \begin{bmatrix} R_1 \\ R_2 \\ R_3 \\ R_4 \end{bmatrix}^{\overset{(k)}{n+1}} \tag{7.117}$$

where the Jacobian zero entries correspond to locations where residual equations have no dependencies on associated unknowns (e.g., residual equation R_1 does not depend on gridblock pressures 3 or 4, and residual equation R_4 does not depend on gridblock pressures 1 or 2).

The Newton–Raphson protocol would proceed as follows:

- *Provide a first approximation (guess) to the solution* ($\overset{(k+1)}{p_1^{n+1}}$, $\overset{(k+1)}{p_2^{n+1}}$, $\overset{(k+1)}{p_3^{n+1}}$, and $\overset{(k+1)}{p_4^{n+1}}$). Typically, initial pressure for a reservoir study.
- *Evaluate the right-hand side of Eq. (7.117):* $-\overset{(k)}{R_1^{n+1}}$, $-\overset{(k)}{R_2^{n+1}}$, $-\overset{(k)}{R_3^{n+1}}$, and $-\overset{(k)}{R_4^{n+1}}$ using Eqs. (7.113) to (7.116) evaluated at current values for $\overset{(k)}{p_1^{n+1}}$, $\overset{(k)}{p_2^{n+1}}$, $\overset{(k)}{p_3^{n+1}}$, and $\overset{(k)}{p_4^{n+1}}$.
- *Calculate the entries of the Jacobian* $\nabla\vec{R}(\vec{p}^k)$. These entries may be calculated analytically. For the case at hand, for example, the entry $\frac{\partial R_2}{\partial p_2}$ would be computed as:

$$\frac{\partial R_2}{\partial p_2} = -\overset{(k)}{E_2^{n+1}} + \left(\overset{(k+1)}{p_3^{n+1}} - \overset{(k+1)}{p_2^{n+1}}\right)\frac{\partial \overset{(k)}{E_2^{n+1}}}{\partial p_2}$$
$$-\overset{(k)}{W_2^{n+1}} + \left(\overset{(k+1)}{p_1^{n+1}} - \overset{(k+1)}{p_2^{n+1}}\right)\frac{\partial \overset{(k)}{W_2^{n+1}}}{\partial p_2} - \overset{(k)}{\Omega_2^{n+1}} + \left(\overset{(k+1)}{p_2^{n+1}} - p_{wf}^{n+1}\right)\frac{\partial \overset{(k)}{\Omega_2^{n+1}}}{\partial p_2}$$
$$-\left(\frac{V_b}{\alpha_c \Delta t}\right)_2^{n+1}\frac{\partial}{\partial p_2}\left[\left(\frac{\phi}{B_g}\right)_2^{\overset{(k+1)}{n+1}}\right] \tag{7.118}$$

Alternatively, if analytical derivatives are cumbersome or undesirable to compute for a particular application, numerical Jacobian derivatives can be calculated instead using forward, backward, or central differences (Ertekin et al., 2001).

- *Solve for* $\overset{(k+1)}{\Delta\vec{p}^{\,n+1}}$ *in Eq. (7.117)*
- *Update the solution using recently obtained* $\Delta\vec{p}^{(k+1)}$:

$$\overset{(k+1)}{\vec{p}^{\,n+1}} = \overset{(k)}{\vec{p}^{\,n+1}} + \overset{(k+1)}{\Delta\vec{p}^{\,n+1}} \tag{7.119}$$

- *Recalculate Jacobian and residuals until convergence is attained.* Iterations stop when the largest absolute value of any newly calculated $\overset{(k+1)}{\Delta\vec{p}^{\,n+1}}$ is less than a prescribed and desirable tolerance (say, 10^{-5} psia). At this point, the desired pressure solution has been found.

7.5 FVM Representations of Multiphase Flow: Black-Oil Equations

7.5.1 Immiscible Liquid/Liquid (Oil/Water) Flow

Chapter 5 presented the governing equations for the simultaneous flow of two immiscible phases (oil and water) in a porous media. For the oil surface component, $I =$ "O", one writes:

$$\nabla \cdot \left(\frac{\overline{\overline{k}} k_{ro}}{\mu_o B_o} \nabla \Phi_o \right) + \frac{q_{osc}}{V_b} = \frac{1}{\alpha_c} \frac{\partial}{\partial t} \left(\phi \frac{S_o}{B_o} \right) \tag{7.120}$$

and for the water surface component, $I =$ "W":

$$\nabla \cdot \left(\frac{\overline{\overline{k}} k_{rw}}{\mu_w B_w} \nabla \Phi_w \right) + \frac{q_{wsc}}{V_b} = \frac{1}{\alpha_c} \frac{\partial}{\partial t} \left(\phi \frac{S_w}{B_w} \right) \tag{7.121}$$

Up to this point in this chapter, the only unknown of interest was just pressure since we were dealing with single-phase scenarios. In this multiphase (oil and water) flow scenario, however, one seeks to numerically solve these multiphase equations in terms of four unknowns: p_o, p_w, S_o, S_w, everywhere in the domain. To complete the set of four equations to solve for these unknowns, the following saturation and capillary constraints (auxiliary equations) are invoked:

$$S_o = 1 - S_w \tag{7.122}$$

$$p_w = p_o - p_{cow}(S_w) \tag{7.123}$$

Therefore, substituting the auxiliary Eqs. (7.122) and (7.123) into the oil and water component equations, the 3D Cartesian expansion of these equations may be written as:

$$\begin{aligned} &\frac{\partial}{\partial x}\left[\frac{k_x k_{ro}}{\mu_o B_o}\left(\frac{\partial p_o}{\partial x} - \gamma_o \frac{\partial G}{\partial x}\right)\right] + \frac{\partial}{\partial y}\left[\frac{k_y k_{ro}}{\mu_o B_o}\left(\frac{\partial p_o}{\partial y} - \gamma_o \frac{\partial G}{\partial y}\right)\right] \\ &+ \frac{\partial}{\partial z}\left[\frac{k_z k_{ro}}{\mu_o B_o}\left(\frac{\partial p_o}{\partial z} - \gamma_o \frac{\partial G}{\partial z}\right)\right] + \frac{q_{osc}}{V_b} = \frac{1}{\alpha_c}\frac{\partial}{\partial t}\left(\phi \frac{(1 - S_w)}{B_o}\right) \end{aligned} \tag{7.124}$$

$$\begin{aligned} &\frac{\partial}{\partial x}\left[\frac{k_x k_{rw}}{\mu_w B_w}\left(\frac{\partial p_o}{\partial x} - \frac{\partial p_{cow}}{\partial x} - \gamma_w \frac{\partial G}{\partial x}\right)\right] \\ &+ \frac{\partial}{\partial y}\left[\frac{k_y k_{rw}}{\mu_w B_w}\left(\frac{\partial p_o}{\partial y} - \frac{\partial p_{cow}}{\partial y} - \gamma_w \frac{\partial G}{\partial y}\right)\right] \\ &+ \frac{\partial}{\partial z}\left[\frac{k_z k_{rw}}{\mu_w B_w}\left(\frac{\partial p_o}{\partial z} - \frac{\partial p_{cow}}{\partial z} - \gamma_w \frac{\partial G}{\partial z}\right)\right] \\ &+ \frac{q_{wsc}}{V_b} = \frac{1}{\alpha_c}\frac{\partial}{\partial t}\left(\phi \frac{S_w}{B_w}\right) \end{aligned} \tag{7.125}$$

which is known as the p_o, S_w formulation for this problem, since the two auxiliary equations were used to eliminate the other two unknowns (p_g, S_o) from the original equations. The surviving unknowns (p_o, S_w) are referred to as the principal or primary unknowns for this set of equations.

The *FVM* integration of these governing equations leads to:

$$\begin{aligned}&\int_t^{t+\Delta t}\int_{V_b}\frac{\partial}{\partial x}\left[\frac{k_x k_{ro}}{\mu_o B_o}\left(\frac{\partial p_o}{\partial x}-\gamma_o\frac{\partial G}{\partial x}\right)\right]dVdt\\&+\int_t^{t+\Delta t}\int_{V_b}\frac{\partial}{\partial y}\left[\frac{k_y k_{ro}}{\mu_o B_o}\left(\frac{\partial p_o}{\partial y}-\gamma_o\frac{\partial G}{\partial y}\right)\right]dVdt\\&+\int_t^{t+\Delta t}\int_{V_b}\frac{\partial}{\partial z}\left[\frac{k_z k_{ro}}{\mu_o B_o}\left(\frac{\partial p_o}{\partial z}-\gamma_o\frac{\partial G}{\partial z}\right)\right]dVdt\\&+\frac{1}{V_b}\int_t^{t+\Delta t}\int_{V_b}q_{osc}dVdt=\frac{1}{\alpha_c}\int_t^{t+\Delta t}\int_{V_b}\frac{\partial}{\partial t}\left(\phi\frac{(1-S_w)}{B_o}\right)dVdt\end{aligned}\tag{7.126}$$

$$\begin{aligned}&\int_t^{t+\Delta t}\int_{V_b}\frac{\partial}{\partial x}\left[\frac{k_x k_{rw}}{\mu_w B_w}\left(\frac{\partial p_o}{\partial x}-\frac{\partial p_{cow}}{\partial x}-\gamma_w\frac{\partial G}{\partial x}\right)\right]dVdt\\&+\int_t^{t+\Delta t}\int_{V_b}\frac{\partial}{\partial y}\left[\frac{k_y k_{rw}}{\mu_w B_w}\left(\frac{\partial p_o}{\partial y}-\frac{\partial p_{cow}}{\partial y}-\gamma_w\frac{\partial G}{\partial y}\right)\right]dVdt\\&+\int_t^{t+\Delta t}\int_{V_b}\frac{\partial}{\partial z}\left[\frac{k_z k_{rw}}{\mu_w B_w}\left(\frac{\partial p_o}{\partial z}-\frac{\partial p_{cow}}{\partial z}-\gamma_w\frac{\partial G}{\partial z}\right)\right]dVdt\\&+\frac{1}{V_b}\int_t^{t+\Delta t}\int_{V_b}q_{wsc}dVdt=\frac{1}{\alpha_c}\int_t^{t+\Delta t}\int_{V_b}\frac{\partial}{\partial t}\left(\phi\frac{S_w}{B_w}\right)dVdt\end{aligned}\tag{7.127}$$

Following the *FVM* integration procedures discussed in the previous sections, we resolve spatial and time integrations and deploy central-difference approximations for pressure, capillary pressure, and depth gradients to obtain:

$$\begin{aligned}&T_{O,x_e}^m\left(p_{oi+1,j,k}^m-p_{oi,j,k}^m\right)+T_{O,x_w}^m\left(p_{oi-1,j,k}^m-p_{oi,j,k}^m\right)\\&-T_{O,x_e}^m\cdot\gamma_{ox_e}^m\left(G_{i+1,j,k}-G_{i,j,k}\right)-T_{O,x_w}^m\cdot\gamma_{ox_w}^m\left(G_{i-1,j,k}-G_{i,j,k}\right)\\&+T_{O,y_n}^m\left(p_{oi,j+1,k}^m-p_{oi,j,k}^m\right)+T_{O,y_s}^m\left(p_{oi,j-1,k}^m-p_{oi,j,k}^m\right)\\&-T_{O,y_n}^m\cdot\gamma_{oy_n}^m\left(G_{i,j+1,k}-G_{i,j,k}\right)-T_{O,y_s}^m\cdot\gamma_{oy_s}^m\left(G_{i,j-1,k}-G_{i,j,k}\right)\\&+T_{O,z_a}^m\left(p_{oi,j,k+1}^m-p_{oi,j,k}^m\right)+T_{O,z_b}^m\left(p_{oi,j,k-1}^m-p_{oi,j,k}^m\right)\\&-T_{O,z_a}^m\cdot\gamma_{oz_a}^m\left(G_{i,j,k+1}-G_{i,j,k}\right)-T_{O,z_b}^m\cdot\gamma_{oz_b}^m\left(G_{i,j,k-1}-G_{i,j,k}\right)\\&+q_{osci,j,k}^m=\left(\frac{V_b}{\alpha_c\Delta t}\right)_{i,j,k}^{n+1}\left\{\left(\phi\frac{1-S_w}{B_o}\right)_{i,j,k}^{n+1}-\left(\phi\frac{1-S_w}{B_o}\right)_{i,j,k}^{n}\right\}\end{aligned}\tag{7.128}$$

where oil-component transmissibilities in each direction are defined by the expressions:

$$T^m_{O,x_e} = \left[\frac{k_x A_x}{\Delta x}\frac{k_{ro}}{\mu_o B_o}\right]^m_{x_e} = E^m_{O,i,j,k} \tag{7.129}$$

$$T^m_{O,x_w} = \left[\frac{k_x A_x}{\Delta x}\frac{k_{ro}}{\mu_o B_o}\right]^m_{x_w} = W^m_{O,i,j,k} \tag{7.130}$$

$$T^m_{O,y_n} = \left[\frac{k_y A_y}{\Delta y}\frac{k_{ro}}{\mu_o B_o}\right]^m_{y_n} = N^m_{O,i,j,k} \tag{7.131}$$

$$T^m_{O,y_s} = \left[\frac{k_y A_y}{\Delta y}\frac{k_{ro}}{\mu_o B_o}\right]^m_{y_s} = S^m_{O,i,j,k} \tag{7.132}$$

$$T^m_{O,z_a} = \left[\frac{k_z A_z}{\Delta z}\frac{k_{ro}}{\mu_o B_o}\right]^m_{z_a} = A^m_{O,i,j,k} \tag{7.133}$$

$$T^m_{O,z_b} = \left[\frac{k_z A_z}{\Delta z}\frac{k_{ro}}{\mu_o B_o}\right]^m_{z_b} = B^m_{O,i,j,k} \tag{7.134}$$

Performing the same *FVM* integrations for the water component flow equation, one obtains:

$$\begin{aligned}
&T^m_{W,x_e}\left(p^m_{oi+1,j,k} - p^m_{oi,j,k}\right) + T^m_{W,x_w}\left(p^m_{oi-1,j,k} - p^m_{oi,j,k}\right)\\
&- T^m_{W,x_e}\left(p^m_{cowi+1,j,k} - p^m_{cowi,j,k}\right) - T^m_{W,x_w}\left(p^m_{cowi-1,j,k} - p^m_{cowi,j,k}\right)\\
&- T^m_{W,x_e}\cdot\gamma^m_{wx_e}\left(G_{i+1,j,k} - G_{i,j,k}\right) - T^m_{W,x_w}\cdot\gamma^m_{wx_w}\left(G_{i-1,j,k} - G_{i,j,k}\right)\\
&+ T^m_{W,y_n}\left(p^m_{oi,j+1,k} - p^m_{oi,j,k}\right) + T^m_{W,y_s}\left(p^m_{oi,j-1,k} - p^m_{oi,j,k}\right)\\
&- T^m_{W,y_n}\left(p^m_{cowi,j+1,k} - p^m_{cowi,j,k}\right) - T^m_{W,y_s}\left(p^m_{cowi,j-1,k} - p^m_{cowi,j,k}\right)\\
&- T^m_{W,y_n}\cdot\gamma^m_{wy_n}\left(G_{i,j+1,k} - G_{i,j,k}\right) - T^m_{W,y_s}\cdot\gamma^m_{wy_s}\left(G_{i,j-1,k} - G_{i,j,k}\right)\\
&+ T^m_{W,z_a}\left(p^m_{oi,j,k+1} - p^m_{oi,j,k}\right) + T^m_{W,z_b}\left(p^m_{oi,j,k-1} - p^m_{oi,j,k}\right)\\
&- T^m_{W,z_a}\left(p^m_{cowi,j,k+1} - p^m_{cowi,j,k}\right) - T^m_{W,z_b}\left(p^m_{cowi,j,k-1} - p^m_{cowi,j,k}\right)\\
&- T^m_{W,z_a}\cdot\gamma^m_{wz_a}\left(G_{i,j,k+1} - G_{i,j,k}\right) - T^m_{W,z_b}\cdot\gamma^m_{wz_b}\left(G_{i,j,k-1} - G_{i,j,k}\right)\\
&+ q^m_{wsci,j,k} = \left(\frac{V_b}{\alpha_c \Delta t}\right)^{n+1}_{i,j,k}\left\{\left(\frac{\phi S_w}{B_w}\right)^{n+1}_{i,j,k} - \left(\frac{\phi S_w}{B_w}\right)^{n}_{i,j,k}\right\}
\end{aligned} \tag{7.135}$$

where water-component transmissibilities are defined by the expressions:

$$T^m_{W,x_e} = \left[\frac{k_x A_x}{\Delta x}\frac{k_{rw}}{\mu_w B_w}\right]^m_{x_e} = E^m_{W,i,j,k} \tag{7.136}$$

$$T^m_{W,x_w} = \left[\frac{k_x A_x}{\Delta x}\frac{k_{rw}}{\mu_w B_w}\right]^m_{x_w} = W^m_{W,i,j,k} \tag{7.137}$$

$$T^m_{W,y_n} = \left[\frac{k_y A_y}{\Delta y}\frac{k_{rw}}{\mu_w B_w}\right]^m_{y_n} = N^m_{W,i,j,k} \tag{7.138}$$

$$T^m_{W,y_s} = \left[\frac{k_y A_y}{\Delta y}\frac{k_{rw}}{\mu_w B_w}\right]^m_{y_s} = S^m_{W,i,j,k} \tag{7.139}$$

$$T^m_{W,z_a} = \left[\frac{k_z A_z}{\Delta z}\frac{k_{rw}}{\mu_w B_w}\right]^m_{z_a} = A^m_{W,i,j,k} \tag{7.140}$$

$$T^m_{W,z_b} = \left[\frac{k_z A_z}{\Delta z}\frac{k_{rw}}{\mu_w B_w}\right]^m_{z_b} = B^m_{W,i,j,k} \tag{7.141}$$

and where $m = n + 1$ *(new time level)* or n *(old time level)* depending on the degree of implicitness or explicitness of the formulation. Equations (7.128) to (7.134) and Eqs. (7.135) to (7.141) represent the FVM numerical approximations deployed in the solution of the immiscible liquid/liquid (oil/water) flow problem based on the p_o, S_w formulation.

Multiphase Transmissibility Calculations

The type of transmissibility coefficients that appear in discretized multiphase equations contain grid-dependent, pressure-dependent, and *saturation*-dependent variables. The variables within oil transmissibility calculations in the x-direction at $x = x_e$ (east boundary), for example, may thus be grouped in three terms as follows:

$$T^m_{O,x_e} = E^m_{O,i,j,k} = \left[\frac{k_x A_x}{\Delta x}\frac{k_{ro}}{\mu_o B_o}\right]^m_{x_e} = \left.\frac{k_x A_x}{\Delta x}\right|_{x_e} \cdot \left.\frac{1}{\mu_o B_o}\right|^m_{x_e} \cdot \left.k_{ro}\right|^m_{x_e} \tag{7.142}$$

These three groups are: a grid-related transmissibility component $\left(\frac{k_x A_x}{\Delta x}\right)$, a pressure-dependent transmissibility component $\left(\frac{1}{\mu_o B_o}\right)$, and a *saturation*-dependent transmissibility component (k_{ro}). Time-invariant, grid-related transmissibilities will continue to be evaluated using harmonic averages as used in single-phase simulation, and pressure-dependent components are evaluated using volume-averaged boundary pressure between two blocks.

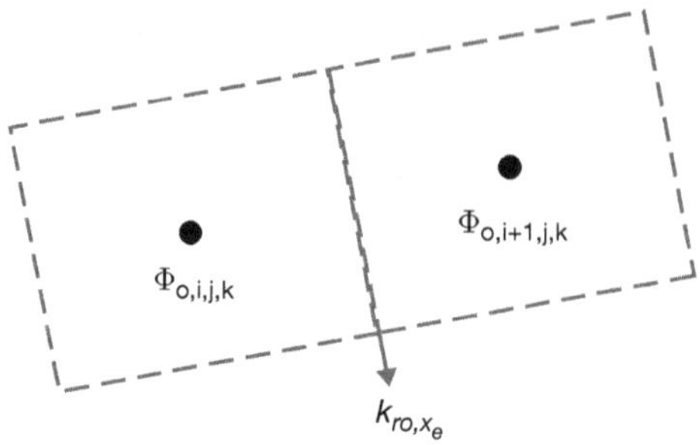

FIGURE 7.7 One-point upstream weighting for k_{ro,x_e}.

Saturation-dependent transmissibility variables, however, represent very strong nonlinearities which cannot be properly captured using arithmetic or volume-averaged values at block boundaries. The customary approach to capture these strong nonlinearities at the interface of blocks is the use of upstream weighting. The simplest and most popular application is the single-point upstream weighting. In this approach, relative permeabilities (or any other strong nonlinearity) are evaluated at the conditions of the gridblock found upstream from the flow. If, for example, one needs to calculate the relative permeability to oil at the interface $x = x_e$ in Fig. 7.7, oil flow potentials are calculated at each gridblock at each side of the interface and the following assignments are made:

If $\Phi_{o,i} > \Phi_{o,i+1}$, *flow of oil phase is from (i,j,k) to (i + 1,j,k) and thus:*

$$k_{ro,x_e} = k_{ro}\left(S_{o,i,j,k}\right) \tag{7.143}$$

If $\Phi_{o,i} < \Phi_{o,i+1}$+, *flow of oil phase is from (i + 1,j,k) to (i,j,k), and:*

$$k_{ro,x_e} = k_{ro}\left(S_{o,i+1,j,k}\right) \tag{7.144}$$

7.5.2 Extended and Standard Black-Oil Formulations

Chapter 5 demonstrated that the governing equations for the flow of three components and three phases using the *extended* black-oil model are:

$$\nabla \cdot \left[\frac{\bar{\bar{k}} k_{ro}}{\mu_o B_o} \nabla \Phi_o + R_v \frac{\bar{\bar{k}} k_{rg}}{\mu_g B_g} \nabla \Phi_g \right] + \frac{1}{V_b}\left(q_{osc} + R_v q_{gsc}\right) = \frac{1}{\alpha_c} \frac{\partial}{\partial t}\left[\phi \left(\frac{S_o}{B_o} + R_v \frac{S_g}{B_g} \right) \right] \tag{7.145}$$

$$\nabla \cdot \left[\frac{\bar{\bar{k}} k_{rg}}{\mu_g B_g} \nabla \Phi_g + R_s \frac{\bar{\bar{k}} k_{ro}}{\mu_o B_o} \nabla \Phi_o \right] + \frac{1}{V_b}\left(q_{gsc} + R_s q_{osc}\right) = \frac{1}{\alpha_c} \frac{\partial}{\partial t}\left[\phi \left(\frac{S_g}{B_g} + R_s \frac{S_o}{B_o} \right) \right] \tag{7.146}$$

$$\nabla \cdot \left[\frac{\bar{\bar{k}} k_{rw}}{\mu_w B_w} \nabla \Phi_w \right] + \frac{q_{wsc}}{V_b} = \frac{1}{\alpha_c} \frac{\partial}{\partial t}\left[\phi \left(\frac{S_w}{B_w} \right) \right] \tag{7.147}$$

As discussed in Example 5.3 (see Chap. 5), this is the most comprehensive black-oil scenario where flows of three phases ($N_p = 3, J =$ "o", "g", and "w") and three (pseudo-) components ($N_c = 3$, $I =$ "O", "G", and "W") are considered. In total, there are six unknowns of interest: all phase pressures, (p_o, p_g, and p_w), and all phase saturations (S_o, S_g and S_w). Three auxiliary equations must be included in this set in order to balance number of equations to that of unknowns: the phase saturation constraint, and the gas/oil and water/oil capillary pressure relationships. These are:

$$S_o + S_g + S_w = 1 \tag{7.148}$$

$$p_{cgo}(S_g) = p_g - p_o \tag{7.149}$$

$$p_{cow}(S_w) = p_o - p_w \tag{7.150}$$

Substitution of auxiliary Eqs. (7.148), (7.149), and (7.150) into the main flow Eqs. (7.145), (7.146), and (7.147) allows reducing this system to one of three equations in three primary unknowns. A typical selection of primary unknowns is one which rewrites this system in terms of the so-called p_o, S_w, *and* S_g formulation, where p_o, S_w, *and* S_g become the principal unknowns. The other saturation (S_o) may be removed from the flow equations via Eq. (7.148). The other pressures (p_g *and* p_w) can be eliminated from the flow equations by substituting Eqs. (7.149) and (7.150). Therefore, Eqs. (7.145), (7.146), and (7.147) can be rewritten as follows in a p_o, S_w, *and* S_g formulation, respectively:

$$\nabla \cdot \left[\frac{\bar{\bar{k}} k_{ro}}{\mu_o B_o} (\nabla p_o - \gamma_o \nabla G) + R_v \frac{\bar{\bar{k}} k_{rg}}{\mu_g B_g} (\nabla p_o + \nabla p_{cgo} - \gamma_g \nabla G) \right] + \frac{1}{V_b} (q_{osc} + R_v q_{gsc}) = \frac{1}{\alpha_c} \frac{\partial}{\partial t} \left[\phi \left(\frac{(1 - S_w - S_g)}{B_o} + R_v \frac{S_g}{B_g} \right) \right] \tag{7.151}$$

$$\nabla \cdot \left[\frac{\bar{\bar{k}} k_{rg}}{\mu_g B_g} (\nabla p_o + \nabla p_{cgo} - \gamma_g \nabla G) + R_s \frac{\bar{\bar{k}} k_{ro}}{\mu_o B_o} (\nabla p_o - \gamma_o \nabla G) \right] + \frac{1}{V_b} (q_{gsc} + R_s q_{osc}) = \frac{1}{\alpha_c} \frac{\partial}{\partial t} \left[\phi \left(\frac{S_g}{B_g} + R_s \frac{(1 - S_w - S_g)}{B_o} \right) \right] \tag{7.152}$$

$$\nabla \cdot \left[\frac{\bar{\bar{k}} k_{rw}}{\mu_w B_w} (\nabla p_o - \nabla p_{cow} - \gamma_w \nabla G) \right] + \frac{q_{wsc}}{V_b} = \frac{1}{\alpha_c} \frac{\partial}{\partial t} \left[\phi \left(\frac{S_w}{B_w} \right) \right] \tag{7.153}$$

where flow potentials have also been rewritten in terms of phase pressure and depth gradients. It should be noted that the *standard* black-oil formulation for oil and gas flow results from forcing $R_v = 0$ in Eqs. (7.145) and (7.151), which would collapse them back to the oil equation shown in the previous section [Eq. (7.120)] for the immiscible oil/water case.

The *FVM* integration of the *3D* Cartesian expansion of the *surface-oil* flow governing equation [Eq. (7.151)] leads to:

$$\int_t^{t+\Delta t}\int_{V_b}\frac{\partial}{\partial x}\left[\frac{k_x k_{ro}}{\mu_o B_o}\left(\frac{\partial p_o}{\partial x}-\gamma_o\frac{\partial G}{\partial x}\right)+R_v\frac{k_x k_{rg}}{\mu_g B_g}\left(\frac{\partial p_o}{\partial x}+\frac{\partial p_{cgo}}{\partial x}-\gamma_g\frac{\partial G}{\partial x}\right)\right]dVdt$$

$$+\int_t^{t+\Delta t}\int_{V_b}\frac{\partial}{\partial y}\left[\frac{k_y k_{ro}}{\mu_o B_o}\left(\frac{\partial p_o}{\partial y}-\gamma_o\frac{\partial G}{\partial y}\right)+R_v\frac{k_y k_{rg}}{\mu_g B_g}\left(\frac{\partial p_o}{\partial y}+\frac{\partial p_{cgo}}{\partial y}-\gamma_g\frac{\partial G}{\partial y}\right)\right]dVdt$$

$$+\int_t^{t+\Delta t}\int_{V_b}\frac{\partial}{\partial z}\left[\frac{k_z k_{ro}}{\mu_o B_o}\left(\frac{\partial p_o}{\partial z}-\gamma_o\frac{\partial G}{\partial z}\right)+R_v\frac{k_z k_{rg}}{\mu_g B_g}\left(\frac{\partial p_o}{\partial z}+\frac{\partial p_{cgo}}{\partial z}-\gamma_g\frac{\partial G}{\partial z}\right)\right]dVdt$$

$$+\frac{1}{V_b}\int_t^{t+\Delta t}\int_{V_b}\left(q_{osc}+R_v q_{gsc}\right)dVdt=\frac{1}{\alpha_c}\int_t^{t+\Delta t}\int_{V_b}\frac{\partial}{\partial t}\left(\phi\left(\frac{\left(1-S_w-S_g\right)}{B_o}+R_v\frac{S_g}{B_g}\right)\right)dVdt \tag{7.154}$$

By resolving the stated spatial and time integrations, and deploying central-difference approximations for the resulting pressure, capillary pressure, and depth gradients, one obtains:

$$T^m_{O,x_e}\left(p^m_{oi+1,j,k}-p^m_{oi,j,k}\right)+T^m_{O,x_w}\left(p^m_{oi-1,j,k}-p^m_{oi,j,k}\right)$$

$$-T^m_{O,x_e}\cdot\gamma^m_{ox_e}\left(G_{i+1,j,k}-G_{i,j,k}\right)-T^m_{O,x_w}\cdot\gamma^m_{ox_w}\left(G_{i-1,j,k}-G_{i,j,k}\right)$$

$$+R^m_{vx_e}\cdot T^m_{G,x_e}\left(p^m_{oi+1,j,k}-p^m_{oi,j,k}\right)+R^m_{vx_w}\cdot T^m_{G,x_w}\left(p^m_{oi-1,j,k}-p^m_{oi,j,k}\right)$$

$$+R^m_{vx_e}\cdot T^m_{G,x_e}\left(p^m_{cgoi+1,j,k}-p^m_{cgoi,j,k}\right)+R^m_{vx_w}\cdot T^m_{G,x_w}\left(p^m_{cgoi-1,j,k}-p^m_{cgoi,j,k}\right)$$

$$-R^m_{vx_e}\cdot T^m_{G,x_e}\cdot\gamma^m_{gx_e}\left(G_{i+1,j,k}-G_{i,j,k}\right)-R^m_{vx_w}\cdot T^m_{G,x_w}\cdot\gamma^m_{gx_w}\left(G_{i-1,j,k}-G_{i,j,k}\right)$$

$$+T^m_{O,y_n}\left(p^m_{oi,j+1,k}-p^m_{oi,j,k}\right)+T^m_{O,y_s}\left(p^m_{oi,j-1,k}-p^m_{oi,j,k}\right)$$

$$-T^m_{O,y_n}\cdot\gamma^m_{oy_n}\left(G_{i,j+1,k}-G_{i,j,k}\right)-T^m_{O,y_s}\cdot\gamma^m_{oy_s}\left(G_{i,j-1,k}-G_{i,j,k}\right)$$

$$+R^m_{vy_n}\cdot T^m_{G,y_n}\left(p^m_{oi,j+1,k}-p^m_{oi,j,k}\right)+R^m_{vy_s}\cdot T^m_{G,y_s}\left(p^m_{oi,j-1,k}-p^m_{oi,j,k}\right)$$

$$+R^m_{vy_n}\cdot T^m_{G,y_n}\left(p^m_{cgoi,j+1,k}-p^m_{cgoi,j,k}\right)+R^m_{vy_s}\cdot T^m_{G,y_s}\left(p^m_{cgoi,j-1,k}-p^m_{cgoi,j,k}\right)$$

$$-R^m_{vy_n}\cdot T^m_{G,y_n}\cdot\gamma^m_{gy_n}\left(G_{i,j+1,k}-G_{i,j,k}\right)-R^m_{vy_s}\cdot T^m_{G,y_s}\cdot\gamma^m_{gy_s}\left(G_{i,j-1,k}-G_{i,j,k}\right)$$

$$+T^m_{O,z_a}\left(p^m_{oi,j,k+1}-p^m_{oi,j,k}\right)+T^m_{O,z_b}\left(p^m_{oi,j,k-1}-p^m_{oi,j,k}\right)$$

$$-T^m_{O,z_a}\cdot\gamma^m_{oz_a}\left(G_{i,j,k+1}-G_{i,j,k}\right)-T^m_{O,z_b}\cdot\gamma^m_{oz_b}\left(G_{i,j,k-1}-G_{i,j,k}\right)$$

$$+R^m_{vz_a}\cdot T^m_{G,z_a}\left(p^m_{oi,j,k+1}-p^m_{oi,j,k}\right)+R^m_{vz_b}\cdot T^m_{G,z_b}\left(p^m_{oi,j,k-1}-p^m_{oi,j,k}\right)$$

$$+R_{vz_a}^{m} \cdot T_{G,z_a}^{m}\left(p_{cgoi,j,k+1}^{m} - p_{cgoi,j,k}^{m}\right) + R_{vz_b}^{m} \cdot T_{G,z_b}^{m}\left(p_{cgoi,j,k-1}^{m} - p_{cgoi,j,k}^{m}\right)$$

$$-R_{vz_a}^{m} \cdot T_{G,z_a}^{m} \cdot \gamma_{gz_a}^{m}\left(G_{i,j,k+1} - G_{i,j,k}\right) - R_{vz_b}^{m} \cdot T_{G,z_b}^{m} \cdot \gamma_{gz_b}^{m}\left(G_{i,j,k-1} - G_{i,j,k}\right)$$

$$+\left(q_{osc} + R_v q_{gsc}\right)_{i,j,k}^{m}$$

$$= \left(\frac{V_b}{\alpha_c \Delta t}\right)_{i,j,k}^{n+1} \left\{\left(\phi\frac{1 - S_w - S_w}{B_o} + R_v\frac{\phi S_g}{B_g}\right)_{i,j,k}^{n+1} - \left(\phi\frac{1 - S_w - S_w}{B_o} + R_v\frac{\phi S_g}{B_g}\right)_{i,j,k}^{n}\right\} \qquad (7.155)$$

The same *FVM* integration and approach applied to the surface-gas Eq. (7.152) yields the following *FVM* approximation for the gas component:

$$T_{G,x_e}^{m}\left(p_{oi+1,j,k}^{m} - p_{oi,j,k}^{m}\right) + T_{G,x_w}^{m}\left(p_{oi-1,j,k}^{m} - p_{oi,j,k}^{m}\right)$$

$$+T_{G,x_e}^{m}\left(p_{cgoi+1,j,k}^{m} - p_{cgoi,j,k}^{m}\right) + T_{G,x_w}^{m}\left(p_{cgoi-1,j,k}^{m} - p_{cgoi,j,k}^{m}\right)$$

$$-T_{G,x_e}^{m} \cdot \gamma_{gx_e}^{m}\left(G_{i+1,j,k} - G_{i,j,k}\right) - T_{G,x_w}^{m} \cdot \gamma_{gx_w}^{m}\left(G_{i-1,j,k} - G_{i,j,k}\right)$$

$$+R_{sx_e}^{m} \cdot T_{O,x_e}^{m}\left(p_{oi+1,j,k}^{m} - p_{oi,j,k}^{m}\right) + R_{sx_w}^{m} \cdot T_{O,x_w}^{m}\left(p_{oi-1,j,k}^{m} - p_{oi,j,k}^{m}\right)$$

$$-R_{sx_e}^{m} \cdot T_{O,x_e}^{m} \cdot \gamma_{ox_e}^{m}\left(G_{i+1,j,k} - G_{i,j,k}\right) - R_{sx_w}^{m} \cdot T_{O,x_w}^{m} \cdot \gamma_{ox_w}^{m}\left(G_{i-1,j,k} - G_{i,j,k}\right)$$

$$+T_{G,y_n}^{m}\left(p_{oi,j+1,k}^{m} - p_{oi,j,k}^{m}\right) + T_{G,y_s}^{m}\left(p_{oi,j-1,k}^{m} - p_{oi,j,k}^{m}\right)$$

$$+T_{G,y_n}^{m}\left(p_{cgoi,j+1,k}^{m} - p_{cgoi,j,k}^{m}\right) + T_{G,y_s}^{m}\left(p_{cgoi,j-1,k}^{m} - p_{cgoi,j,k}^{m}\right)$$

$$-T_{G,y_n}^{m} \cdot \gamma_{gy_n}^{m}\left(G_{i,j+1,k} - G_{i,j,k}\right) - T_{G,y_s}^{m} \cdot \gamma_{gy_s}^{m}\left(G_{i,j-1,k} - G_{i,j,k}\right)$$

$$+R_{sy_n}^{m} \cdot T_{O,y_n}^{m}\left(p_{oi,j+1,k}^{m} - p_{oi,j,k}^{m}\right) + R_{sy_s}^{m} \cdot T_{O,y_s}^{m}\left(p_{oi,j-1,k}^{m} - p_{oi,j,k}^{m}\right)$$

$$R_{sy_n}^{m} \cdot T_{O,y_n}^{m} \cdot \gamma_{oy_n}^{m}\left(G_{i,j+1,k} - G_{i,j,k}\right) - R_{sy_s}^{m} \cdot T_{O,y_s}^{m} \cdot \gamma_{oy_s}^{m}\left(G_{i,j-1,k} - G_{i,j,k}\right)$$

$$+T_{G,z_a}^{m}\left(p_{oi,j,k+1}^{m} - p_{oi,j,k}^{m}\right) + T_{G,z_b}^{m}\left(p_{oi,j,k-1}^{m} - p_{oi,j,k}^{m}\right)$$

$$+T_{G,z_a}^{m}\left(p_{cgoi,j,k+1}^{m} - p_{cgoi,j,k}^{m}\right) + T_{G,z_b}^{m}\left(p_{cgoi,j,k-1}^{m} - p_{cgoi,j,k}^{m}\right)$$

$$-T_{G,z_a}^{m} \cdot \gamma_{gz_a}^{m}\left(G_{i,j,k+1} - G_{i,j,k}\right) - T_{G,z_b}^{m} \cdot \gamma_{gz_b}^{m}\left(G_{i,j,k-1} - G_{i,j,k}\right)$$

$$+R_{sz_a}^{m} \cdot T_{O,z_a}^{m}\left(p_{oi,j,k+1}^{m} - p_{oi,j,k}^{m}\right) + R_{sz_b}^{m} \cdot T_{O,z_b}^{m}\left(p_{oi,j,k-1}^{m} - p_{oi,j,k}^{m}\right)$$

$$-R_{sz_a}^{m} \cdot T_{O,z_a}^{m} \cdot \gamma_{oz_a}^{m}\left(G_{i,j,k+1} - G_{i,j,k}\right) - R_{sz_b}^{m} \cdot T_{O,z_b}^{m} \cdot \gamma_{oz_b}^{m}\left(G_{i,j,k-1} - G_{i,j,k}\right)$$

$$+\left(q_{gsc} + R_s q_{osc}\right)_{i,j,k}^{m}$$

$$= \left(\frac{V_b}{\alpha_c \Delta t}\right)_{i,j,k}^{n+1} \left\{\left(\frac{\phi S_g}{B_g} + R_s\phi\frac{1 - S_w - S_w}{B_o}\right)_{i,j,k}^{n+1} - \left(\frac{\phi S_g}{B_g} + R_s\phi\frac{1 - S_w - S_w}{B_o}\right)_{i,j,k}^{n}\right\} \qquad (7.156)$$

and where $m = n + 1$ *(new time level)* or n *(old time level)* depending on the degree of implicitness or explicitness of the formulation. Oil component transmissibilities were defined by Eqs. (7.129) to (7.134), and gas-component transmissibilities are analogously given by:

$$T^{m}_{G,x_e} = \left[\frac{k_x A_x}{\Delta x}\frac{k_{rg}}{\mu_g B_g}\right]^{m}_{x_e} = E^{m}_{G,i,j,k} \tag{7.157}$$

$$T^{m}_{G,x_w} = \left[\frac{k_x A_x}{\Delta x}\frac{k_{rg}}{\mu_g B_g}\right]^{m}_{x_w} = W^{m}_{G,i,j,k} \tag{7.158}$$

$$T^{m}_{G,y_n} = \left[\frac{k_y A_y}{\Delta y}\frac{k_{rg}}{\mu_g B_g}\right]^{m}_{y_n} = N^{m}_{G,i,j,k} \tag{7.159}$$

$$T^{m}_{G,y_s} = \left[\frac{k_y A_y}{\Delta y}\frac{k_{rg}}{\mu_g B_g}\right]^{m}_{y_s} = S^{m}_{G,i,j,k} \tag{7.160}$$

$$T^{m}_{G,z_a} = \left[\frac{k_z A_z}{\Delta z}\frac{k_{rg}}{\mu_g B_g}\right]^{m}_{z_a} = A^{m}_{G,i,j,k} \tag{7.161}$$

$$T^{m}_{G,z_b} = \left[\frac{k_z A_z}{\Delta z}\frac{k_{rg}}{\mu_g B_g}\right]^{m}_{z_b} = B^{m}_{G,i,j,k} \tag{7.162}$$

In all these multiphase transmissibility calculations at each gridblock interface, saturation-dependent variables such as relative permeabilities are evaluated using the one-point upstream weighting technique described in the previous sections for most cases. The rest of the coefficients next to transmissibility coefficients (e.g., R_s, R_v, γ_o, γ_g) are treated as weaker (pressure-dependent) nonlinearities and thus are calculated at the volumetrically averaged pressure prevailing at the interface.

For the water component Eq. (7.153), the *FVM* numerical approximation remains identical to the one presented in the preceding section as Eq. (7.135). Altogether, Eqs. (7.155), (7.156), and (7.135) represent the *FVM* numerical approximations deployed in the solution of the three-phase (oil/gas/water) flow problem using the *extended* black-oil model and the p_o, S_w, S_g formulation. It should again be noted that the *standard* black-oil formulation would result from forcing $R_v = 0$ in Eq. (7.155). Equations (7.156) and (7.135) would remain the same.

7.6 Solution of Black-Oil Multiphase Numerical Equations

Once the multiphase numerical equations have been formulated, the problem left is that of their simultaneous solution. In here, we will discuss the most widely used methods deployed for solving equations of multiphase flow problems in hydrocarbon reservoir modeling. These are:

- *IMPES* **(Implicit Pressure, Explicit Saturation):** A partially explicit solution technique. Nodal pressures are solved *implicitly* but saturations are solved *explicitly*. Its basic premise is *elimination of non-pressure time-derivatives from the set of conservation equations and obtaining a single pressure equation.*

- **Generalized Newton–Raphson Procedure (*NR*):** A versatile and fully implicit solution technique, covered in Section 7.4 for the case of solving compressible fluid nonlinear equations. We will revisit it here for the case of multiphase equations.

7.6.1. Implicit Pressure, Explicit Saturation *(IMPES)* Method

Without a doubt, the least computationally expensive and faster multiphase solution method is the *IMPES* approach. For a reservoir flow problem with "*N*" total gridblocks, the *IMPES* approach solves a set of "*N*" equations simultaneously (one unknown, pressure, per block). The Newton–Raphson procedure, however, solves a set of "N_c*N" equations simultaneously, where N_c is the number of principal unknowns per block which equals the number of pseudo-components in the fluid mixture.

The main idea behind *IMPES* is collapsing the set of multiphase flow equations to a *single pressure-equation for each numerical gridblock,* and this is done by a linear combination of flow equations that allows the elimination of saturation unknowns. After pressure has been solved for implicitly and advanced in time, saturations are updated explicitly by substituting the available pressures at time level $n + 1$. This is illustrated in Example 7.3 for the case of a simple *1D* waterflooding project.

Example 7.2: 1D Waterflooding project—*IMPES* solution: *Let us consider the 1D waterflooding (oil/water flow) project into a horizontal reservoir depicted in Fig. 7.8. One water injector and one producer are placed in the system as depicted. All external boundaries are no-flow. Derive the pressure and saturation solution equation for the IMPES method with the assumption of zero compressibility for both fluids and rock, zero capillary pressure, and no depth gradients.*

For this problem, Eqs. (7.120) and (7.121) are significantly simplified on the basis of the stated assumptions. They hence become:

$$\frac{\partial}{\partial x}\left[\frac{k_x k_{ro}}{\mu_o B_o}\frac{\partial p}{\partial x}\right] + \frac{q_{osc}}{V_b} = \frac{1}{\alpha_c}\frac{\phi}{B_o}\frac{\partial S_o}{\partial t} \tag{7.163}$$

$$\frac{\partial}{\partial x}\left[\frac{k_x k_{rw}}{\mu_w B_w}\frac{\partial p}{\partial x}\right] + \frac{q_{wsc}}{V_b} = \frac{1}{\alpha_c}\frac{\phi}{B_w}\frac{\partial S_w}{\partial t} \tag{7.164}$$

The IMPES method aims to eliminate the saturation unknowns in the accumulation term (right-hand side) of these equations. It seeks to identify a multiplier, say "*A*",

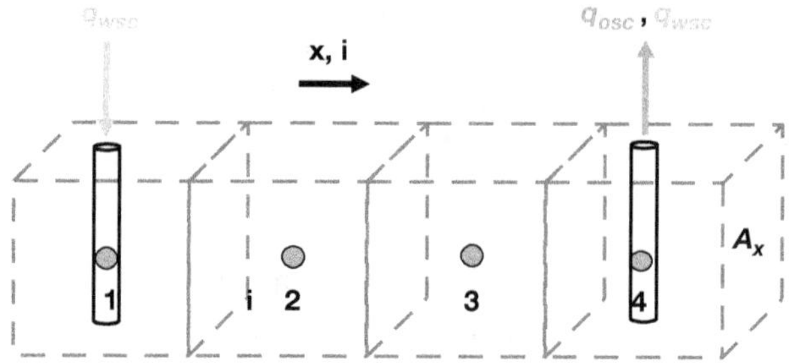

FIGURE 7.8 1D waterflooding project into a reservoir.

that once applied to one of the equations (say, the oil equation) would make saturation terms disappear once both equations are added. In other words, we seek a value of "A" such that:

$$A \cdot \frac{\phi}{B_o}\frac{\partial S_o}{\partial t} + \frac{\phi}{B_w}\frac{\partial S_w}{\partial t} = 0 \qquad (7.165)$$

or, since $\frac{\partial S_w}{\partial t} = -\frac{\partial S_o}{\partial t}$ given the saturation constraint in this problem:

$$\left[A \cdot \frac{\phi}{B_o} - \frac{\phi}{B_w}\right]\frac{\partial S_o}{\partial t} = 0 \qquad (7.166)$$

therefore, by solving for which value of "A" makes this true, $A = \frac{B_o}{B_w}$ is the multiplier we seek. By multiplying the oil equation in (7.163) by $\frac{B_o}{B_w}$ and adding it to the water equation in (7.164),

$$\frac{B_o}{B_w} \cdot \frac{\partial}{\partial x}\left[\frac{k_x k_{ro}}{\mu_o B_o}\frac{\partial p}{\partial x}\right] + \frac{\partial}{\partial x}\left[\frac{k_x k_{rw}}{\mu_w B_w}\frac{\partial p}{\partial x}\right] + \frac{B_o}{B_w} \cdot \frac{q_{osc}}{V_b} + \frac{q_{wsc}}{V_b} =$$

$$\frac{1}{\alpha_c}\overbrace{\left\{\frac{B_o}{B_w} \cdot \frac{\phi}{B_o}\frac{\partial S_o}{\partial t} + + \frac{\phi}{B_w}\frac{\partial S_w}{\partial t}\right\}}^{0} = 0 \qquad (7.167)$$

Equation (7.167) is the *IMPES* pressure equation, where all saturation time derivatives have been eliminated. By integrating this pressure equation over the control volume, the *FVM* approximation for the *IMPES* pressure equation may be written as:

$$\left(\frac{B_o}{B_w}\right)_i^m \left\{T_{O,x_e}^m\left(p_{i+1,}^{n+1}-p_i^{n+1}\right)+T_{O,x_w}^m\left(p_{i-1}^{n+1}-p_i^{n+1}\right)\right\}$$

$$+T_{W,x_e}^m\left(p_{i+1,}^{n+1}-p_i^{n+1}\right)+T_{W,x_w}^m\left(p_{i-1}^{n+1}-p_i^{n+1}\right)$$

$$+\left(\frac{B_o}{B_w}\right)_i^m \left(q_{osc}\right)_i^m+\left(q_{wsc}\right)_i^m=0 \tag{7.168}$$

which is *implicit* in pressure in view of the implicit nature of the pressure solution in *IMPES*. Saturation-dependent terms are always evaluated explicitly (time step n), the rest of the terms may be evaluated at time step $m = n$ (explicitly) to avoid iterations or $m =$ "$n + 1, k$" (one iteration level behind). Equation (7.168) can be rewritten in terms of a three-point numerical stencil as follows to form the pressure matrix that solves this problem:

$$W_i^m p_{i-1}^{n+1}+C_i^m p_i^{n+1}+E_i^m p_{i+1}^{n+1}=Q_i^m \tag{7.169}$$

where

$$W_i^m=\left(\frac{B_o}{B_w}\right)_i^m T_{O,x_w}^m+T_{W,x_w}^m \tag{7.170}$$

$$E_i^m=\left(\frac{B_o}{B_w}\right)_i^m T_{O,x_e}^m+T_{W,x_e}^m \tag{7.171}$$

$$C_i^m=-\left(W_i^m+E_i^m\right) \tag{7.172}$$

$$Q_i^m=-\left(\frac{B_o}{B_w}\right)_i^m \left(q_{osc}\right)_i^m-\left(q_{wsc}\right)_i^m \tag{7.173}$$

Since all external boundaries are no flow, $W_1 = E_4 = 0$ (zero transmissibility at the external boundaries) in this problem. For this problem, the application of Eq. (7.169) to each gridblock in the system will yield to the following system of four algebraic equations that need to be solved simultaneously for all four pressures (p_1, p_2, p_3, p_4):

$$\begin{bmatrix} C_1 & E_1 & 0 & 0 \\ W_2 & C_2 & E_2 & 0 \\ 0 & W_3 & C_3 & E_3 \\ 0 & 0 & W_4 & C_4 \end{bmatrix}^m \begin{bmatrix} p_1 \\ p_2 \\ p_3 \\ p_4 \end{bmatrix}^{n+1} = -\begin{bmatrix} Q_1 \\ Q_2 \\ Q_3 \\ Q_4 \end{bmatrix}^m \tag{7.174}$$

Upon the calculation of all stated transmissibilities (matrix coefficients), all pressures can be solved for simultaneously (implicitly).

Once pressures have been calculated implicitly, **saturations are updated explicitly.** One may select any of the original flow equations (oil or water) for this purpose. For example, if one selects the oil equation, the *FVM* approximation is:

$$T^{m}_{O,x_e}\left(p^{n+1}_{i+1,}-p^{n+1}_{i}\right)+T^{m}_{O,x_w}\left(p^{n+1}_{i-1}-p^{n+1}_{i}\right)+\left(q_{osc}\right)^{m}_{i}$$
$$=\left(\frac{V_b\phi}{\alpha_c B_o \Delta t}\right)_i\left(S_{oi}^{\,n+1}-S_{oi}^{\,n}\right) \tag{7.175}$$

Therefore, the *explicit IMPES saturation update* may be written as:

$$S_{oi}^{\,n+1}=S_{oi}^{\,n}+\left(\frac{\alpha_c B_o \Delta t}{V_b\phi}\right)_i\left[T^{m}_{O,x_e}\left(p^{n+1}_{i+1,}-p^{n+1}_{i}\right)+T^{m}_{O,x_w}\left(p^{n+1}_{i-1}-p^{n+1}_{i}\right)+\left(q_{osc}\right)^{m}_{i}\right] \tag{7.176}$$

Because of the explicit treatment of saturation and saturation-dependent variables (such as relative permeabilities and capillary pressures), the *IMPES* method is faster per time step than other multiphase solvers such as Newton–Raphson method. However, such explicitness also implies that the method is less numerically stable and requires taking much smaller time steps when compared to Newton–Raphson method. Time-step control is needed to prevent unstable solutions. Ertekin et al. (2001) presents more details about *IMPES* and its implementation when additional effects, such as capillary pressure, pore compressibility, and multiple dimensions, are taken into account.

7.6.2 Generalized Newton-Raphson Procedure

As first introduced in Section 7.4, the Newton–Raphson (*NR*) protocol is one of the most popular methods for the simultaneous solution of nonlinear equations in multiple variables. In a typical Newton-Raphson implementation, "N" functional relationships (or "residuals") must be zeroed and the final goal of the method is to find the combination of *N*-independent variables that satisfies such condition.

A well-known drawback of this method is that it demands a good initialization of the unknowns or independent variables (close to the expected solution) for satisfactory performance. If the initial approximation is good enough, the updating sequence would converge rapidly (quadratically) to the solution. The method, however, has the unfortunate tendency of diverging hopelessly if the initial guess is not sufficiently close to the root. As a consequence, unaided Newton–Raphson methods are said to have *local convergence* because convergence is only guaranteed to occur in some neighborhood of the actual solution. The maximum allowable difference between initial guesses and the actual solution that would still guarantee convergence is known as the "convergence radius," and thus any initialization is required to be within said radius of the system of equations for convergence to be possible. When no restriction is placed on how close the initialization needs to be to the actual solution, then the convergence is said to be *global*, with a theoretical radius of convergence of infinity. There is no general way of knowing *apriori*, however, if any given initialization would be found within the convergence radius until a given combination is tried out. Obtaining the best-possible first approximation using any physical or knowledge available about the problem is always a must for a locally convergent Newton–Raphson method. Fortunately, for most reservoir engineering problems, using initial pressures and saturations usually suffices as good Newton–Raphson initializations.

In the typical Newton–Raphson sequence, the "N" functional relationships (or "residuals") must be zeroed may be denoted as:

$$\vec{R} = (R_1, R_2, R_3, R_4, \ldots, R_N) = 0 \tag{7.177}$$

and the N-independent variables or unknowns are given by:

$$\vec{x} = (x_1, x_2, x_3, x_4, \ldots, x_N) \tag{7.178}$$

The Jacobian of the system is represented by its gradient vector whose components can be written in terms of a matrix containing all first-order partial derivatives of the residual function $\vec{R}$ with respect to each component of the vector of unknowns $\vec{x}$:

$$\nabla\vec{R}(\vec{x}) = \frac{\partial\vec{R}}{\partial\vec{x}} = \begin{bmatrix} \frac{\partial R_1}{\partial x_1} & \frac{\partial R_1}{\partial x_2} & \frac{\partial R_1}{\partial x_3} & \frac{\partial R_1}{\partial x_4} & \cdots & \frac{\partial R_1}{\partial x_N} \\ \frac{\partial R_2}{\partial x_1} & \frac{\partial R_2}{\partial x_2} & \frac{\partial R_2}{\partial x_3} & \frac{\partial R_2}{\partial x_4} & \cdots & \frac{\partial R_2}{\partial x_N} \\ \frac{\partial R_3}{\partial x_1} & \frac{\partial R_3}{\partial x_2} & \frac{\partial R_3}{\partial x_3} & \frac{\partial R_3}{\partial x_4} & \cdots & \frac{\partial R_3}{\partial x_N} \\ \frac{\partial R_4}{\partial x_1} & \frac{\partial R_4}{\partial x_2} & \frac{\partial R_4}{\partial x_3} & \frac{\partial R_4}{\partial x_4} & \cdots & \frac{\partial R_4}{\partial x_N} \\ \cdots & \cdots & \cdots & \cdots & \cdots & \cdots \\ \frac{\partial R_N}{\partial x_1} & \frac{\partial R_N}{\partial x_2} & \frac{\partial R_N}{\partial x_3} & \frac{\partial R_N}{\partial x_4} & \cdots & \frac{\partial R_N}{\partial x_N} \end{bmatrix} \tag{7.179}$$

If a good initialization for $\vec{x}$ is available ($\Delta\vec{x}^{\text{old}}$), and the value of the Jacobian is known at such location ($\nabla\vec{R}(\vec{x}^{old})$), the *multivariate Newton–Raphson formulation* indicates that a better approximation for $\vec{x}$ is given by:

$$\nabla\vec{R}(\vec{x}^{old})\Delta\vec{x}^{new} = -\vec{R}(\vec{x}^{old}) \tag{7.180}$$

from where

$$\vec{x}^{new} = \vec{x}^{old} + \Delta\vec{x}^{new} \tag{7.181}$$

where the final goal of the method is to find the combination of N-independent variables in $\vec{x}$ that satisfies the zeroed-residual condition in Eq. (7.177).

Example 7.3: 1D Waterflooding project—Newton solution: *Let us go back to the 1D waterflooding (oil/water flow) project depicted in Fig. 7.8. Derive the Newton-Raphson scheme for the simultaneous solution of pressure and saturation. Again, all external boundaries are no-flow but please do not assume zero compressibility for either fluids or rock, and incorporate capillary pressure effects.*

Equations (7.120) and (7.121) represent the applicable governing equations for the 1D waterflooding scenario. Using a p_o-S_w formulation, we write:

$$\frac{\partial}{\partial x}\left[\frac{k_x k_{ro}}{\mu_o B_o}\left(\frac{\partial p_o}{\partial x}-\gamma_o\frac{\partial G}{\partial x}\right)\right]+\frac{q_{osc}}{V_b}=\frac{1}{\alpha_c}\frac{\partial}{\partial t}\left(\phi\frac{(1-S_w)}{B_o}\right) \tag{7.182}$$

$$\frac{\partial}{\partial x}\left[\frac{k_x k_{rw}}{\mu_w B_w}\left(\frac{\partial p_o}{\partial x}-\frac{\partial p_{cow}}{\partial x}-\gamma_w\frac{\partial G}{\partial x}\right)\right]+\frac{q_{wsc}}{V_b}=\frac{1}{\alpha_c}\frac{\partial}{\partial t}\left(\phi\frac{S_w}{B_w}\right) \tag{7.183}$$

Their associated *FVM* approximations are found in Eqs. (7.128) and (7.135). Primary unknowns in this formulation are p_o and S_w. Thus, for the system of four gridblocks under study in Fig. 7.8, a total of eight unknowns must be simultaneously solved using these *FVM* equations, two unknowns per block ($p_{o1}^{n+1}, S_{w1}^{n+1}; p_{o2}^{n+1}, S_{w2}^{n+1}; p_{o3}^{n+1}, S_{w3}^{n+1}; p_{o4}^{n+1}, S_{w4}^{n+1}$). Eight residual equations are thus needed, which are the two *FVM* equations [Eqs. (7.128) and (7.135)] need to be written in residual form (equaled to zero) for each of the gridblocks. The oil residual ($R_{O,i}^{n+1}$) and water residual ($R_{W,i}^{n+1}$) equations written for this scenario directly follow from Eqs. (7.128) and (7.135), respectively:

$$\begin{aligned}R_{Oi}^{n+1} &= E_{O,i}^{n+1}\left(p_{oi+1}^{n+1}-p_{oi}^{n+1}\right)+W_{O,i}^{n+1}\left(p_{oi-1}^{n+1}-p_{oi}^{n+1}\right)\\ &\quad +q_{osci,}^{\;n+1}-\left(\frac{V_b}{\alpha_c\Delta t}\right)_i^{n+1}\left\{\left(\phi\frac{1-S_w}{B_o}\right)_i^{n+1}-\left(\phi\frac{1-S_w}{B_o}\right)_i^{n}\right\}=0\end{aligned} \tag{7.184}$$

$$\begin{aligned}R_{Wi}^{n+1} &= E_{W,i}^{n+1}\left(p_{oi+1}^{n+1}-p_{oi}^{n+1}\right)+W_{W,i}^{m}\left(p_{oi-1}^{n+1}-p_{oi}^{n+1}\right)\\ &\quad -E_{W,i}^{n+1}\left(p_{cowi+1}^{n+1}-p_{cowi}^{n+1}\right)-W_{W,i}^{n+1}\left(p_{cowi-1}^{n+1}-p_{cowi}^{n+1}\right)\\ &\quad +q_{wsci}^{\;n+1}-\left(\frac{V_b}{\alpha_c\Delta t}\right)_i^{n+1}\left\{\left(\frac{\phi S_w}{B_w}\right)_i^{n+1}-\left(\frac{\phi S_w}{B_w}\right)_i^{n}\right\}=0\end{aligned} \tag{7.185}$$

where $E_{O,i}^{n+1}$ and $E_{W,i}^{n+1}$ represent the east transmissibility of the oil and water components, respectively, for block "i" (i.e., $T_{O,xe}^{n+1}$ and $T_{W,xe}^{n+1}$); and $W_{O,i}^{n+1}$ and $W_{W,i}^{n+1}$ represent the west transmissibility of oil and water, respectively, for block "i" (i.e., $T_{O,xe}^{n+1}$ and $T_{W,xe}^{n+1}$). No depth gradients are considered in these equations, and the fully implicit scheme ($m = n + 1$) was implemented.

Application of these *FVM* governing flow equations to each gridblock yields, keeping in mind that due to the presence of external no-flow boundaries, $W_{O1} = W_{W1} = E_{O4} = E_{W4} = 0$:

Block $i = 1$:

$$R_{O1}^{n+1} = E_{O1}^{n+1}\left(p_{o2}^{\ n+1} - p_{o1}^{\ n+1}\right) - \left(\frac{V_b}{\alpha_c \Delta t}\right)_1^{n+1} \left\{\left(\phi \frac{1-S_w}{B_o}\right)_1^{n+1} - \left(\phi \frac{1-S_w}{B_o}\right)_1^{n}\right\} = 0 \quad (7.186)$$

$$R_{W1}^{n+1} = E_{W1}^{n+1}\left(p_{o2}^{\ n+1} - p_{o1}^{\ n+1}\right) - E_{W1}^{n+1}\left(p_{cow2}^{\ \ n+1} - p_{cow1}^{\ \ n+1}\right)$$

$$+q_{wsc1}^{\ \ n+1} - \left(\frac{V_b}{\alpha_c \Delta t}\right)_1^{n+1} \left\{\left(\frac{\phi S_w}{B_w}\right)_1^{n+1} - \left(\frac{\phi S_w}{B_w}\right)_1^{n}\right\} = 0 \quad (7.187)$$

Block $i = 2$:

$$R_{O2}^{n+1} = E_{O2}^{n+1}\left(p_{o3}^{\ n+1} - p_{o2}^{\ n+1}\right) + W_{O2}^{n+1}\left(p_{o1}^{\ n+1} - p_{o2}^{\ n+1}\right)$$

$$-\left(\frac{V_b}{\alpha_c \Delta t}\right)_2^{n+1} \left\{\left(\phi \frac{1-S_w}{B_o}\right)_2^{n+1} - \left(\phi \frac{1-S_w}{B_o}\right)_2^{n}\right\} = 0 \quad (7.188)$$

$$R_{W2}^{n+1} = E_{W2}^{n+1}\left(p_{o3}^{\ n+1} - p_{o2}^{\ n+1}\right) + W_{W2}^{n+1}\left(p_{o1}^{\ n+1} - p_{o2}^{\ n+1}\right)$$

$$-E_{W2}^{n+1}\left(p_{cow3}^{\ \ n+1} - p_{cow2}^{\ \ n+1}\right) - W_{W2}^{n+1}\left(p_{cow1}^{\ \ n+1} - p_{cow2}^{\ \ n+1}\right)$$

$$-\left(\frac{V_b}{\alpha_c \Delta t}\right)_2^{n+1} \left\{\left(\frac{\phi S_w}{B_w}\right)_2^{n+1} - \left(\frac{\phi S_w}{B_w}\right)_2^{n}\right\} = 0 \quad (7.189)$$

Block $i = 3$:

$$R_{O3}^{n+1} = E_{O3}^{n+1}\left(p_{o4}^{\ n+1} - p_{o3}^{\ n+1}\right) + W_{O3}^{n+1}\left(p_{o2}^{\ n+1} - p_{o3}^{\ n+1}\right)$$

$$-\left(\frac{V_b}{\alpha_c \Delta t}\right)_3^{n+1} \left\{\left(\phi \frac{1-S_w}{B_o}\right)_3^{n+1} - \left(\phi \frac{1-S_w}{B_o}\right)_3^{n}\right\} = 0 \quad (7.190)$$

$$R_{W3}^{n+1} = E_{W3}^{n+1}\left(p_{o4}^{\ n+1} - p_{o3}^{\ n+1}\right) + W_{W3}^{n+1}\left(p_{o2}^{\ n+1} - p_{o3}^{\ n+1}\right)$$

$$-E_{W3}^{n+1}\left(p_{cow4}^{\ \ n+1} - p_{cow3}^{\ \ n+1}\right) - W_{W3}^{n+1}\left(p_{cow2}^{\ \ n+1} - p_{cow3}^{\ \ n+1}\right)$$

$$-\left(\frac{V_b}{\alpha_c \Delta t}\right)_3^{n+1}\left\{\left(\frac{\phi S_w}{B_w}\right)_3^{n+1}-\left(\frac{\phi S_w}{B_w}\right)_3^{n}\right\}=0 \tag{7.191}$$

Block $i = 4$:

$$R_{O4}^{n+1}=W_{O4}^{n+1}\left(p_{o3}^{n+1}-p_{o4}^{n+1}\right)$$
$$-\left(\frac{V_b}{\alpha_c \Delta t}\right)_4^{n+1}\left\{\left(\phi\frac{1-S_w}{B_o}\right)_4^{n+1}-\left(\phi\frac{1-S_w}{B_o}\right)_4^{n}\right\}=0 \tag{7.192}$$

$$R_{W4}^{n+1}=W_{W4}^{n+1}\left(p_{o3}^{n+1}-p_{o4}^{n+1}\right)-W_{W4}^{n+1}\left(p_{cow3}^{n+1}-p_{cow4}^{n+1}\right)$$
$$-\left(\frac{V_b}{\alpha_c \Delta t}\right)_4^{n+1}\left\{\left(\frac{\phi S_w}{B_w}\right)_4^{n+1}-\left(\frac{\phi S_w}{B_w}\right)_4^{n}\right\}=0 \tag{7.193}$$

These eight residual equations are to be used to iteratively solve for the eight principal unknowns: $p_{o1}^{n+1}, S_{w1}^{n+1}, p_{o2}^{n+1}, S_{w2}^{n+1}, p_{o3}^{n+1}, S_{w3}^{n+1}, p_{o4}^{n+1}, S_{w4}^{n+1}$. On the basis of these residual equations, the Newton–Raphson iteration (k-iteration level) protocol would then be written as:

$$
\begin{array}{ll}
\text{Block 1} & \begin{array}{l}\text{Oil Res 1}\\ \text{Wat Res 1}\end{array}\\
\text{Block 2} & \begin{array}{l}\text{Oil Res 2}\\ \text{Wat Res 2}\end{array}\\
\text{Block 3} & \begin{array}{l}\text{Oil Res 3}\\ \text{Wat Res 3}\end{array}\\
\text{Block 4} & \begin{array}{l}\text{Oil Res 4}\\ \text{Wat Res 4}\end{array}
\end{array}
\left(\begin{array}{cccccccc}
\frac{\partial R_{o1}}{\partial p_{o1}} & \frac{\partial R_{o1}}{\partial S_{w1}} & \frac{\partial R_{o1}}{\partial p_{o2}} & \frac{\partial R_{o1}}{\partial Sw_2} & \mathbf{0} & \mathbf{0} & \mathbf{0} & \mathbf{0} \\
\frac{\partial R_{w1}}{\partial p_{o1}} & \frac{\partial R_{w1}}{\partial S_{w1}} & \frac{\partial R_{w1}}{\partial p_{o2}} & \frac{\partial R_{w1}}{\partial Sw_2} & \mathbf{0} & \mathbf{0} & \mathbf{0} & \mathbf{0} \\
\hdashline
\frac{\partial R_{o2}}{\partial p_{o1}} & \frac{\partial R_{o2}}{\partial S_{w1}} & \frac{\partial R_{o2}}{\partial p_{o2}} & \frac{\partial R_{o2}}{\partial S_{w2}} & \frac{\partial R_{o2}}{\partial p_{o3}} & \frac{\partial R_{o2}}{\partial_{w3}} & \mathbf{0} & \mathbf{0} \\
\frac{\partial R_{w2}}{\partial p_{o1}} & \frac{\partial R_{w2}}{\partial S_{w1}} & \frac{\partial R_{w2}}{\partial p_{o2}} & \frac{\partial R_{w2}}{\partial S_{w2}} & \frac{\partial R_{w2}}{\partial p_{o3}} & \frac{\partial R_{w2}}{\partial S_{w3}} & \mathbf{0} & \mathbf{0} \\
\hdashline
\mathbf{0} & \mathbf{0} & \frac{\partial R_{o3}}{\partial p_{o2}} & \frac{\partial R_{o3}}{\partial S_{w2}} & \frac{\partial R_{o3}}{\partial p_{o3}} & \frac{\partial R_{o3}}{\partial S_{w3}} & \frac{\partial R_{o3}}{\partial p_{o4}} & \frac{\partial R_{o3}}{\partial S_{w4}} \\
\mathbf{0} & \mathbf{0} & \frac{\partial R_{w3}}{\partial p_{o2}} & \frac{\partial R_{w3}}{\partial S_{w2}} & \frac{\partial R_{w3}}{\partial p_{o3}} & \frac{\partial R_{w3}}{\partial S_{w3}} & \frac{\partial R_{w3}}{\partial p_{o4}} & \frac{\partial R_{w3}}{\partial S_{w4}} \\
\hdashline
\mathbf{0} & \mathbf{0} & \mathbf{0} & \mathbf{0} & \frac{\partial R_{o4}}{\partial p_{o3}} & \frac{\partial R_{o4}}{\partial S_{w3}} & \frac{\partial R_{o4}}{\partial p_{o4}} & \frac{\partial R_{o4}}{\partial S_{w4}} \\
\mathbf{0} & \mathbf{0} & \mathbf{0} & \mathbf{0} & \frac{\partial R_{w4}}{\partial p_{o3}} & \frac{\partial R_{w4}}{\partial S_{w3}} & \frac{\partial R_{w4}}{\partial p_{o4}} & \frac{\partial R_{w4}}{\partial S_{w4}}
\end{array}\right)^{\substack{(k)\\ n+1}}
\left(\begin{array}{c}
\Delta p_{o1}\\ \Delta S_{w1}\\ \hdashline \Delta p_{o2}\\ \Delta S_{w2}\\ \hdashline \Delta p_{o3}\\ \Delta S_{w3}\\ \hdashline \Delta p_{o4}\\ \Delta S_{w4}
\end{array}\right)^{\substack{(k)\\ n+1}}
=
\left(\begin{array}{c}
-R_{o1}\\ -R_{w1}\\ \hdashline -R_{o2}\\ -R_{w2}\\ \hdashline -R_{o3}\\ -R_{w3}\\ \hdashline -R_{o4}\\ -R_{w4}
\end{array}\right)^{\substack{(k)\\ n+1}}
\tag{7.194}
$$

where the Jacobian matrix zero entries correspond to locations where residual equations have no dependencies on associated unknowns (e.g., residual equation R_{o1} and R_{w1} do not depend on gridblock pressures or saturations in blocks 3 or 4, and residual equations R_{o4} and R_{w4} do not depend on gridblock pressures or saturations in blocks 1 or 2).

The Newton–Raphson protocol would then proceed as follows:

- *Provide a first approximation (initialization or guess) to the solution* ($p_{o1}^{\stackrel{(k)}{n+1}}$, $p_{o2}^{\stackrel{(k)}{n+1}}$, $p_{o3}^{\stackrel{(k)}{n+1}}$, and $p_{o4}^{\stackrel{(k)}{n+1}}$ for pressure, $S_{w1}^{\stackrel{(k)}{n+1}}$, $S_{w2}^{\stackrel{(k)}{n+1}}$, $S_{w3}^{\stackrel{(k)}{n+1}}$, and $S_{w4}^{\stackrel{(k)}{n+1}}$ for saturation). Initial pressure and saturations are used for the first time step in a reservoir study. For subsequent time steps, initializations are equal to the values of the unknowns in the previous time step.
- *Evaluate the right-hand side of Eq. (7.194):* $-R_{o1}^{\stackrel{(k)}{n+1}}$, $-R_{w1}^{\stackrel{(k)}{n+1}}$, $-R_{o2}^{\stackrel{(k)}{n+1}}$, $-R_{w2}^{\stackrel{(k)}{n+1}}$, $-R_{o3}^{\stackrel{(k)}{n+1}}$, $-R_{w3}^{\stackrel{(k)}{n+1}}$, $-R_{o4}^{\stackrel{(k)}{n+1}}$, and $-R_{w4}^{\stackrel{(k)}{n+1}}$, using Eqs. (7.186) to (7.193) evaluated at current values for $p_{o1}^{\stackrel{(k)}{n+1}}$, $p_{o2}^{\stackrel{(k)}{n+1}}$, $p_{o3}^{\stackrel{(k)}{n+1}}$, and $p_{o4}^{\stackrel{(k)}{n+1}}$ and $S_{w1}^{\stackrel{(k)}{n+1}}$, $S_{w2}^{\stackrel{(k)}{n+1}}$, $S_{w3}^{\stackrel{(k)}{n+1}}$, and $S_{w4}^{\stackrel{(k)}{n+1}}$.
- *Calculate the entries of the Jacobian* $\vec{\mathbf{R}}$*:* These entries may be calculated analytically or numerically. For multiphase cases, these derivatives are most often evaluated numerically using forward, backward, or central differences (Ertekin et al., 2001).
- *Solve for* $\Delta p_{o,i}^{\stackrel{(k+1)}{n+1}}$ *and* $\Delta S_{w,i}^{\stackrel{(k+1)}{n+1}}$ *(*$i = 1, 2, 3$ *and 4) in Eq. (7.194).* This entails solving the stated system of simultaneous algebraic equations.
- *Update the solution using recently obtained* $\Delta p_{o,i}^{\stackrel{(k+1)}{n+1}}$ *and* $\Delta S_{w,i}^{\stackrel{(k+1)}{n+1}}$:

$$p_{o,i}^{\stackrel{(k+1)}{n+1}} = p_{o,i}^{\stackrel{(k)}{n+1}} + \Delta p_{o,i}^{\stackrel{(k+1)}{n+1}} \tag{7.195}$$

$$S_{w,i}^{\stackrel{(k+1)}{n+1}} = S_{w,i}^{\stackrel{(k)}{n+1}} + \Delta S_{w,i}^{\stackrel{(k+1)}{n+1}} \tag{7.196}$$

  for all gridblocks $i = 1, 2, 3, and\ 4$.
- *Recalculate residuals and Jacobian entries until convergence is attained.* Convergence is attained when the largest absolute value of $\Delta p_{o,i}^{\stackrel{(k+1)}{n+1}}$ and $\Delta S_{w,i}^{\stackrel{(k+1)}{n+1}}$ are both found below a prescribed tolerance (say, 10^{-5} psia for pressure, and 10^{-9} for saturation). At this point, the multiphase solution s calculated for the given time step.

7.7 FVM Representations of Multiphase Flow: Compositional Equations

Compositional reservoir modeling studies the most general flow scenario where N_c—chemical species (*fluid components*) are found flowing within the reservoir domain. Each of the fluid components co-exist in *various phases* within the reservoir system. In total, the model considers up to "N_c" species ($I = 1, 2, \ldots, N_c$) or components distributed among N_p reservoir phases ($J = 1, 2, \ldots, N_p$). In compositional reservoir modeling, each numerical gridblock is treated as a thermodynamic system with well-defined thermodynamic properties, and the generalized flow equation [Eq. (5.30) in Chap. 5] is implemented without any further simplifications:

$$\nabla \cdot \left[\sum_{J=1}^{N_p} \left(\omega_{IJ} \rho_J \frac{\bar{\bar{k}} k_{rJ}}{\mu_J} \nabla \Phi_J \right) \right] + \frac{1}{V_b} \sum_{J=1}^{N_p} \left(\omega_{IJ} \rho_J q_J \right) = \frac{1}{\alpha_c} \frac{\partial}{\partial t} \left[\phi \sum_{J=1}^{N_p} S_J \, \omega_{IJ} \, \rho_J \right]$$

$$I = 1, 2, \ldots, N_c \tag{7.197}$$

As studied in Example 5.2 in Chap. 5, compositional flow modeling investigates flows of N_p-phases and N_c-components. The ultimate goal is being able to solve for all phase pressures (for all N_p-phases), phase saturations (N_p-phases), and phase compositions (N_p*N_c compositions) at every point of the domain and at every time step. Total numbers of target unknowns are 2 $N_p + N_p$*N_c, namely:

- *Phase pressures,* p_J ($J = 1, 2, \ldots, N_p$)
- *Phase saturations,* S_J ($J = 1, 2, \ldots, N_p$)
- *Phase compositions,* ω_{IJ} (N_p*N_c compositions)

To solve for these 2 $N_p + N_p$*N_c unknowns, Example 5.2 showed that we indeed have 2 $N_p + N_p$*N_c equations available for such purpose. Beyond the N_c-flow main equations readily available in Eq. (7.197), we have the following auxiliary equations:

- The phase saturation constraint (all phase saturation must add up to one):

$$\sum_{J=1}^{N_p} S_J = 1 \tag{7.198}$$

- N_p-phase composition constraints (one per phase in the system):

$$\sum_{I=1}^{N_c} \omega_{IJ} = 1 \tag{7.199}$$

- $N_p - 1$ capillary pressure constraints (written for each fluid interface in the system):

$$p_c(S_J) = p_{non-wet} - p_{wet} \tag{7.200}$$

- N_c * ($N_p - 1$) equilibrium thermodynamic constraints, written in terms of equilibrium ratios or fugacity equilibrium (see Chap. 4).

For illustration purposes, one of the simplest applications of compositional equations is deployed when the model is used to analyze *simultaneous flow of gas and condensate in a gas-condensate reservoir*. For such a system, assuming nonmobile (irreducible) water, N_c could take a variety of values depending on the available fluid characterization but $N_p = 2$ (J = "o" or "g"). Equation (7.197) is now written as follows for a gas/condensate compositional system:

$$\nabla \cdot \left[\omega_{Io} \rho_o \frac{\bar{\bar{k}} k_{ro}}{\mu_o} \nabla \Phi_o + \omega_{Ig} \rho_g \frac{\bar{\bar{k}} k_{rg}}{\mu_g} \nabla \Phi_g \right] + \frac{1}{V_b} \left(\omega_{Io} \rho_o q_o + \omega_{Ig} \rho_g q_g \right)$$

$$= \frac{1}{\alpha_c}\frac{\partial}{\partial t}\left[\phi\left(S_o\,\omega_{Io}\,\rho_o + S_g\,\omega_{Ig}\,\rho_g\right)\right]$$

$$I = 1, 2, \ldots, N_c \tag{7.201}$$

The 3D Cartesian expansion of these compositional equations would be written as:

$$\frac{\partial}{\partial x}\left[\omega_{Io}\rho_o\frac{k_x k_{ro}}{\mu_o}\left(\frac{\partial p_o}{\partial x} - \gamma_o\frac{\partial G}{\partial x}\right) + \omega_{Ig}\rho_g\frac{k_x k_{rg}}{\mu_g}\left(\frac{\partial p_g}{\partial x} - \gamma_g\frac{\partial G}{\partial x}\right)\right]$$
$$+\frac{\partial}{\partial y}\left[\omega_{Io}\rho_o\frac{k_y k_{ro}}{\mu_o}\left(\frac{\partial p_o}{\partial y} - \gamma_o\frac{\partial G}{\partial y}\right) + \omega_{Ig}\rho_g\frac{k_y k_{rg}}{\mu_g}\left(\frac{\partial p_g}{\partial y} - \gamma_g\frac{\partial G}{\partial y}\right)\right]$$
$$+\frac{\partial}{\partial z}\left[\omega_{Io}\rho_o\frac{k_z k_{ro}}{\mu_o}\left(\frac{\partial p_o}{\partial z} - \gamma_o\frac{\partial G}{\partial z}\right) + \omega_{Ig}\rho_g\frac{k_z k_{rg}}{\mu_g}\left(\frac{\partial p_g}{\partial z} - \gamma_g\frac{\partial G}{\partial z}\right)\right]$$
$$+\frac{1}{V_b}\left(\omega_{Io}\rho_o q_o + \omega_{Ig}\rho_g q_g\right) = \frac{1}{\alpha_c}\frac{\partial}{\partial t}\left[\phi\left(S_o\,\omega_{Io}\,\rho_o + S_g\,\omega_{Ig}\,\rho_g\right)\right]$$

$$I = 1, 2, \ldots, N_c \tag{7.202}$$

In these equations, one phase pressure is always selected as one of the primary unknowns, which is also referred to as the reference phase pressure. p_g would be the best choice for a gas-dominated system such as a gas-condensate fluid, and p_o would be a better choice for an oil-dominated system such as a volatile oil fluid. The other phase pressures would be solved for using appropriate capillary pressure relationship [Eq. (7.200)]. The rest of the primary unknowns in compositional models are always compositions or composition-related variables. Choices include overall in situ composition, composition of one the phases in the system, or normalized number of moles of each component in the system. Phase saturations are seldom selected as primary unknowns in compositional systems because they become directly dependent on prevailing pressure and in situ composition of each gridblock, through the compositional flash calculations studied in Chap. 4.

FVM development of these compositional equations revisits familiar grounds. One first integrates the compositional equation of interest over the entire domain and within the time step of interest. For Eq. (7.202), this leads to, after substituting the applicable capillary pressure relationship:

$$\int_t^{t+\Delta t}\int_{V_b}\frac{\partial}{\partial x}\left[\omega_{Io}\rho_o\frac{k_x k_{ro}}{\mu_o}\left(\frac{\partial p_g}{\partial x} - \frac{\partial p_{cgo}}{\partial x} - \gamma_o\frac{\partial G}{\partial x}\right) + \omega_{Ig}\rho_g\frac{k_x k_{rg}}{\mu_g}\left(\frac{\partial p_g}{\partial x} - \gamma_g\frac{\partial G}{\partial x}\right)\right]dVdt$$
$$+\int_t^{t+\Delta t}\int_{V_b}\frac{\partial}{\partial y}\left[\omega_{Io}\rho_o\frac{k_y k_{ro}}{\mu_o}\left(\frac{\partial p_g}{\partial y} - \frac{\partial p_{cgo}}{\partial y} - \gamma_o\frac{\partial G}{\partial y}\right) + \omega_{Ig}\rho_g\frac{k_y k_{rg}}{\mu_g}\left(\frac{\partial p_g}{\partial y} - \gamma_g\frac{\partial G}{\partial y}\right)\right]dVdt$$
$$+\int_t^{t+\Delta t}\int_{V_b}\frac{\partial}{\partial z}\left[\omega_{Io}\rho_o\frac{k_z k_{ro}}{\mu_o}\left(\frac{\partial p_g}{\partial z} - \frac{\partial p_{cgo}}{\partial z} - \gamma_o\frac{\partial G}{\partial z}\right) + \omega_{Ig}\rho_g\frac{k_z k_{rg}}{\mu_g}\left(\frac{\partial p_g}{\partial z} - \gamma_g\frac{\partial G}{\partial z}\right)\right]dVdt$$
$$+\frac{1}{V_b}\int_t^{t+\Delta t}\int_{V_b}\left(\omega_{Io}\rho_o q_o + \omega_{Ig}\rho_g q_g\right)dVdt = \frac{1}{\alpha_c}\int_t^{t+\Delta t}\int_{V_b}\frac{\partial}{\partial t}\left(\phi\left(S_o\,\omega_{Io}\,\rho_o + S_g\,\omega_{Ig}\,\rho_g\right)\right)dVdt$$

$$I = 1, 2, \ldots, N_c \tag{7.203}$$

Performing the specified spatial and time integrations, and deploying central-difference approximations for pressure, capillary pressure, and depth gradients, one obtains the following *FVM* approximation for this compositional system:

$$
\begin{aligned}
& T_{Io,x_e}^{m}\left(p_{gi+1,j,k}^{\ m}-p_{gi,j,k}^{\ m}\right)+T_{Io,x_w}^{m}\left(p_{gi-1,j,k}^{\ m}-p_{gi,j,k}^{\ m}\right) \\
& -T_{Io,x_e}^{m}\left(p_{cgoi+1,j,k}^{\ m}-p_{cgoi,j,k}^{\ m}\right)-T_{Io,x_w}^{m}\left(p_{cgoi-1,j,k}^{\ m}-p_{cgoi,j,k}^{\ m}\right) \\
& -T_{Io,x_e}^{m}\cdot\gamma_{ox_e}^{\ m}\left(G_{i+1,j,k}-G_{i,j,k}\right)-T_{Io,x_w}^{m}\cdot\gamma_{ox_w}^{\ m}\left(G_{i-1,j,k}-G_{i,j,k}\right) \\
& +T_{Ig,x_e}^{m}\left(p_{gi+1,j,k}^{\ m}-p_{gi,j,k}^{\ m}\right)+T_{Ig,x_w}^{m}\left(p_{gi-1,j,k}^{\ m}-p_{gi,j,k}^{\ m}\right) \\
& -T_{Ig,x_e}^{m}\cdot\gamma_{gx_e}^{\ m}\left(G_{i+1,j,k}-G_{i,j,k}\right)-T_{Ig,x_w}^{m}\cdot\gamma_{gx_w}^{\ m}\left(G_{i-1,j,k}-G_{i,j,k}\right) \\
& +T_{Io,y_n}^{m}\left(p_{gi,j+1,k}^{\ m}-p_{gi,j,k}^{\ m}\right)+T_{Io,y_s}^{m}\left(p_{gi,j-1,k}^{\ m}-p_{gi,j,k}^{\ m}\right) \\
& -T_{Io,y_n}^{m}\left(p_{cgoi,j+1,k}^{\ m}-p_{cgoi,j,k}^{\ m}\right)-T_{Io,y_s}^{m}\left(p_{cgoi,j-1,k}^{\ m}-p_{cgoi,j,k}^{\ m}\right) \\
& -T_{Io,y_n}^{m}\cdot\gamma_{oy_n}^{\ m}\left(G_{i,j+1,k}-G_{i,j,k}\right)-T_{Io,y_s}^{m}\cdot\gamma_{oy_s}^{\ m}\left(G_{i,j-1,k}-G_{i,j,k}\right) \\
& +T_{Ig,y_n}^{m}\left(p_{gi,j+1,k}^{\ m}-p_{gi,j,k}^{\ m}\right)+T_{Ig,y_s}^{m}\left(p_{gi,j-1,k}^{\ m}-p_{gi,j,k}^{\ m}\right) \\
& -T_{Ig,y_n}^{m}\cdot\gamma_{gy_n}^{\ m}\left(G_{i,j+1,k}-G_{i,j,k}\right)-T_{Ig,y_s}^{m}\cdot\gamma_{gy_s}^{\ m}\left(G_{i,j-1,k}-G_{i,j,k}\right) \\
& +T_{Io,z_a}^{m}\left(p_{gi,j,k+1}^{\ m}-p_{gi,j,k}^{\ m}\right)+T_{Io,z_b}^{m}\left(p_{gi,j,k-1}^{\ m}-p_{gi,j,k}^{\ m}\right) \\
& -T_{Io,z_a}^{m}\left(p_{cgoi,j,k+1}^{\ m}-p_{cgoi,j,k}^{\ m}\right)-T_{Io,z_b}^{m}\left(p_{cgoi,j,k-1}^{\ m}-p_{cgoi,j,k}^{\ m}\right) \\
& -T_{Io,z_a}^{m}\cdot\gamma_{oz_a}^{\ m}\left(G_{i,j,k+1}-G_{i,j,k}\right)-T_{Io,z_b}^{m}\cdot\gamma_{oz_b}^{\ m}\left(G_{i,j,k-1}-G_{i,j,k}\right) \\
& +T_{Ig,z_a}^{m}\left(p_{gi,j,k+1}^{\ m}-p_{gi,j,k}^{\ m}\right)+T_{Ig,z_b}^{m}\left(p_{gi,j,k-1}^{\ m}-p_{gi,j,k}^{\ m}\right) \\
& -T_{Ig,z_a}^{m}\cdot\gamma_{gz_a}^{\ m}\left(G_{i,j,k+1}-G_{i,j,k}\right)-T_{Ig,z_b}^{m}\cdot\gamma_{gz_b}^{\ m}\left(G_{i,j,k-1}-G_{i,j,k}\right) \\
& +\left(\omega_{Io}\rho_o q_o+\omega_{Ig}\rho_g q_g\right)_{i,j,k}^{m} \\
& =\left(\frac{V_b}{\alpha_c\Delta t}\right)_{i,j,k}^{n+1}\left\{\left(\phi S_o\omega_{Io}\rho_o+\phi S_g\omega_{Ig}\rho_g\right)_{i,j,k}^{n+1}-\left(\phi S_o\omega_{Io}\rho_o+\phi S_g\omega_{Ig}\rho_g\right)_{i,j,k}^{n}\right\}
\end{aligned}
$$

$$I=1,2,\ldots,N_c \tag{7.204}$$

where p_g has been deployed as the principal pressure unknown and the applicable capillary pressure relationship [Eq. (7.149)] has been used to substitute for p_o in terms of p_g. Also, depending on the degree of implicitness or explicitness desired for the formulation, "m" may be selected to be $n+1$ (new time level) or n (old time level).

Compositional transmissibilities in Eq. (7.204) are now defined as:

$$T_{IJ,x_e}^{m}=\left[\omega_{IJ}k_{rJ}\frac{\rho_j}{\mu_j}\frac{k_xA_x}{\Delta x}\right]_{x_e}^{m}=E_{IJ,i,j,k}^{m} \tag{7.205}$$

$$T^m_{IJ,x_w} = \left[\omega_{IJ} k_{rJ} \frac{\rho_j}{\mu_j} \frac{k_x A_x}{\Delta x}\right]^m_{x_w} = W^m_{IJ,i,j,k} \tag{7.206}$$

$$T^m_{IJ,y_n} = \left[\omega_{IJ} k_{rJ} \frac{\rho_j}{\mu_j} \frac{k_y A_y}{\Delta y}\right]^m_{y_n} = N^m_{IJ,i,j,k} \tag{7.207}$$

$$T^m_{IJ,y_s} = \left[\omega_{IJ} k_{rJ} \frac{\rho_j}{\mu_j} \frac{k_y A_y}{\Delta y}\right]^m_{y_s} = S^m_{IJ,i,j,k} \tag{7.208}$$

$$T^m_{IJ,z_a} = \left[\omega_{IJ} k_{rJ} \frac{\rho_j}{\mu_j} \frac{k_z A_z}{\Delta z}\right]^m_{z_a} = A^m_{IJ,i,j,k} \tag{7.209}$$

$$T^m_{IJ,z_b} = \left[\omega_{IJ} k_{rJ} \frac{\rho_j}{\mu_j} \frac{k_z A_z}{\Delta z}\right]^m_{z_b} = B^m_{IJ,i,j,k} \tag{7.210}$$

Compositional Transmissibility Calculations

The different types of transmissibility coefficients that now appear in discretized *FVM* multiphase compositional equations contain three types of variables which require different treatment. As discussed earlier, there are grid-dependent and pressure-dependent variables, but also compositional and saturation-dependent variables. For example, the compositional x-direction transmissibility at $x = x_e$ (east boundary) may be decomposed in three terms:

$$T^m_{IJ,x_e} = E^m_{IJ,i,j,k} = \left.\frac{k_x A_x}{\Delta x}\right|_{x_e} \cdot \left.\frac{\rho_J}{\mu_J}\right|^m_{x_e} \cdot \left(\omega_{IJ} k_{rJ}\right)\Big|^m_{x_e} \tag{7.211}$$

where the familiar grid-related transmissibility component $\left(\frac{k_x A_x}{\Delta x}\right)$ is readily recognized, and also the largely pressure-dependent transmissibility component $\left(\frac{\rho_j}{\mu_j}\right)$. However, the third coefficient $(\omega_{IJ} k_{rJ})$ represents a much stronger nonlinearity, which is a *saturation-* and *composition-*dependent variable. Time-invariant, grid-related transmissibilities $\frac{k_x A_x}{\Delta x}$ continue to be evaluated using harmonic averages (as done in single-phase simulation), and pressure-dependent components are evaluated using volume-averaged boundary pressure between two blocks as done before for weaker nonlinearities.

For strong nonlinearities such as $\omega_{IJ} k_{rJ}$, neither harmonic nor arithmetic or volume-averaged values at block boundaries are suggested in multiphase simulation (Ertekin et al., 2001). For these very strong nonlinearities in transmissibility terms, the implementation of the simple and popular upstream weighting is recommended as previously done for the case of black-oil simulation. As previously discussed, the value of these nonlinearities at block interfaces are each to be evaluated using in situ conditions at gridblock located upstream to the boundary of interest.

7.8 Solution of Compositional Multiphase Numerical Equations

7.8.1 Pressure-Overall *I*-th Composition Solutions

N_c-intensive variables may be selected as *principal unknowns* in order to solve the system of isothermal compositional equations: *one* phase pressure (p_J) and N_c – *1* overall mass compositions ($z_I, I = 1,2,\ldots,N_c - 1$) (Nghiem et al. 1983; Ayala, 2004; Ayala et al. 2006; Zhang and Ayala, 2018; Batista et al., 2018). Thermodynamically, intensive variables are those which are independent of the extent or amount of mass in the system. Selecting *one* phase pressure and $N_c - 1$ overall z_I compositions ($I = 1,2,\ldots,N_c - 1$) as intensive variables of choice and *principal unknowns* constitutes a most natural choice in terms of coupling with thermodynamic flash calculations. As discussed in Chap. 4, Rachford–Rice equations are the workhorse behind all flash calculations in compositional reservoir simulation. In flash calculations, multi-component fluid mixture of known compositions (z_I) are flashed or separated at a given pressure and temperature into a number of fluid phases in equilibrium and their amounts (β_J's) and distinct compositions (ω_{IJ}'s) are calculated. As a result, once this set of principal unknowns is estimated, all remaining terms in the compositional equations can be evaluated using the following protocol:

- *Molar fraction of N_c-th component is calculated using the composition constraint* $z_{N_c} = 1 - \sum_1^{N_c-1} z_I$;
- *Phase mass fractions (β_J's), resulting phase compositions (ω_{IJ}'s), and phase densities (ρ_J's), are obtained via flash thermodynamic equilibrium calculations at prevailing pressure, temperature, and overall composition using a cubic equation of state (see Chap. 4). Phase viscosities (μ_J's) are calculated with standard correlations in terms of phase densities and compositions (e.g., Lorentz et al., 1964). This flash calculation step must honor all N_p-phase composition constrains $\left(\sum_{I=1}^{N_c} \omega_{IJ} = 1; J = 1,2,\ldots,N_P\right)$ and all N_c * ($N_P - 1$) equilibrium thermodynamic constraints (see Chap. 4).*
- *Phase saturations (S_J's) can be calculated as a function of phase mass fraction (β_J's) and phase densities (ρ_J's) (see Chap. 4). Phase saturation calculations must honor the $\sum_{J=1}^{N_p} S_J = 1$ constraint.*
- *The remaining $N_P - 1$ phase saturations are calculated in terms of the phase pressure selected as the principal unknown and phase saturation using capillary pressure constraints (written for each fluid interface in the system):*

$$p_c(S_J) = p_{non-wet} - p_{wet} \tag{7.212}$$

The proposed system has N_c principal unknowns, which can be obtained by simultaneous solution of the N_c-compositional flow equations implied by Eq. (7.197). Such simultaneous solution may be accomplished via Newton–Raphson method as shown in previous section for black-oil multiphase simulation. Example 7.4 illustrates the application of the Newton–Raphson procedure using the p_g-z_I compositional formulation for a gas-cycling project in a very simple reservoir system.

Example 7.4: 1D gas cycling project in near-critical gas reservoir: ***Let us consider the case of 1D Cartesian two-phase (gas-condensate) flow presented in Fig. 7.9. Dry gas injection takes place in gridblock i = 1 and gas and condensate are being produced at i = 2. Present the Newton–Raphson scheme for the simultaneous solution of the compositional equations using pressure and overall composition as primary unknowns. All external boundaries are no-flow.***

Equations (7.202) represent the applicable governing equations for compositional flow analysis of two phases (gas and condensate). Rewriting for 1D Cartesian flow, neglecting depth gradients, and selecting p_g as the reference phase pressure of choice yield:

$$\frac{\partial}{\partial x}\left[\omega_{Io}\rho_o\frac{k_x k_{ro}}{\mu_o}\left(\frac{\partial p_g}{\partial x}-\frac{\partial p_{cgo}}{\partial x}\right)+\omega_{Ig}\rho_g\frac{k_x k_{rg}}{\mu_g}\frac{\partial p_g}{\partial x}\right]$$

$$+\frac{1}{V_b}\left(\omega_{Io}\rho_o q_o+\omega_{Ig}\rho_g q_g\right)=\frac{1}{\alpha_c}\frac{\partial}{\partial t}\left[\phi\left(S_o\,\omega_{Io}\,\rho_o+S_g\,\omega_{Ig}\,\rho_g\right)\right]$$

$$I=1,2,\ldots,N_c \tag{7.213}$$

The associated FVM approximation was presented in Eq. (7.204). Rewriting for this case in a residual, implicit ($m = n + 1$) form (for Newton–Raphson purposes) and using East and West transmissibility notation in Eqs. (7.205) and (7.206) lead to:

$$R_{I,i}^{n+1}=E_{Io,i}^{n+1}\left(p_{gi+1}^{\;n+1}-p_{gi}^{\;n+1}\right)+W_{Io,i}^{n+1}\left(p_{gi-1}^{\;n+1}-p_{gi}^{\;n+1}\right)$$

$$-E_{Io,i}^{n+1}\left(p_{cgoi+1}^{\;n+1}-p_{cgoi}^{\;n+1}\right)-W_{Io,i}^{n+1}\left(p_{cgoi-1}^{\;n+1}-p_{cgoi}^{\;n+1}\right)$$

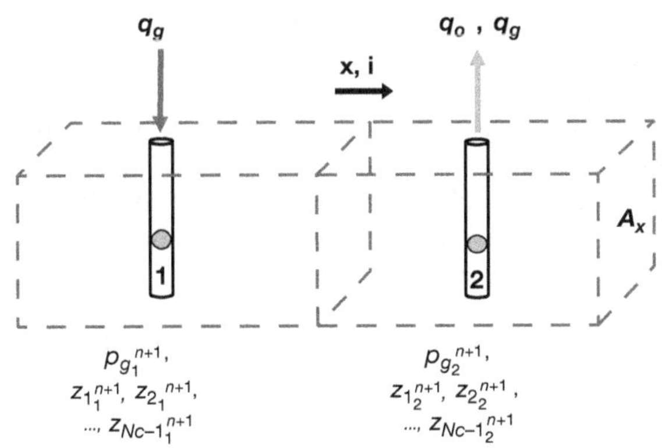

FIGURE 7.9 1D gas cycling project into a near-critical gas reservoir.

$$+E_{Ig,i}^{n+1}\left(p_{gi+1}^{\ n+1}-p_{gi}^{\ n+1}\right)+W_{Ig,i}^{n+1}\left(p_{gi-1}^{\ n+1}-p_{gi}^{\ n+1}\right)$$

$$+\left(\omega_{Io}\rho_o q_o+\omega_{Ig}\rho_g q_g\right)_i^{n+1}$$

$$-\left(\frac{V_b}{\alpha_c \Delta t}\right)_i^{n+1}\begin{Bmatrix}\left(\phi S_o\,\omega_{Io}\,\rho_o+\phi S_g\,\omega_{Ig}\,\rho_g\right)_i^{n+1}\\-\left(\phi S_o\,\omega_{Io}\,\rho_o+\phi S_g\,\omega_{Ig}\,\rho_g\right)_i^{n}\end{Bmatrix}=0$$

$$I=1,2,\ldots,N_c \tag{7.214}$$

where $E_{Io,i}^{n+1}$ and $E_{Io,i}^{n+1}$ represent the east transmissibility of the I-th component in the oil and gas phases, respectively, for block "i"; and $W_{Io,i}^{n+1}$ and $W_{Ig,i}^{n+1}$ represent the west transmissibility of the I-th component in the oil and gas phases, respectively, for block "i". They are calculated as:

$$E_{Io,i}^{n+1}=\left.\frac{k_x A_x}{\Delta x}\right|_{x_e}\cdot\left.\frac{\rho_o}{\mu_o}\right|_{x_e}^{n+1}\cdot\left.(\omega_{Io}k_{ro})\right|_{x_e}^{n+1} \tag{7.215}$$

$$E_{Ig,i}^{n+1}=\left.\frac{k_x A_x}{\Delta x}\right|_{x_e}\cdot\left.\frac{\rho_g}{\mu_g}\right|_{x_e}^{n+1}\cdot\left.(\omega_{Ig}k_{rg})\right|_{x_e}^{n+1} \tag{7.216}$$

$$W_{Io,i}^{n+1}=\left.\frac{k_x A_x}{\Delta x}\right|_{x_w}\cdot\left.\frac{\rho_o}{\mu_o}\right|_{x_w}^{n+1}\cdot\left.(\omega_{Io}k_{ro})\right|_{x_w}^{n+1} \tag{7.217}$$

$$W_{Ig,i}^{n+1}=\left.\frac{k_x A_x}{\Delta x}\right|_{x_w}\cdot\left.\frac{\rho_g}{\mu_g}\right|_{x_w}^{n+1}\cdot\left.(\omega_{Ig}k_{rg})\right|_{x_w}^{n+1} \tag{7.218}$$

Primary unknowns in this formulation are p_g and z_I ($I = 1,2,\ldots,N_c - 1$). For the two-gridblock system in Fig. 7.9, a total of $2\cdot N_C$ unknowns are to be simultaneously solved using these N_C equations at each gridblock (for a total of $2\cdot N_C$ equations). The dependency of these residual equations on pressure is clear through all the pressure terms present in these equations, in addition to the pressure effect on transmissibilities and source and accumulation terms. The dependency of these residual equations on overall composition (z_I) may seem less evident by mere visual inspection, but quite strong due to the direct coupling of the variables in these equations with flash calculations. Phase compositions (ω_{IJ}'s), phase properties (ρ_J's and μ_J's), and saturations (S_J's) [mass fractions (β_J's)] are directly dependent on prevailing overall composition (z_I) through flash calculations (see Chap. 4).

Application of the *FVM* residual equations to each gridblock yields, after enforcing $W_{IJ,1} = W_{IJ,1} = E_{IJ,2} = E_{IJ,2} = 0$ due to the presence of external no-flow boundaries:

<u>Block $i = 1$:</u>

$$R_{I,1}^{n+1}=E_{Io,1}^{n+1}\left(p_{g2}^{\ n+1}-p_{g1}^{\ n+1}\right)-E_{Io,i}^{n+1}\left(p_{cgo2}^{\ n+1}-p_{cgo1}^{\ n+1}\right)$$

$$
\begin{aligned}
&+E_{Ig,i}^{n+1}\left(p_{g2}^{\;n+1}-p_{g1}^{\;n+1}\right)+\left(\omega_{Ig}\rho_g q_g\right)_1^{n+1} \\
&-\left(\frac{V_b}{\alpha_c\Delta t}\right)_1^{n+1}\left\{\begin{matrix}\left(\phi S_o\,\omega_{Io}\,\rho_o+\phi S_g\,\omega_{Ig}\,\rho_g\right)_1^{n+1} \\ -\left(\phi S_o\,\omega_{Io}\,\rho_o+\phi S_g\,\omega_{Ig}\,\rho_g\right)_1^{n}\end{matrix}\right\}=0 \\
&I=1,2,\ldots,N_c
\end{aligned}
\tag{7.219}
$$

<u>Block $i=2$</u>:

$$
\begin{aligned}
R_{I,2}^{n+1}&=W_{Io,2}^{n+1}\left(p_{g1}^{\;n+1}-p_{g2}^{\;n+1}\right)-W_{Io,2}^{n+1}\left(p_{cgo1}^{\;\;n+1}-p_{cgo2}^{\;\;n+1}\right) \\
&+W_{Ig,2}^{n+1}\left(p_{g1}^{\;n+1}-p_{g2}^{\;n+1}\right)+\left(\omega_{Io}\rho_o q_o+\omega_{Ig}\rho_g q_g\right)_2^{n+1} \\
&-\left(\frac{V_b}{\alpha_c\Delta t}\right)_2^{n+1}\left\{\begin{matrix}\left(\phi S_o\,\omega_{Io}\,\rho_o+\phi S_g\,\omega_{Ig}\,\rho_g\right)_2^{n+1} \\ -\left(\phi S_o\,\omega_{Io}\,\rho_o+\phi S_g\,\omega_{Ig}\,\rho_g\right)_2^{n}\end{matrix}\right\}=0 \\
&I=1,2,\ldots,N_c
\end{aligned}
\tag{7.220}
$$

These $2\cdot N_c$ residual equations are to be deployed to be iteratively solved in terms of the $2\cdot N_c$ principal unknowns $p_{g1}^{\;n+1},\left(z_{I1}^{\;n+1},I=1,2,\ldots,N_c-1\right)$ and $p_{g2}^{\;n+1},\left(z_{I2}^{\;n+1},I=1,2,\ldots,N_c-1\right)$, as depicted in Fig. 7.9. On the basis of these residual equations, the Newton Raphson iteration (which represents the *k-iteration* level) protocol is expressed as:

$$
\begin{array}{ll}
\text{Block } i=1 & \text{Comp } l=1 \\
& \text{Comp } l=2 \\
& (\ldots) \\
& \text{Comp } l=\mathrm{N_c} \\
\text{Block } i=2 & \text{Comp } l=1 \\
& \text{Comp } l=2 \\
& (\ldots) \\
& \text{Comp } l=\mathrm{N_c}
\end{array}
\left(\begin{array}{ccccc|ccccc}
\frac{\partial R_{1,1}}{\partial p_{g1}} & \frac{\partial R_{1,1}}{\partial z_{1,1}} & \frac{\partial R_{1,1}}{\partial z_{2,1}} & (\ldots) & \frac{\partial R_{1,1}}{\partial z_{Nc-1,1}} & \frac{\partial R_{1,1}}{\partial p_{g2}} & \frac{\partial R_{1,1}}{\partial z_{1,2}} & \frac{\partial R_{1,1}}{\partial z_{2,2}} & (\ldots) & \frac{\partial R_{1,1}}{\partial z_{Nc-1,2}} \\
\frac{\partial R_{2,1}}{\partial p_{g1}} & \frac{\partial R_{2,1}}{\partial z_{1,1}} & \frac{\partial R_{2,1}}{\partial z_{2,1}} & (\ldots) & \frac{\partial R_{2,1}}{\partial z_{Nc-1,1}} & \frac{\partial R_{2,1}}{\partial p_{g2}} & \frac{\partial R_{2,1}}{\partial z_{1,2}} & \frac{\partial R_{2,1}}{\partial z_{2,2}} & (\ldots) & \frac{\partial R_{2,1}}{\partial z_{Nc-1,2}} \\
(\ldots) & (\ldots) & (\ldots) & (\ldots) & (\ldots) & (\ldots) & (\ldots) & (\ldots) & (\ldots) & (\ldots) \\
\frac{\partial R_{Nc,1}}{\partial p_{g1}} & \frac{\partial R_{Nc,1}}{\partial z_{1,1}} & \frac{\partial R_{Nc,1}}{\partial z_{2,1}} & (\ldots) & \frac{\partial R_{Nc,1}}{\partial z_{Nc-1,1}} & \frac{\partial R_{Nc,1}}{\partial p_{g2}} & \frac{\partial R_{Nc,1}}{\partial z_{1,2}} & \frac{\partial R_{Nc,1}}{\partial z_{2,2}} & (\ldots) & \frac{\partial R_{Nc,1}}{\partial z_{Nc-1,2}} \\
\hline
\frac{\partial R_{1,2}}{\partial p_{g1}} & \frac{\partial R_{1,2}}{\partial z_{1,1}} & \frac{\partial R_{1,2}}{\partial z_{2,1}} & (\ldots) & \frac{\partial R_{1,2}}{\partial z_{Nc-1,1}} & \frac{\partial R_{1,2}}{\partial p_{g2}} & \frac{\partial R_{1,2}}{\partial z_{1,2}} & \frac{\partial R_{1,2}}{\partial z_{2,2}} & (\ldots) & \frac{\partial R_{1,2}}{\partial z_{Nc-1,2}} \\
\frac{\partial R_{2,2}}{\partial p_{g1}} & \frac{\partial R_{2,2}}{\partial z_{1,1}} & \frac{\partial R_{2,2}}{\partial z_{2,1}} & (\ldots) & \frac{\partial R_{2,2}}{\partial z_{Nc-1,1}} & \frac{\partial R_{2,2}}{\partial p_{g2}} & \frac{\partial R_{2,2}}{\partial z_{1,2}} & \frac{\partial R_{2,2}}{\partial z_{2,2}} & (\ldots) & \frac{\partial R_{2,2}}{\partial z_{Nc-1,2}} \\
(\ldots) & (\ldots) & (\ldots) & (\ldots) & (\ldots) & (\ldots) & (\ldots) & (\ldots) & (\ldots) & (\ldots) \\
\frac{\partial R_{Nc,2}}{\partial p_{g1}} & \frac{\partial R_{Nc,2}}{\partial z_{1,1}} & \frac{\partial R_{Nc,2}}{\partial z_{2,1}} & (\ldots) & \frac{\partial R_{Nc,2}}{\partial z_{Nc-1,1}} & \frac{\partial R_{Nc,2}}{\partial p_{g2}} & \frac{\partial R_{Nc,2}}{\partial z_{1,2}} & \frac{\partial R_{Nc,2}}{\partial z_{2,2}} & (\ldots) & \frac{\partial R_{Nc,2}}{\partial z_{Nc-1,2}}
\end{array}\right)_{n+1}^{(k)}
\left(\begin{array}{c}
\Delta pg_{i=1} \\ \Delta z_{1,1} \\ (\ldots) \\ \Delta z_{Nc-1,1} \\ \hline \Delta pg_{i=2} \\ \Delta z_{1,2} \\ (\ldots) \\ \Delta z_{Nc-1,2}
\end{array}\right)_{n+1}^{(k+1)}
=
\left(\begin{array}{c}
-R_{1,1} \\ -R_{2,1} \\ (\ldots) \\ -R_{Nc,1} \\ \hline -R_{1,2} \\ -R_{2,2} \\ (\ldots) \\ -R_{Nc,2}
\end{array}\right)^{(k)}
\tag{7.221}
$$

The Newton–Raphson protocol would then proceed as follows:

- *Provide a first approximation (initialization or guess) to the solution* $\left[\overset{(k)}{p_{g1}^{\;n+1}},\left(\overset{(k)}{z_{I1}^{\;n+1}},I=1,2,\ldots,N_c-1\right)\text{ and }\overset{(k)}{p_{g2}^{\;n+1}},\left(\overset{(k)}{z_{I2}^{\;n+1}},I=1,2,\ldots,N_c-1\right)\right]$. Initial pressure and initial fluid compositions are used for the first time step in a reservoir study. For subsequent time steps, initializations are equal to the values of the unknowns found in the previous time step.

- *Evaluate the right-hand side of Eq. (7.221):* $-\overset{(k)}{R_{1,1}^{n+1}}, -\overset{(k)}{R_{1,1}^{n+1}}, \ldots, -\overset{(k)}{R_{N_c,1}^{n+1}}$, and $-\overset{(k)}{R_{1,2}^{n+1}}$, $-\overset{(k)}{R_{1,2}^{n+1}}, \ldots, -\overset{(k)}{R_{N_c,2}^{n+1}}$, using Eqs. (7.219) and (7.220) evaluated at current values for $\overset{(k)}{p_{g1}^{n+1}}, \left(\overset{(k)}{\Delta z_{I1}^{n+1}}, I = 1, 2, \ldots, N_c - 1\right)$ and $\overset{(k)}{p_{g2}^{n+1}}, \left(\overset{(k)}{\Delta z_{I2}^{n+1}}, I = 1, 2, \ldots, N_c - 1\right)$. Flash calculations are required for the calculation of phase properties as a function of pressure and in-situ overall composition at the prevailing reservoir temperature.
- *Calculate Jacobian matrix entries:* These compositional entries may be calculated numerically while coupling these residuals to flash calculations. In estimating these Jacobian entries numerically, care should be taken to account for changes in the composition of N_c-th component when evaluating the partial derivatives with respect to compositional changes of other components ($z_I, I = 1, 2, \ldots, N_c - 1$). Overall composition constraint should be preserved at all times when perturbing residuals [i.e., $R_{I,1}(z_{1,1} + \varepsilon) = R_{I,1}(p_{g,1}, z_{1,1} + \varepsilon, z_{2,1}, \ldots z_{N_c-1}, z_{N_c} - \varepsilon)$ as shown in Ayala et al., 2006].
- *Solve for* $\overset{(k+1)}{\Delta p_{g1}^{n+1}}, (\overset{(k+1)}{\Delta z_{I1}^{n+1}}, I = 1, 2, \ldots, N_c - 1)$ ($\boldsymbol{i} = 1$ and 2) *in Eq. (7.221).* This entails solving the resulting system of simultaneous algebraic equations.
- *Update the solution using* $\overset{(k+1)}{\Delta p_{g,i}^{n+1}}$ *and* $\overset{(k+1)}{\Delta z_{I,i}^{n+1}}$:

$$\overset{(k+1)}{p_{g,i}^{n+1}} = \overset{(k)}{p_{g,i}^{n+1}} + \overset{k+1}{\Delta p_{g,i}^{n+1}} \tag{7.222}$$

$$\overset{(k+1)}{z_{I,i}^{n+1}} = \overset{(k)}{z_{I,i}^{n+1}} + \overset{(k+1)}{\Delta z_{I,i}^{n+1}}; \; I = 1, 2, \ldots, N_c - 1 \tag{7.223}$$

for all gridblocks ($i = 1$ *and* 2) and $z_{N_c} = 1 - \sum_1^{N_c-1} z_I$. Flash calculations must be performed to update all secondary variables.
- *Recalculate residuals and Jacobian entries until convergence is attained.* Iterations stop when the largest absolute value of $\overset{(k+1)}{\Delta p_{g,i}^{n+1}}$ and $\overset{(k+1)}{\Delta z_{I,i}^{n+1}}$ both are found below a prescribed tolerance (say, 10^{-5} psia for pressure, and 10^{-9} for composition).

7.8.2 Pressure-Overall *I*-th Mass Solutions

N_c-extensive variables plus *one* intensive variable may be selected as *principal unknowns* to solve the isothermal compositional reservoir flow modeling problem: *one* phase pressure and N_c-overall masses for each of the *I*-th components (M_I; I = 1, 2, …, N_c). This is also a popular selection of principal unknowns in compositional reservoir simulation (Chang 1990; Chang *et al.*, 1990; Collins et al., 1992; Qiao, 2015; Qiao et al., 2017; Zhang and Ayala, 2018).

Given its use of extensive variables, this formulation solves for a total of $N_c + 1$ principal unknowns; i.e., it employs one more variable than the previous pressure-overall composition formulation. In this approach, the M_I variable is typically defined as it appears on the right-hand side or accumulation term of the compositional flow equations—normalized per unit bulk volume as conservations equations are written. In the case of Eq. (7.201), e.g., the total mass of I-th component per unit bulk volume is:

$$M_I = \phi S_o \omega_{Io} \rho_o + \phi S_g \omega_{Ig} \rho_g \tag{7.224}$$

Thus, Eq. (7.201) is rewritten as:

$$\frac{\partial M_I}{\partial t} = \alpha_c \nabla \cdot \left[\omega_{Io}\rho_o \frac{\bar{\bar{k}} k_{ro}}{\mu_o} \nabla \Phi_o + \omega_{Ig}\rho_g \frac{\bar{\bar{k}} k_{rg}}{\mu_g} \nabla \Phi_g \right] + \frac{\alpha_c}{V_b}\left(\omega_{Io}\rho_o q_o + \omega_{Ig}\rho_g q_g\right)$$

$$I = 1, 2, \ldots, N_c \tag{7.225}$$

By using one phase pressure and all M_I's as principal unknowns ($N_c + 1$ principal unknowns), remaining terms in the compositional equations are evaluated as follows:

- *Overall in situ composition of the mixture is calculated directly as the ratio of mass of each component to total mass:*

$$z_I = \frac{M_I}{\sum_I^{N_c} M_I}; \; I = 1, 2, \ldots, N_c \tag{7.226}$$

- *Resulting phase mass fractions (β_J's), phase compositions (ω_{IJ}'s) and phase densities (ρ_J's), are obtained via flash thermodynamic equilibrium calculations at prevailing pressure, temperature, and overall composition using an appropriate equation of state (see Chap. 4). Phase viscosities (μ_J's) are calculated with standard correlations in terms of phase densities and compositions (e.g., Lorentz et al., 1964). Flash calculations must honor all N_p-phase composition constrains $\left(\sum_{I=1}^{N_c} \omega_{IJ} = 1; J = 1, 2, \ldots, N_P\right)$ and all $N_c * (N_P - 1)$ equilibrium thermodynamic constraints (see Chap. 4).*
- *Phase saturations (S_J's) can be calculated as a function of phase mass fraction (β_J's) and phase densities (ρ_J's) [Eq. (4.22)]. Phase saturation calculations must honor the $\sum_{J=1}^{N_p} S_J = 1$ constraint.*
- *The remaining $N_p - 1$ phase saturations are calculated in terms of the phase pressure selected as the principal unknown and phase saturation using capillary pressure constraints, written for each fluid interface in the system:*

$$p_c(S_j) = p_{non-wet} - p_{wet} \tag{7.227}$$

It should be noted, however, that in this scenario of $N_c + 1$ unknowns, the set of N_c-compositional flow equations becomes insufficient by 1. To complete the equation set, the volume (balance) preservation principle is the additional equation routinely used (Ács et al., 1985; Watts, 1986; Wong et al., 1990). The principle of preservation of volume enforces that the total fluid volume occupied by all phases (in aggregate) must equal the available pore volume in the rock at any time. This volume balance equation may be written as follows:

$$V_t\left(p, M_1, \ldots, M_{N_c}\right) = V_p(p) \tag{7.228}$$

where V_t corresponds to the addition of the volume occupied by every phase $\left(V_t = \sum_{J=1}^{N_P} V_J\right)$. In Eq. (7.228), pore volume calculations must take into account any pore volume change with pressure:

$$V_p(p) = V_b\phi_o\left[1 + c_\phi\left(p - p_{ref}\right)\right] \tag{7.229}$$

Note that because of the dependency of total fluid volume with respect to the $N_c + 1$ unknowns $[V_t(p, M_1, \ldots, M_{N_c})]$, its total derivative is given by:

$$dV_t = \left(\frac{\partial V_t}{\partial p}\right)_M dp + \left(\frac{\partial V_t}{\partial M_1}\right)_{p,M_{I\neq 1}} dM_1 + \ldots + \left(\frac{\partial V_t}{\partial M_{N_c}}\right)_{p,M_{I\neq N_c}} dM_{N_c} \tag{7.230}$$

where the coefficient $\left(\frac{\partial V_t}{\partial p}\right)_M$ is directly related to fluid isothermal compressibilities and $\left(\frac{\partial V_t}{\partial M_j}\right)_{p,M_{I\neq j}}$ to thermodynamic partial molar volumes.

A fully implicit simulator that uses Newton–Raphson iteration and solves all compositional flow and volume balance equations ($N_c + 1$ equations) simultaneously in terms of *one phase pressure* and *overall mass of I-th component* (M_I) ($N_c + 1$ principal unknowns) at each numerical gridblock would be a suitable choice for this problem. When pressure (p) and overall mass of the i-th component (M_I) are used as primary unknowns, the volume balance equation must be used as part of the system of equations.

In addition, compositional, multiphase, multicomponent numerical simulators have also been proposed based on Implicit Pressure Explicit Composition (*IMPEC*) approaches (e.g., Chang, 1990 and Chang et al., 1990). In these *IMPEC* formulations, the compositional flow equations are written in terms of M_I [e.g., see Eq. (7.225) for the case of oil and gas flow] and *pressure* is solved implicitly in the first step. The compositional pressure-implicit *IMPEC* equation is constructed by enforcing the volume balance constraint and bringing the compositional flow equations into it. For the case of oil and gas flow, the time-derivative of the volume balance constraint in Eq. (7.228), taking into account the dependencies in Eqs (7.229) and (7.230), yields:

$$\left(\frac{\partial V_t}{\partial p}\right)_M \frac{\partial p}{\partial t} + \sum_{I=1}^{N_c}\left(\frac{\partial V_t}{\partial M_I}\right)_{p,M_{k\neq I}} \frac{\partial M_I}{\partial t} = V_b\phi_o c_\phi \frac{\partial p}{\partial t} \tag{7.231}$$

where $\frac{\partial M_l}{\partial t}$ may be eliminated by substituting Eq. (7.225) into it:

$$\sum_{l=1}^{N_c}\left(\frac{\partial V_t}{\partial M_l}\right)_{p,M_{k\neq l}}\left\{\begin{array}{c}\alpha_c\nabla\cdot\left[\omega_{lo}\rho_o\frac{\bar{\bar{k}}k_{ro}}{\mu_o}\nabla\Phi_o+\omega_{lg}\rho_g\frac{\bar{\bar{k}}k_{rg}}{\mu_g}\nabla\Phi_g\right]\\+\frac{\alpha_c}{V_b}\left(\omega_{lo}\rho_o q_o+\omega_{lg}\rho_g q_g\right)\end{array}\right\}$$

$$=\left[V_b\phi_o c_\phi-\left(\frac{\partial V_t}{\partial p}\right)_M\right]\frac{\partial p}{\partial t} \tag{7.232}$$

Equation (7.232) is the typical compositional pressure equation of an *IMPEC* methodology. After pressure is solved for implicitly, resulting masses of species (M_l) in the gridblock can be then evaluated via explicit treatment of Eq. (7.225) and overall in situ composition (z_l) would follow. Due to this explicit treatment of composition, *IMPEC* approaches could be significantly limited in handling large time-steps. One way of increasing the degree of implicitness to handle larger time steps is considering a second sequential implicit step for saturation calculations into the formulation (Watts, 1986), which is at times referred to as the *IMPSAT* (Implicit Pressure and Saturation) methodology. For increased stability and the ability to handle even larger time steps, a fully implicit Newton–Raphson methodology would need to be deployed—at the expense of increased computational costs.

References

Ács, G., S. Doleschall, and E. Farkas, "General Purpose Compositional Model," *SPE J.* 25(4): 543–553, 1985.

Ayala, L.F., *Compositional Modeling of Naturally-Fractured Gas-Condensate Reservoirs in Multi-Mechanistic Flow Domains*, PhD dissertation, The Pennsylvania State University, University Park, PA, 2004.

Ayala, L.F., T. Ertekin, and M. Adewumi, "Compositional Modeling of Gas-Condensate Reservoirs in Multimechanistic Flow Domains," SPE 94856, *SPE J.* 11(4): 480–487, December 2006.

Batista, B.R., F. Marcondes, and K. Sepehrnoori, "Development of a Fully Implicit Approach with Intensive Variables for Compositional Reservoir Simulation," *J Pet Sc Eng.*, 169: 317–336, October 2018.

Chang, Y. B., G.A. Pope, and K. Sepehrnoori, "A Higher-Order Finite Difference Compositional Simulator," *J Pet Sc Eng.* 5(1):35–50, 1990.

Chang, Y.B., *Development and Application of an Equation of State Compositional Simulator*. PhD dissertation. University of Texas at Austin, 1990.

Collins, D. A., L. X. Nghiem, Y.-K. Li, and J.E. Grabonstotter, "An Efficient Approach to Adaptive-Implicit Compositional Simulation with an Equation of State," *SPE Res Engg J.* 7(02):259–264, 1992.

Ertekin, T., J. Abou-Kassem, and G. King, *Basic Applied Reservoir Simulation*, SPE Textbook Series Vol. 7, Richardson, TX, 2001.

Lorentz, J., B. Bray, and C. Clark, "Calculating Viscosities of Reservoir Fluids from their Composition," SPE Paper 915, *J Petrol Technol.* 1171–1176, October 1964.

Mazumder, S., *Numerical Methods for Partial Differential Equations: Finite Difference and Finite Volume Methods*, Academic Press, Amsterdam, 2016.

Nghiem, L. X., D.K. Fong, and K. Aziz, "Compositional Modeling with an Equation of State," *SPE J.* 21(06):687–698, 1983.

Qiao, C., *General Purpose Compositional Simulation for Multiphase Reactive Flow with a Fast Linear Solver,* PhD dissertation, The Pennsylvania State University, University Park, PA, 2015.

Qiao, C., S. Khorsandi, and R. Johns, "A General Purpose Simulation Framework for Multiphase Multicomponent Reactive Fluids." SPE Paper 182715, *Proceedings of the SPE Reservoir Simulation Conference*, Montgomery, Texas, 20–22, February 2017.

Versteeg H.K., and W. Malalasekera, *Computational Fluid Dynamics: The Finite Volume Method*, 2nd ed., Pearson Education Limited, England, 2007.

Watts, J.W., "A Compositional Formulation Based on Pressure and Saturation Equations," *SPE Res Engg J*, 1(3):243–252, 1986.

Wong, T.W., A. Firoozabadi, and K. Aziz, "Relationship of the Volume-Balance Method of Compositional Simulation to the Newton–Raphson Method," SPE Paper 18424, *SPE Res Engg.* 415–422, August 1990.

Zhang, M., and Ayala, L.F., "A Semi-Analytical Solution to Compositional Flow in Liquid-Rich Gas Plays," *Fuels* 212:274–292, January 2018.

Zikanov, O. *Essential Computational Fluid Dynamics*, Wiley, New Jersey, 2010.

CHAPTER 8

Proxy and Hybrid Models in Reservoir Engineering

Most of the reservoir engineering problems at the beginning can be categorized as ill-defined problems. This is because there is a good degree of uncertainty not only in spatial distribution of reservoir parameters but also in many occasions, uncertainties extend toward the description of processes and process controlling mechanisms in place.

Models that are classified under intelligent systems are typically knowledge-based systems. These knowledge-based systems mimic the human intelligence by learning, making adaptation, memorization, etc., so that they can make decisions and associated reasoning. The principal goal is to exploit uncertainty to achieve tractable, robust, and low-energy solutions. These knowledge-based models include artificial neural networks, genetic algorithms, fuzzy logic, case-based reasoning, and model-based reasoning. In this chapter, we will focus on artificial neural network applications in reservoir engineering.

8.1 Use of Artificial Intelligence-Based Models in the Upstream Petroleum Industry

In today's practices of the upstream petroleum engineering technologies, reservoir engineering principles and fundamentals are instrumental in making sound decisions toward the development and production of a hydrocarbon reservoir. Most of the time these decisions are needed to be made in the presence of limited data.

Fundamentally, the basic relation for reservoir evaluation for a given process using a deterministic protocol can be expressed as:

$$q(t)[or\ p(t)] = \Omega(D, R) \tag{8.1}$$

In Eq. (8.1), we can identify the following terms:

$q_{x,y,t}(t)$: *production rate at point x, y, z at time t*

$p_{x,y,t}(t)$: *pressure at point x, y, z at time t*

$\Omega(D, R)$: *mathematical flow model with relevant physics and thermodynamics in place*

D: flow domain characteristics

R: some production mechanism and/or recovery process

In "deterministic" modeling as we have seen in the previous chapters, the entries on the right-hand side of Eq. (8.1) (Ω, D and R) are all fully known. In "stochastic" modeling, the same entries on the right-hand side (Ω, D and R) are partially known. In "proxy" (surrogate) models the right-hand side of the equation does not exist and the information to structure the right-hand side comes in the form of field data. The field data carries important signatures about the influence of existing physical, chemical and thermodynamical processes which determine the magnitude and behavior of the data collected, while the conventional modeling protocols use bivalent logic numerical analysis probability functions and differential equations; the proxy models dwell on fuzzy logic, neuro-computing, probabilistic reasoning, genetic algorithms and pattern recognition. Accordingly, conventional computing generates quantitative, precise and formal results, whereas in artificial intelligence-based models the principal goal is to exploit imprecision and uncertainty to achieve tractable, robust, and low-cost solutions. Figure 8.1 shows the general work flow for artificial intelligence-based models.

In reservoir engineering applications, one can treat the artificial-neural-network application as systems analysis approach which can be expressed similar to Eq. (8.1) but in the form of:

$$\begin{bmatrix} Reservoir \\ Parameters \end{bmatrix} \times \begin{bmatrix} Project \\ Design\ Parameters \end{bmatrix} = \begin{bmatrix} Response\ function \\ (Pressure\ or\ flow\ rate) \end{bmatrix} \tag{8.2}$$

In Eq. (8.2), if reservoir properties and project design parameters are known, solving the equation in the forward direction generates prediction-based results. Therefore, a forward implementation can be considered as process output predictor (forecasting).

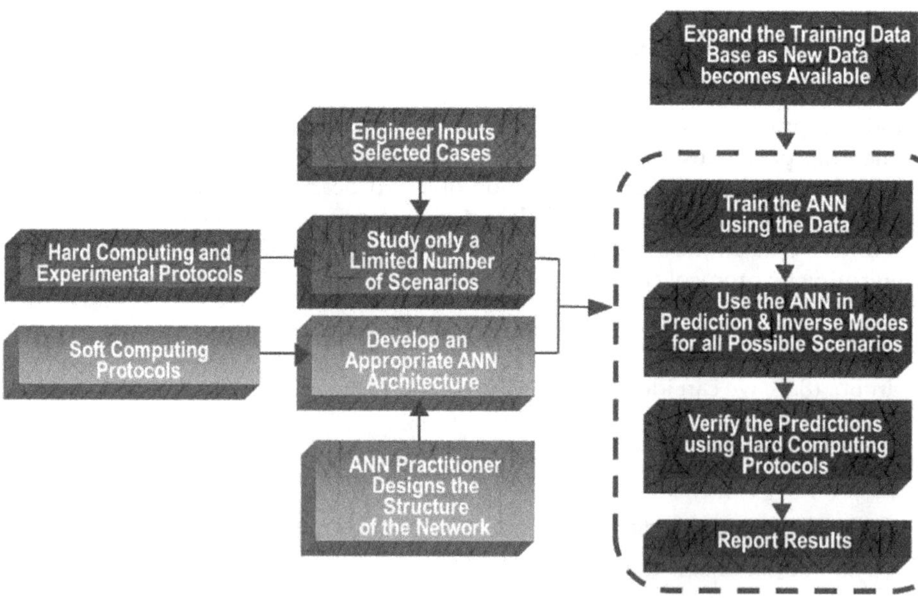

FIGURE 8.1 General workflow for a typical new dimensional protocol.

Equation (8.2) can also be studied in two inverse forms. The first inverse form involves solving for project design parameters [Eq. (8.3)]:

$$[Project\ Design\ Parameters] = \frac{[Response\ Function]}{[Reservoir\ Parameters]} \tag{8.3}$$

Equation (8.3) indicates project design parameters for the process being studied in a reservoir (as defined by reservoir properties) and constrained with response function (constrained by an economic analysis). Accordingly, the implementation of this inverse model can be interpreted as a project design tool.

The second inverse application of Eq. (8.2) is as:

$$[Reservoir\ Parameters] = \frac{[Response\ Function]}{[Project\ Design\ Parameters]} \tag{8.4}$$

What is indicated by Eq. (8.4) is simply a short-hand notation for describing a history-matching protocol.

It should be noted that Eqs. (8.2), (8.3), and (8.4) can be solved using the conventional formalisms and associated solution protocols. However, with the introduction of proxy models and/or hybrid models it will be possible to reduce the manpower needs as well as the computational overhead. With the rapid evaluation options offered by virtual intelligence technology it will be possible to solve some reservoir engineering and production engineering problems at a much rapid rate.

Virtual intelligence technology offers a different paradigm for computing, as biologically inspired computing is different from conventional computing. For example, one can indicate that neural networks are rough models for mental processes. Because of their massive parallelism, they can process information rapidly and carry out the solution protocols almost simultaneously in an intuitive manner. Accordingly, some interactive nonlinear thermodynamical, chemical, or physical processes that we are not able to identify through the diligent use of physical processes can be practically solved using virtual intelligence. It should be recognized that in putting the architecture of the virtual expert system (model) the fundamentals of the problem under consideration are captured in an implicit fashion by the observed data (data from the field of data sets generated by fundamental models) and transmitted to the virtual tool during its training session.

It is also possible to generate hybrid models. In hybrid models, the data-driven models and models based on fundamentals are coupled to each other. This modeling approach provides an avenue to integrate all available information and knowledge in the forms of data and fundamentals. Figure 8.2 shows a computational platform architecture which is built for hybrid models applications. In this platform, the data-driven models shown on the right side of the control unit can be standalone units (tool boxes). In the same way, the collection of modes based on fundamentals which are displayed on the left side of the control unit can also be utilized as standalone models. When data-driven models and high-fidelity models are coupled to each other through an effective hand-shaking protocol a hybrid model is born.

In general, the coupling of data-driven and high-fidelity models can be achieved in two different way. In one case, a history matching protocol results that are accomplished by implementing and artificial expert model can be sent directly to the

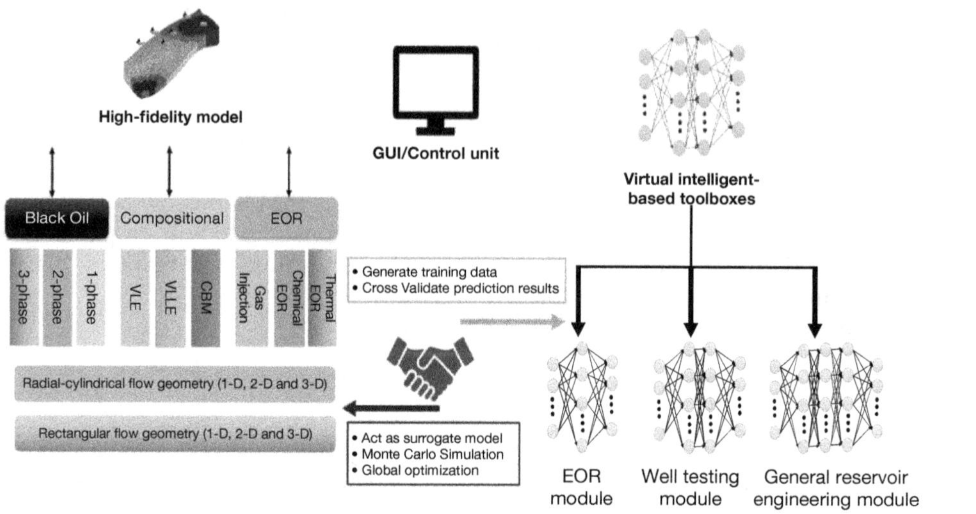

FIGURE 8.2 A schematic representation of computational architecture of a hybrid model.

relevant high-fidelity model for checking the accuracy levels of the history-matched parameters. It should also be noted that the same high-fidelity model has been used to generate some training data for the artificial intelligence-based history-matching tool. As a second option, coupling of these two radically different models can be achieved by embedding some of the data-driven models directly into the physics-driven models. For example, the following implementation describes such a coupling. Consider, the simulation of an enhanced gas recovery project of a coalbed methane reservoir by CO_2 injection. In such a study, typically, a compositional coalbed methane simulator is utilized. There will be two gas components including CH_4 and CO_2. A data-driven model (data is generated using the ideal adsorbate solution) is first built as an expert system which calculates the partitioning of CO_2 and CH_4 molecules between the adsorbed gas phase and free gas phase. As the reservoir model carries the flow calculations at every block and at each iteration, it will need to execute the ideal-adsorbate solution calculations which are tedious and slow (similar to vapor-liquid-equilibrium calculations). Therefore, transferring that part of the calculations to the expert system potentially increases the computational speed by approximately three orders of magnitude.

Figure 8.3 shows how a hybrid model for integrated reservoir modeling studies can be constructed. In such an integrated model one boundary condition is at the physical limits and the other extreme boundary condition is at the pipeline pump station. All other flow domains such as reservoir and well systems, and separator and pipeline network systems also have internal boundary conditions. The reservoir system model which is formulated using the fundamentals of transport in porous media is connected to a wellbore hydraulics expert model. Furthermore another interface between the well system and separator/pipeline network system connects the last link of these integrated models. It should be mentioned that all of these connected reservoir

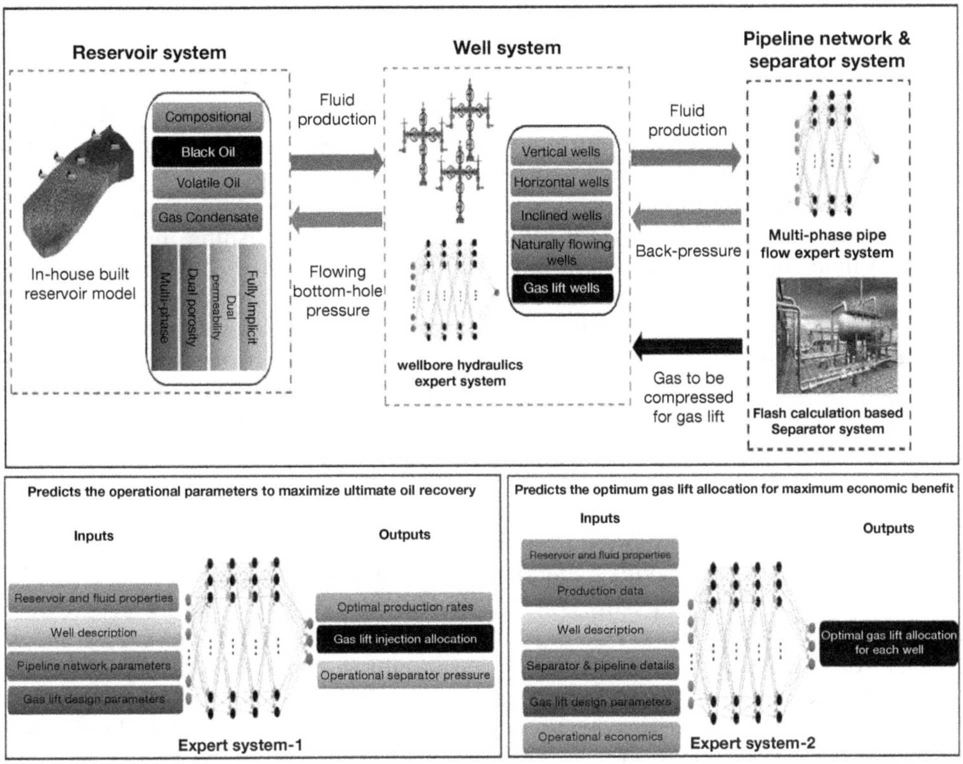

FIGURE 8.3 Field scale production optimization (Putcha, 2007).

models will be operating at the same iteration level at any time step computations. As shown in Fig. 8.3, such an integrated model can be instrumental in designing gas lift operations.

The lower section of Fig. 8.3 suggests two additional expert systems. While the first expert system is proposed to predict the operational parameters to maximize ultimate oil recovery, the second expert system is structured to predict the optimum gas lift allocation for maximum economic benefit from each well.

8.2 ANN-Based Toolboxes for Decision Making in Reservoir Engineering

In this section, some of the toolboxes equipped with ANN-based neurosimulation tools applicable to three different classes of reservoir engineering applications are discussed. In constructing these toolboxes, full advantages offered by deterministic methods (in terms of the accurate description and capture of physical processing power of the artificial neural network technology) are taken into account to ensure the solution integrity and computational speed in the overall analysis and decision processes.

8.2.1 Enhanced Oil Recovery Screening Toolbox

The suitability and application of an enhanced oil recovery (EOR) process depend on reservoir characteristics, fluid properties, reservoir pressure, and reservoir architecture. Large spatial variations in reservoir depositional geology, petrophysical properties, fluid properties, and formation pressure bring in challenges for evaluation design and subsequent applications of the most optimal improved oil recovery technologies for a given field under consideration. Assessing and designing suitable recovery processes are certainly considered as a critical analysis step in establishing the economic success of field deployment of a process. Such direct assessment of potential EOR technologies for possible deployment plans for increased oil production and recovery will require detailed data analysis, integrated modeling, and exhaustive simulation studies. Accordingly, a sound and technically based screening and performance predictive tool is highly promising to fulfill the need for identification of applicable EOR processes and their expected recovery performances. A toolbox as described here promises to be more powerful than the existing guidelines as it provides information not only about the expected reservoir performance but also suggests a set of project design parameters that can be used toward the optimal development of the process decided. The development and implementation of a knowledge-based neurosimulation model arms the reservoir engineers to provide general screening and design criteria for potential IOR/EOR processes for a field with given reservoir characteristics. At the same time, the neurosimulation model can be used to introduce some quantification of the anticipated oil production to form a basis to assess the economic feasibility of the projects envisaged. As shown in Fig. 8.4, the forward-looking tool will allow a reservoir engineer to assess the applicability of various EOR processes to varying reservoir and fluid characteristics of a given formation to identify the most promising process. Figure 8.5 shows the basic framework of the project design tool which can be used to investigate design and operational considerations for the optimized execution of the EOR process under consideration.

The tool described in Fig. 8.6 basically describes a history matching tool that can be used in analyzing the reservoir performance during the pilot testing of the EOR process. With the help of such a tool a final check and analysis can be placed on the reservoir properties that play an important role on the project design parameters. Such a check will be important in guiding the design team in extending the pilot test to a full-field scale implementation.

8.2.2 Well Test Analysis Toolbox

As discussed in Chap. 6, the overall goal of well testing is to collect information about flow conditions in the well around the immediate vicinity of the well, in the virgin portions of the reservoir not influenced by the drilling and well completion operations and information about the location and nature of the boundaries of the reservoir. The measurement of variation of pressure with time provides a pressure transient data which then can be analyzed to determine the formation parameters that characterize the existing flow conditions in the reservoir/well system.

In reservoir engineering, finding the storage and transport characteristics of the hydrocarbon reservoirs is of paramount importance. These reservoir properties help the reservoir engineer in the process of reservoir evaluation and development, reservoir

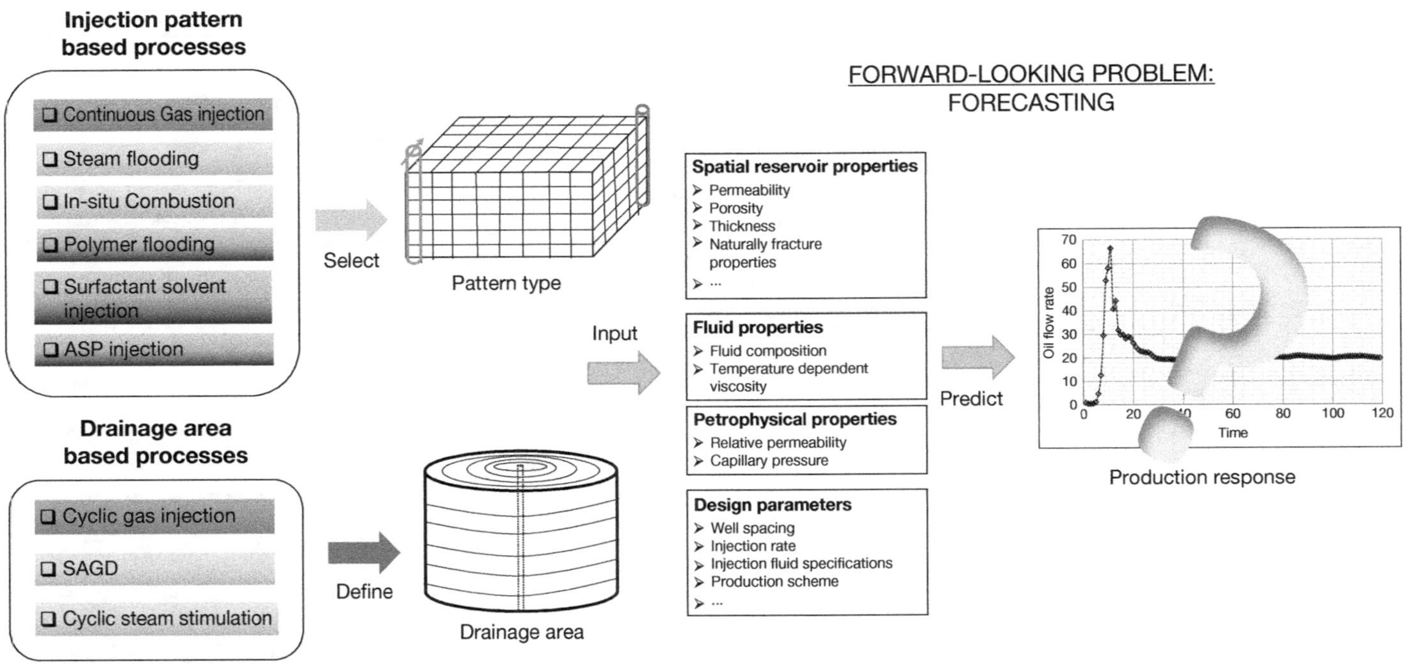

FIGURE.8.4 EOR project screening tool (Sun, 2017).

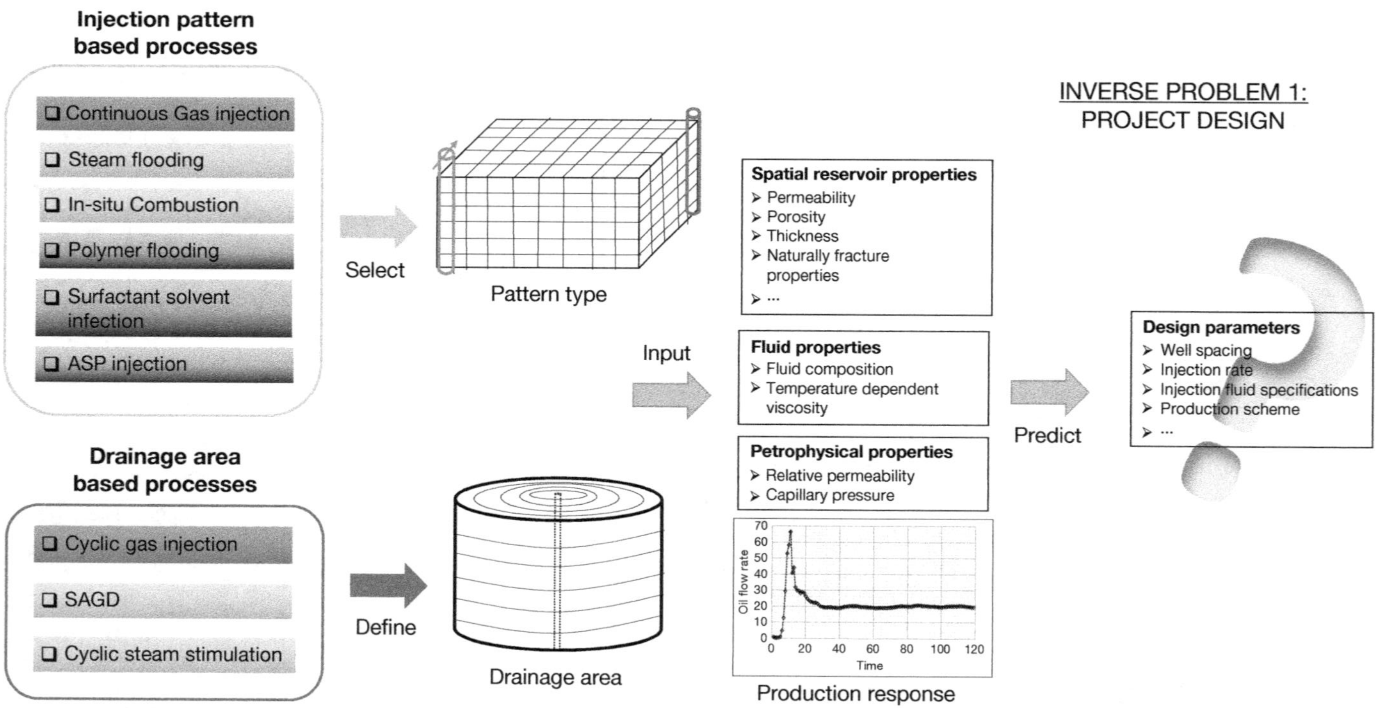

FIGURE 8.5 Basic framework for an EOR process design tool (Sun, 2017).

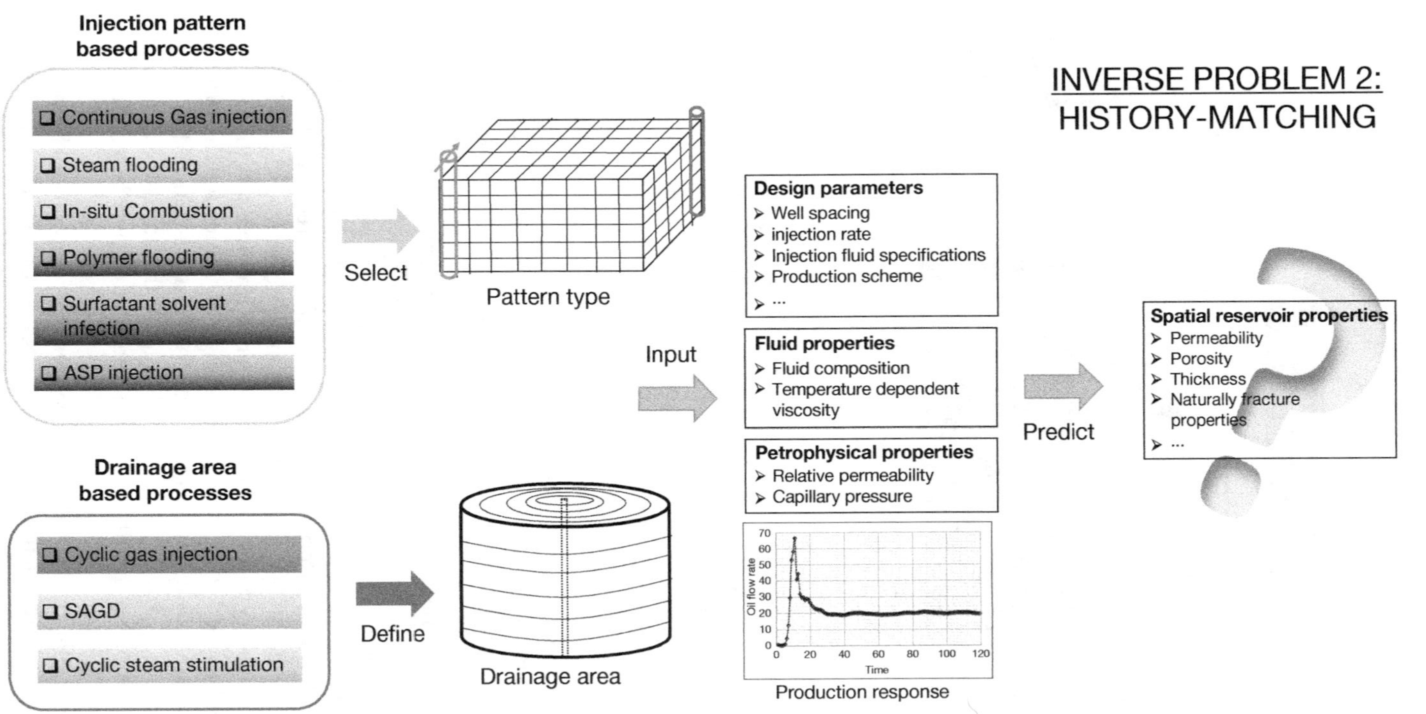

FIGURE 8.6 EOR process history matching tool used during the pilot test application (Sun, 2017).

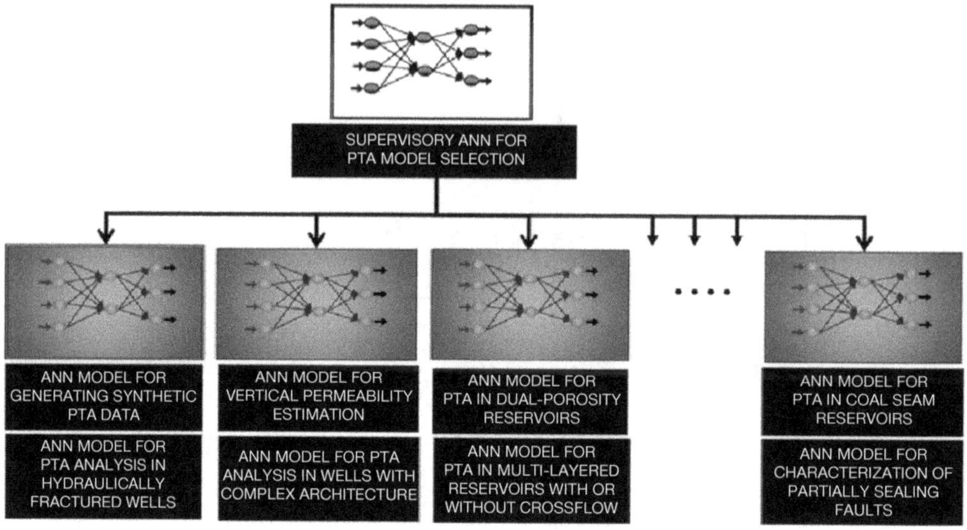

Figure 8.7 Computational toolbox for well test analysis (Ertekin and Sun, 2017).

pressure maintenance, and reservoir recovery optimization to name a few. Well test analysis can be considered as a systems analysis technique. The principles governing the analysis of well tests can be more easily visualized when one considers well test interpretation as a special pattern recognition technique in the background of the overall framework. In Fig. 8.7, a well test analysis toolbox is presented. Each individual tool in the toolbox represents an artificial intelligence-based expert system. In well test analysis perhaps the most critical step is the identification of a relevant model that will be used in interpreting the test data. Typically, the model identification is achieved using an iterative protocol. In other words, a model is selected and analysis is performed if the results best match the data collected, then analysis is reported. If the quality of the match is not found to be satisfactory, another model is selected and analysis is repeated until a satisfactory match is achieved. The supervisory expert system shown on the top of Fig. 8.7 functions solely to direct the incoming requests for pressure transient analysis to a specific expert with the relevant domain expertise. The number of tools in the toolbox can be increased as new tools are developed and introduced into the toolbox.

The expert systems (tools) shown in Fig. 8.7 have received their training using the pressure transient data generated using the analytical forward-looking solutions and/or numerical models. Typically, a large number of training sets (in the order of hundreds to thousands) is used to ensure a wide range of values for the properties that determine the pressure transient response of the systems. Since the learning is the end product of training, any expert system has to go through an extensive amount of testing and validation so that it becomes a qualified expert before it is assigned a place in the toolbox.

As an example, an artificial expert system developed for characterizing a partially sealing fault is briefly discussed here. Table 8.1 gives a summary of the ranges of properties used in generating training data for this example application.

Property (units)	Range
Fluid viscosity (cp)	$1 \leq \mu \leq 50$
Initial pressure (psi)	$500 \leq p_c \leq 8{,}000$
Formation permeability (md)	$2 \leq k \leq 1{,}300$
Formation porosity (%)	$10 \leq \phi \leq 50$
Formation thickness (ft)	$10 \leq h \leq 200$
Total formation compressibility (psi^{-1})	$10^{-6} \leq C_t \leq 10^{-4}$
Distance to the fault (ft)	$98 \leq L_f \leq 1{,}060$
Orientation of the fault* (degrees)	$0 \leq \theta \leq 80$
Sealing capacity of the fault (%)	$0 \leq SC \leq 40$
Flow rate (STB/d)	$20 \leq q \leq 2{,}000$

*With respect to principal flow direction.

Table 8.1 Range of Parameters Used in Training Sets (Aydinoglu et al., 2002).

Figure 8.8 shows the architecture of expert system built for characterizing a partially sealing fault. The ANN topology consists of four hidden layers and one input and one output layer. There are 22 neurons on the input layer including some functional links. The hidden layers have a rather large number of neurons (100, 100, 60, and 30). The output layer has 15 neurons and 6 of them (k_x, k_y, l, SC, ϕ and θ) are the principal parameters that the expert system is asked to determine. The remaining nine neurons are some functional links assigned to the output layer. Figure 8.9 gives a summary of the results of 30 blind tests that are conducted to understand the level of accuracy in the estimated values of the six principal parameters.

8.2.3 General Reservoir and Production Engineering Analysis Toolbox

In Fig. 8.10, we see a generalized reservoir and production engineering toolbox containing several expert systems (tools). Growing demand of oil and natural gas and depletion of conventional reserves led to the achievement of technology to economically produce oil and natural gas from the unconventional reservoirs. These unconventional reservoirs include shale reservoirs, tight sands, heavy oil reservoirs, coalbed methane reservoirs, and methane hydrate systems.

Each of these reservoirs has their own intricacies and in the analysis of this class of reservoirs, we often see that the classical reservoir and production engineering analysis tools are not sufficiently robust or well developed to effectively tackle most of the reservoir engineering problems. There is a definite need for some advanced and fast techniques and analysis protocols to sustain an economic level of production from these reservoirs. For example, multilateral wells are one of the new alternatives of field development in this class of reservoirs as they can increase the productivity per well, reduce the cost of field development, reduce the footprint at the surface, and minimize the need for hydraulic fracturing. Furthermore, as some new advanced technologies are developed their applications to some conventional reservoirs for increasing the

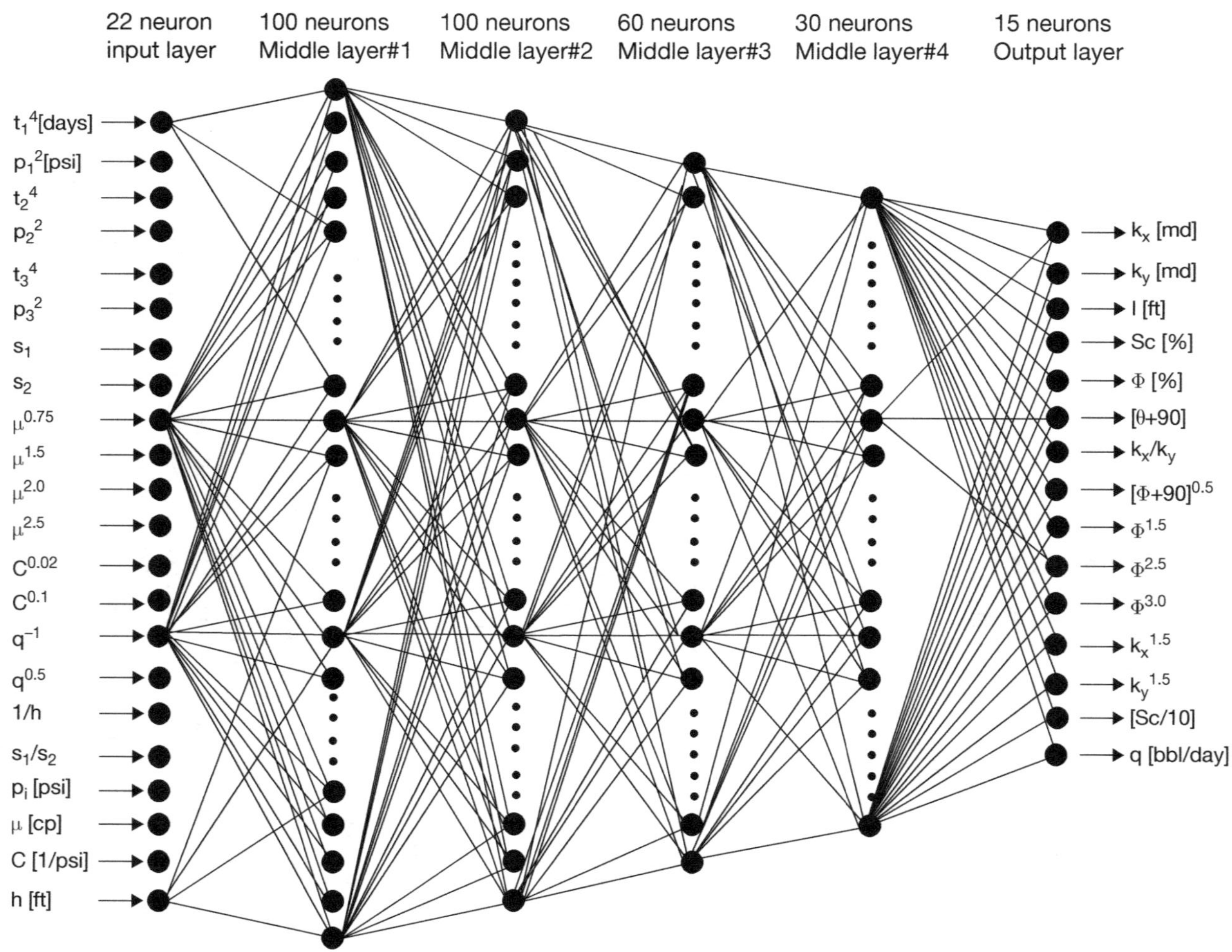

FIGURE 8.8 Architecture of the ANN-based expert system built for fault characterization (Aydinoglu et al. 2002).

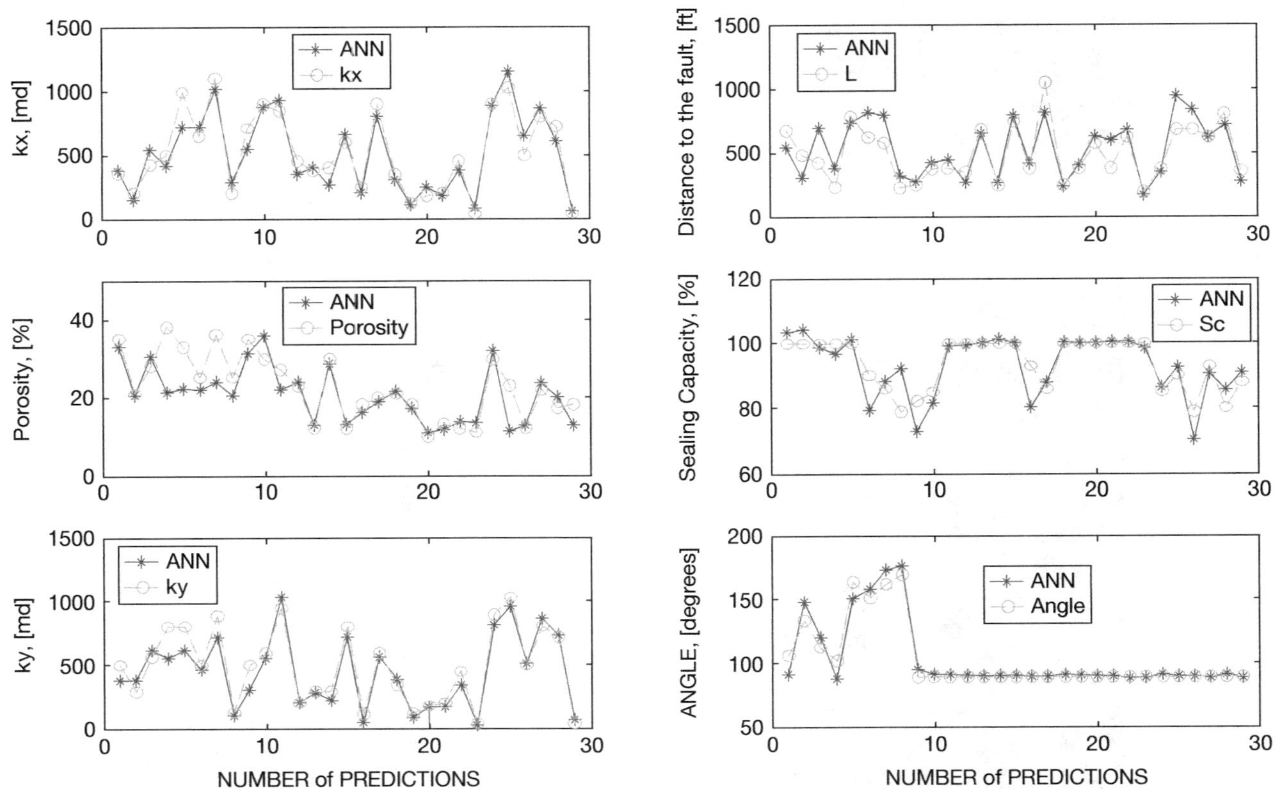

FIGURE 8.9 Accuracy of the values of the six principal unknowns as predicted by the expert system (Aydinoglu et al., 2002).

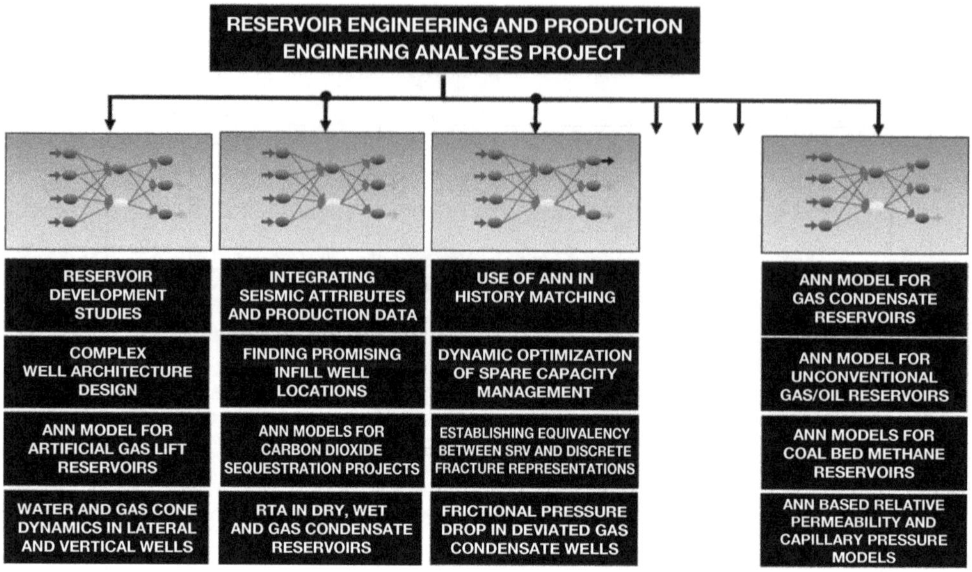

FIGURE 8.10 Overall structure of a general reservoir and production engineering toolbox (Ertekin and Sun, 2017).

efficiency and the comprehensive nature of the solution can become a possibility as well. This is why in the toolbox displayed in Fig. 8.10, tools such as "Assisted History Matching"and "Reservoir Development" are included.

Example 8.1: Case Study: Structuring an integrated reservoir characterization and field development protocol utilizing artificial intelligence.

In this case study, artificial intelligence is utilized to characterize and optimize the field development strategy of a conventional oil reservoir in North America. In structuring such a strategy two sets of ANN-based expert systems are developed. The first set of ANN tools generates synthetic well logs and the second set of ANN tools is developed to generate (predict) productivity maps to be utilized in determining the infill locations.

Figure 8.11 displays the physical boundaries of the reservoir and the locations of the existing wells. More than 40 wells had been developed.

First set of tools addresses the need for synthetic well logs. This tool set is capable of generating five different types of synthetic well logs at any location within the seismic

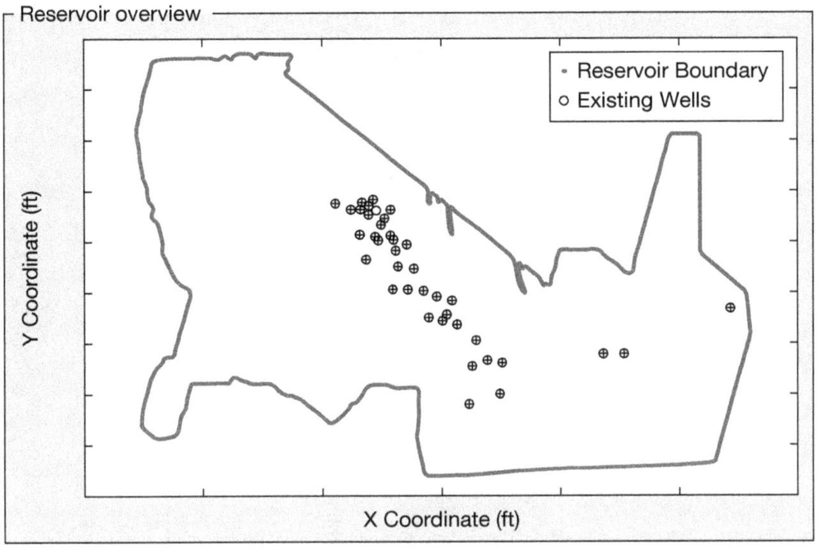

FIGURE 8.11 Reservoir boundaries and existing well locations (Ozdemir et al., 2016).

boundaries. These synthetic well logs can also be used for wells with advanced trajectories such as slanted wells and horizontal wells. Seismic data is the crucial input in reservoir characterization studies. Even though it provides a low-resolution information compared to well logs and core analysis data, seismic data has the most useful information in terms of 3D coverage over a large area. In this study 39 different seismic attributes are utilized. The attributes are extracted by using 100 ft depth intervals within a 3500 ft section. In other words, each seismic attribute is extracted within uniformly divided 35 intervals.

Well logs are primarily used for developing the synthetic well log generation tool. In addition, for certain wells when a certain type of actual log does not exist, synthetic well logs are generated. In this way, at every well location five different types of logs become available for further studies. At the well locations following well logs are available (at each of the 48 wells):

- Gamma ray log (18 actual, 30 synthetic)
- Induction resistivity log (35 actual, 17 synthetic)
- Induction conductivity (29 actual, 19 synthetic)
- Short normal resistivity (24 actual, 24 synthetic)
- Spontaneous potential (27 actual, 21 synthetic)

Figure 8.12 shows the workflow implemented during the synthetic well log generation phase of the study. It should be noted that with the help of the well log cube

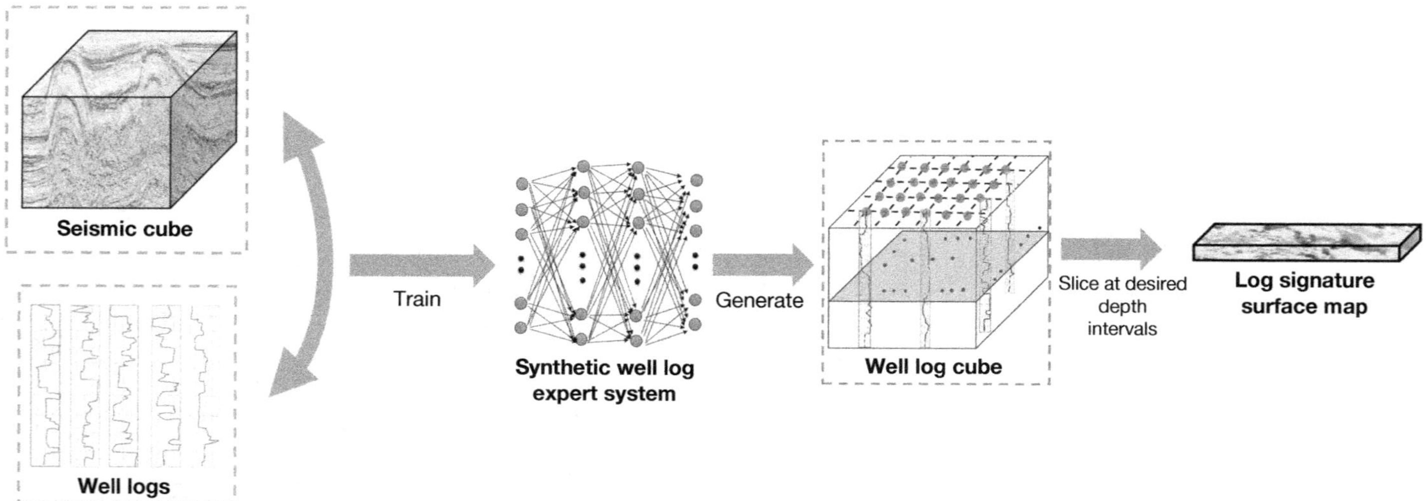

Figure.8.12 Workflow for synthetic well log generation (Ertekin and Sun, 2017).

generated, it will be possible to look at the spatial distribution of a particular well log signature over the entire reservoir in three dimensions. In this way, by taking slices at different intervals, it will be possible to locate zones that should be computed and/or location of the planes in which multilateral wells should be drilled. This is shown in Fig. 8.13.

In the second stage of the study using the existing three-year oil production history in conjunction with the available seismic and well log signature cubes, an expert system which is capable to predict the production performance (productivity) cube is developed. For the purpose of reducing the data dimension, a curve-fitting scheme is implemented to the field production history. During this process the well production histories are decomposed into a plateau and decline period, as shown in Fig. 8.14.

Two curve-fitting equations are applied to the plateau and decline regions. A plateau flow rate which is calculated as the arithmetic average of the flow rates within the plateau region is used as the fitting parameter. Also, the duration of the plateau period needs to be recorded, the production rate of the decline period is fitted using the hyperbolic equation given in Fig. 8.14. Accordingly, only five curve-fitting parameters are needed to be estimated by the expert system. These five parameters include plateau production rate ($q_{plateau}$), duration of the plateau period ($t_{plateau}$), hyperbolic coefficients initial flow rate (q_i), a and b. Figure 8.15 shows the overall workflow for this second stage of the analysis during which production performance curves at different location are predicted.

Figure 8.16 shows a testing case for a well which was not shown to the network during training (blind testing). Promising matches were obtained both for oil production rate and cumulative oil production. The average absolute errors for this testing case are 10% and 6% in predicting oil flow rate and cumulative production as a function of time, respectively.

The final objective of the analysis is to generate a heat map showing the expected brown-stage productivity distribution over the entire field. In this way it will be possible to identify the potential sweet spots for infill drilling purposes. This is schematically represented as the last step of the second-stage workflow in Fig. 8.15. As shown in Fig. 8.17, when such an analysis is conducted it will be comforting to see the heat zones that match the regions where the existing wells are located. High-permeability channels and new sweet spots can be easily identified from the heat maps, as shown in Fig. 8.17. It is interesting to note that there are a number of areas other than the developed section which are identified as potential sweet spots for infill drilling. Also, it needs to be emphasized that to generate predictions using the developed expert system we do not need any information from structural map expert and petrophysical property distributions as they are implicitly conveyed to the network through seismic data, well logs, and production data.

This case study offers an innovative protocol to characterize oil reservoirs in terms of their productivities. What makes the entire protocol more attractive is the high accuracy levels in prediction are achieved at high speeds.

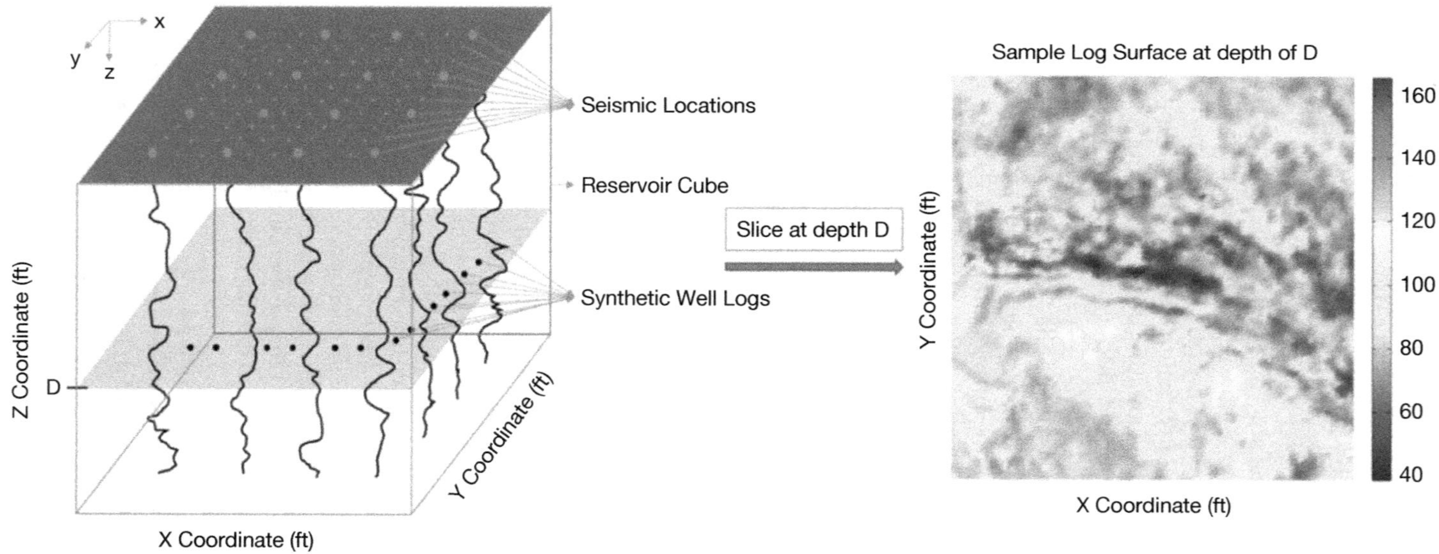

FIGURE 8.13 Generation of log signature surface map at a selected depth (Ertekin and Sun, 2017).

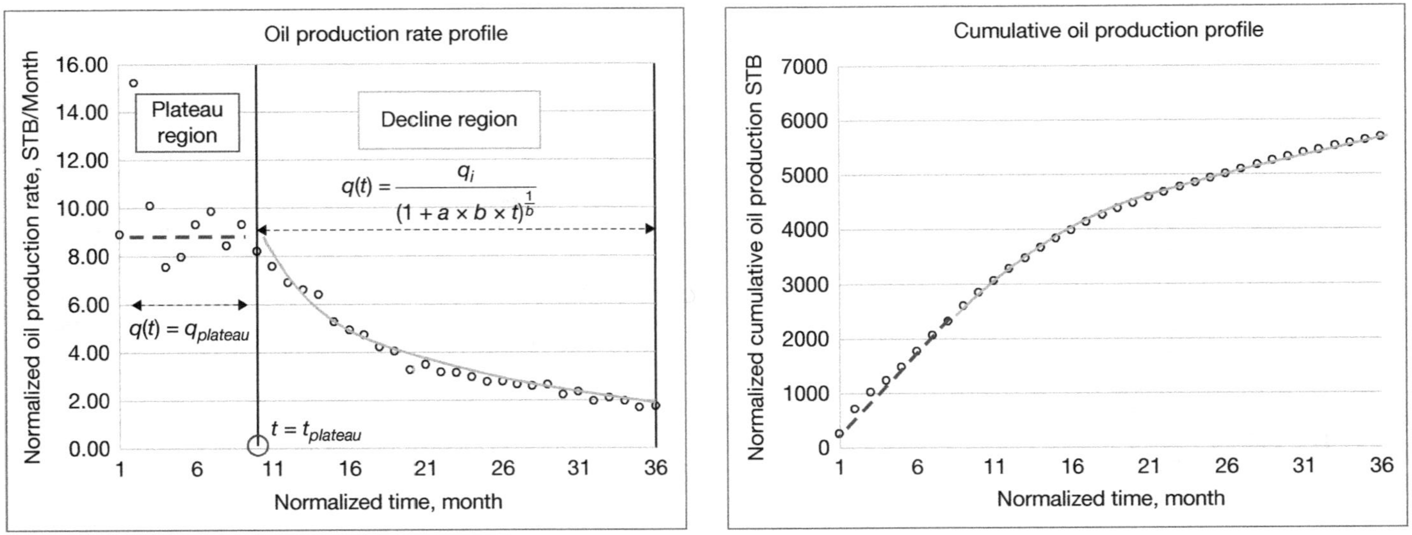

Figure.8.14 Curve-fitting scheme applied to the production histories (Ozdemir et al., 2016).

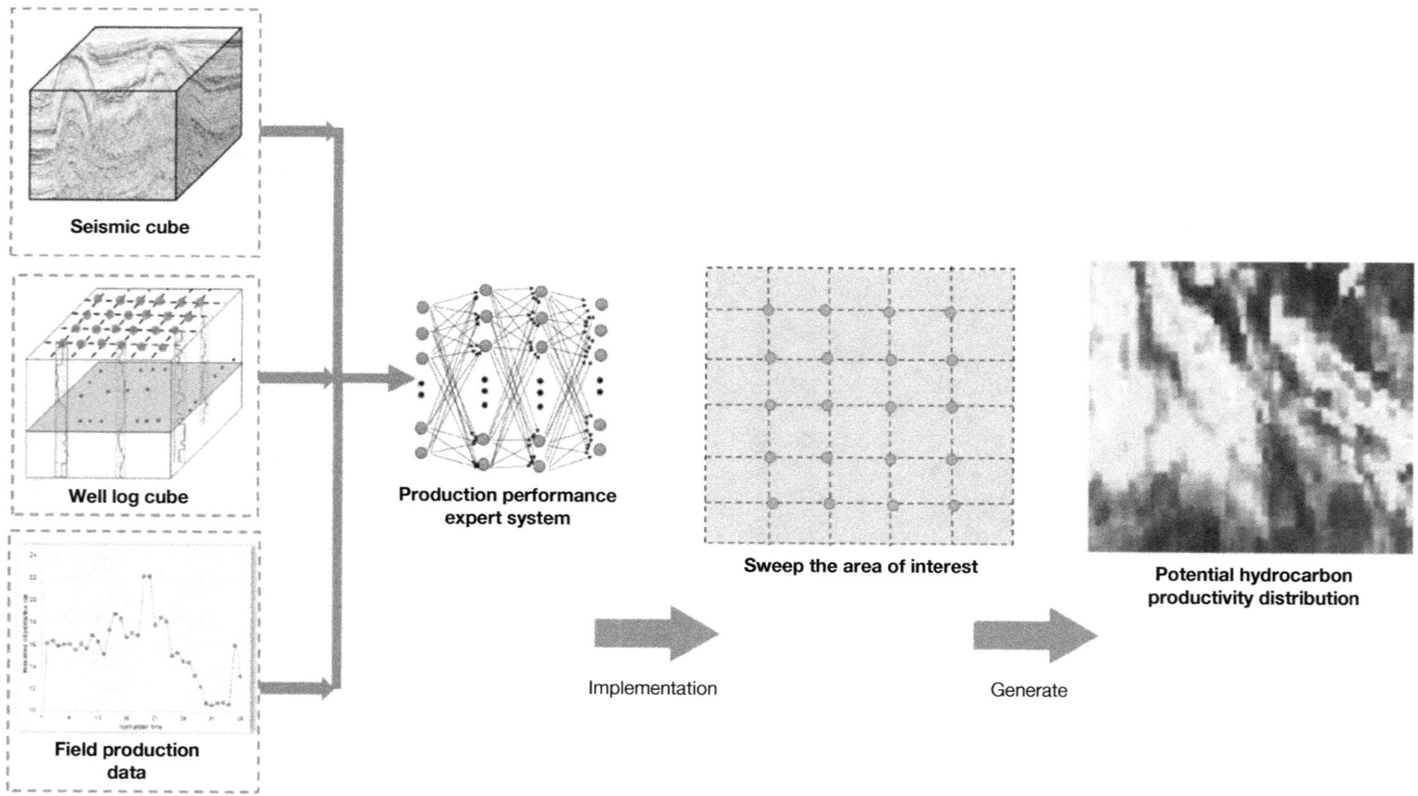

FIGURE.8.15 Workflow for second and final stages of analysis (Ertekin and Sun, 2017).

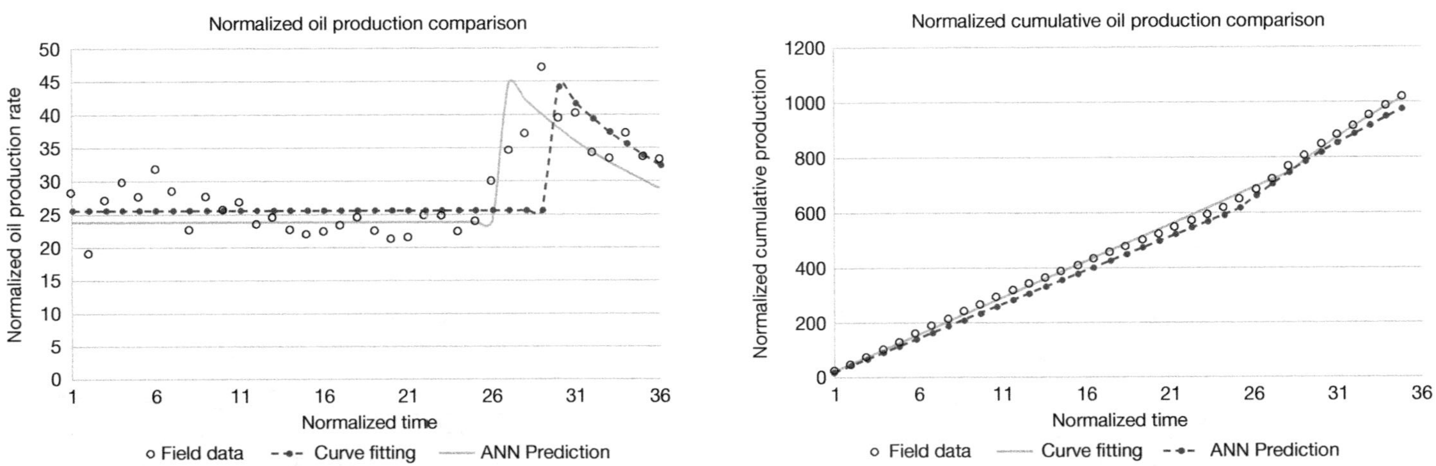

Figure 8.16 Testing of the production performance predictor for one of the testing wells (Ozdemir et al. 2016).

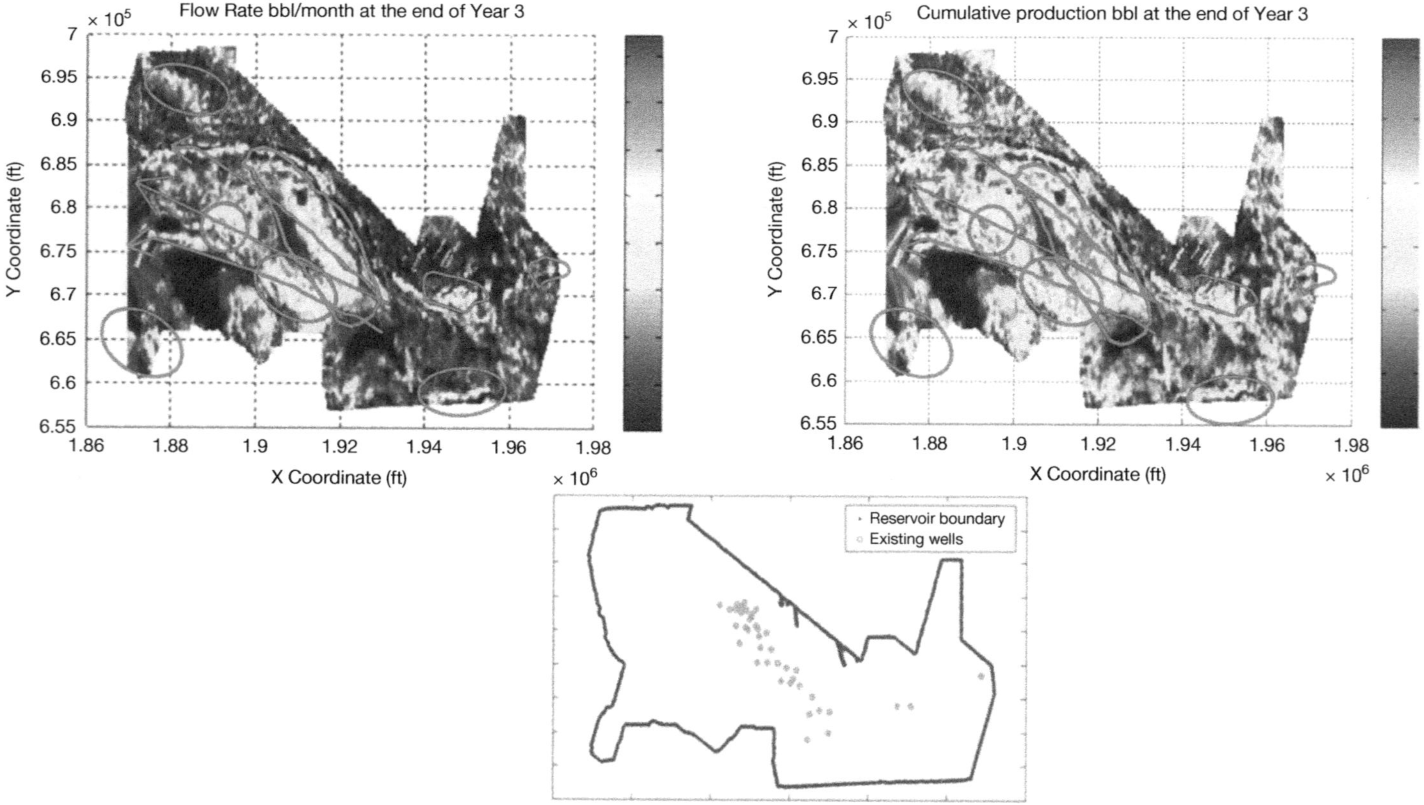

Figure 8.17 Identification of sweet spot and permeability channels for infill drilling process (Ozdemir et al., 2016).

The purpose of this chapter was to provide some examples on how artificial expert systems can be effectively utilized in addressing some of the long-standing reservoir engineering problems. It is hoped that by virtue of artificial intelligence techniques one will be able to assemble expert tools that will enable the reservoir engineers in implementing better controls on decision systems, optimization studies, and information management and in making smart inferences concerning operations.

In developing and assembling tools as described in this chapter there will be opportunities to examine challenges. The barriers to implementing intelligence system-based solutions are not the cost, lack of benefits, or technical risk, but possibly other issues such as the resistance to change. It is also recognized that new technologies may require changes to established workflows which practitioners may resist; however, it is generally assumed that practitioners are not resisting to specific technologies but creation of a discontinuity in technology adaptation, but it is generally recognized that it is this very same discontinuity which sets the level of betterment of our understanding and thinking on a new trajectory in technology adaptation. This is why the creation of workflows similar to the ones described in this chapter will have immediate applications.

8.3 Overview of Mathematical Manipulations in ANNs

In the previous sections of this chapter, we have focused on some possible applications of artificial intelligence in the petroleum industry without discussing the mathematical backdrop of the process. There are volumes of material in the literature about the mathematics of artificial intelligence protocols. Here we will provide a brief review of what kind of simple mathematics is taking place in the development of artificial neural networks.

An artificial neural network is defined as an information processing system. It is a computational model inspired by human brain. It comprises of a number of simple and highly interconnected processors called neurons, which are simplified analogs of the biological cells in the brain. In a typical ANN, each node is connected to every other neuron by numbers called weights. The magnitude of these weights dictates the value of the ANN's output for a particular input. Neural networks basically serve for the following purpose: They are trained with known input-output data following which they are asked to predict the corresponding output for the input that the network has not been trained with. In other words, one can think of neural networks as a highly nonlinear multidimensional interpolators.

There are different types of ANNs that have been used in various petroleum engineering applications. A synopsis of some of the most commonly used neural networks is presented below:

- Back-Propagation Networks (BPNs): The BPN is probably the most commonly used network due to its ease of use and its speed of training. BPNs are used widely in engineering applications when multiple data sets exist.
- Kohonen Networks (KNs): This class of networks can be widely used in areas such as pattern recognition and voice recognition. Their structures differ from BPNs as they fall into the self-supervising class of artificial neural networks.
- Recurrent Networks (RNs): RNs can be considered as offshoots of the BPN. RNs have the important advantage of feedback between neurons, which make it ideal to study problems in control systems where the system is highly dynamic. Their structures are similar to that of the BPN.

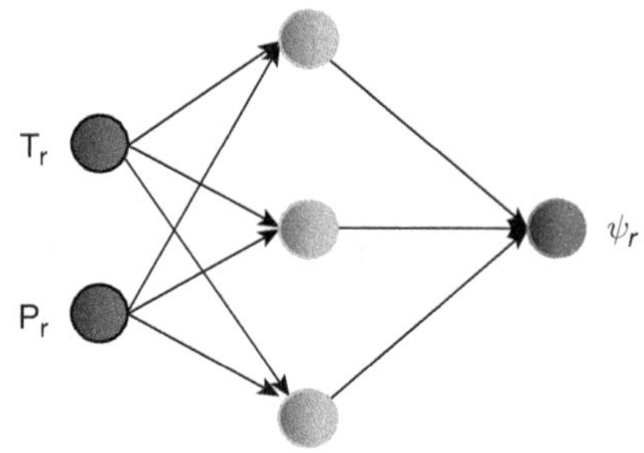

FIGURE 8.18 Structure of the BPN used in the numerical example (Doraisamy, 1998).

Training is defined as the process of determining the magnitude of the weights that yield the desired output for the corresponding input. A fully trained network consists of a set of weights that give the desired output. A fully trained network should be able to generalize and make reasonably accurate estimates of the unknown output for a given input.

The process of training is different for each type of network. In the following sections, we will focus on the back-propagation network as the BPN is one of the most versatile networks and is quite widely utilized for engineering applications.

The BPN consists of a minimum of three layers: the input, middle (hidden) and the output layer. There can exist more than one hidden layer in the middle layers. Each neuron in a particular layer is fully connected to every other neuron in the layer following it (a fully interconnected ANN). A schematic diagram of a BPN is presented in Fig. 8.18. It can be seen from Fig. 8.18 that there are connections between neurons in the input and middle layers and between the middle and the output layers. The number of middle layers is determined by the network designer. The number of neurons is the inputs that are required to obtain an output set. In a similar manner, the number of output layer neurons is equal to the number of output values that exist for a set of input data. The number of middle-layer neurons is not specified, though it is usually greater than the number of input and output layer neurons for maximum efficiency.

Example 8.2: Training a BPN

Consider the following simple problem: We have values of ψ_r as a function of T_R and P_R. It is desired to obtain the value of ψ_r for an unknown value of T_R and P_R which translates into a simple interpolation problem. (Note: This example involves a well-known protocol related to conversion of reduced pressure P_R to reduced pseudo pressure real-gas potential at a specific T_R. In order to show the long-hand implementation of the BPN training only a limited number of training sets (4) are used.

The structure of the BPN used to solve the above problem is presented in Fig. 8.18. There are two inputs (T_R and P_R) that give one output (ψ). Thus, the BPN has a structure with two neurons in the input layer and one neuron in the output layer. The number of middle-layer neurons is selected to be three for this problem.

For a BPN with one middle layer, there exist two weight matrices: between the input and middle layers and between the middle and output layers. For convention, the matrix associated with the input-middle-layer neurons is labeled V while the matrix for the middle-output neurons is labeled W. In this problem, V is a 3×2 matrix and W is a 1×3 matrix.

Step 1: The first step in training a BPN is to initialize the weight matrices to random values. This is done using a random number generator supported on most platforms. Thus, if a network is trained for two cases which are identical in all aspects but the initial random weights, one would expect to see a similarity in the weights as training progresses. Assume that the initial randomly generated values are:

$$V = \begin{bmatrix} 0.1 & 0.1 \\ 0.2 & 0.2 \\ 0.3 & 0.4 \end{bmatrix}$$

and

$$W = \begin{bmatrix} 0.1 & 0.2 & 0.3 \end{bmatrix}$$

The training sets can be arranged as follows:

Pattern #	T_R	P_R	ψ
1	1.3	2.6	3.406
2	1.3	2.7	3.6367
3	1.5	2.6	3.2672
4	1.5	2.7	3.5251

The input and output values are normalized to a value between 0 and 1. The above input is normalized by dividing the input by 3 and output by 5 leading to the following input-output data:

Pattern #	T_R	P_R	ψ
1	0.433	0.866	0.6812
2	0.433	0.9	0.7273
3	0.5	0.866	0.6534
4	0.5	0.9	0.7050

The choice of the divisors (3 and 5 in this case) to normalize the input and output is purely arbitrary. The goal here is to secure the stability of the network. When values of the input are greater than 1, it will lead to increasing magnitudes of numbers when multiplied with the weight matrices. The output needs to be normalized between 0 and 1

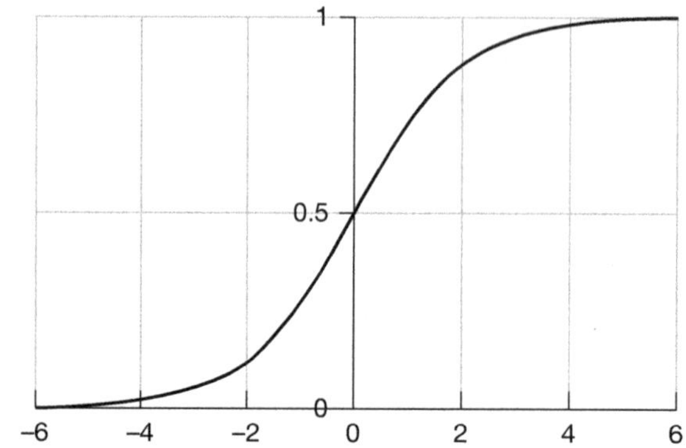

FIGURE 8.19 The sigmoid function.

because the output of the sigmoid function (Fig. 8.19) is bounded between 0 and –1 or between –1 and 1 depending on the choice of the sigmoid function. Thus, it is important to know the upper and lower bounds of the input and the output in the normalization so that input/output values greater than 1 are not encountered.

Step 2: Consider pattern #1:

The input matrix ***I*** is defined as:

$$I = \begin{bmatrix} 0.433 \\ 0.866 \end{bmatrix}$$

The desired output matrix for this input is:

$$o = \begin{bmatrix} 0.6812 \end{bmatrix}$$

Step 3: Multiply the weight matrix ***V*** with the input matrix ***I***. The above multiplication yields:

$$\begin{bmatrix} 0.1 & 0.1 \\ 0.2 & 0.2 \\ 0.3 & 0.4 \end{bmatrix} \times \begin{bmatrix} 0.433 \\ 0.866 \end{bmatrix} = \begin{bmatrix} 0.1299 \\ 0.2598 \\ 0.4763 \end{bmatrix}$$

A transfer function is then applied to the resulting matrix. The transfer function, also called an activation or "squashing" function, can generally be any S-shaped function, but is mostly the sigmoid function:

$$f(x) = \frac{1}{1 + e^{-x}}$$

The shape of the sigmoid function is shown in Fig. 8.19. A key point is that $f(x)$ has a minimum value of 0 and a maximum value of 1 and is differentiable everywhere with a positive slope. The above function is called unipolar sigmoid function. At times, a bipolar sigmoid function can be used, which has an output that varies between –1 and 1. The current example makes use of the unipolar sigmoid function. Apply the transfer function on the resulting matrix following output from the middle-layer neurons.

$$y = \begin{bmatrix} \dfrac{1}{1+e^{-0.1299}} \\ \dfrac{1}{1+e^{-0.2598}} \\ \dfrac{1}{1+e^{-0.4763}} \end{bmatrix} = \begin{bmatrix} 0.5324 \\ 0.5645 \\ 0.6168 \end{bmatrix}$$

Step 4: The next step is to multiply the ***W*** matrix with the y matrix to obtain the following matrix:

$$\begin{bmatrix} 0.1 & 0.2 & 0.3 \end{bmatrix} \times \begin{bmatrix} 0.5324 \\ 0.5645 \\ 0.6168 \end{bmatrix} = \begin{bmatrix} 0.35118 \end{bmatrix}$$

The transfer function when applied to this output provides an output with a magnitude of 0.5869. Recall that output of this input is 0.6812. Therefore, the weights need to be updated.

Step 5: The errors propagated from each layer need to be calculated. To do this, the first-order derivative of the transfer function is utilized.

$$f'(x) = \frac{d}{dx}[f(x)] = \frac{d}{dx}\left(\frac{1}{1+e^{-x}}\right) = f(x)[1 - f(x)]$$

Thus, the first-order derivative for the output layer is calculated.

$$f_0'(x) = (0.5869)[1 - 0.05869] = 0.24244$$

Similarly, the first order derivative of the middle layer will be:

$$f_m'(x) = \begin{bmatrix} 0.5324 \times (1 - 0.5324) \\ 0.5645 \times (1 - 0.5645) \\ 0.6168 \times (1 - 0.6168) \end{bmatrix} = \begin{bmatrix} 0.2489 \\ 0.2458 \\ 0.2363 \end{bmatrix}$$

Step 6: The errors are computed using the following equation:

$$\delta_\theta = \delta_{output\ layer} = (d_k - o_k) f_\theta'(output\ layer)$$

and similarly,

$$\delta_{yj} = \delta_{middle\ layer} = f_{m,f}'(middle\ layer) \sum_{k=1}^{number\ of\ outputs} \delta_{\theta,k} W_{kj}$$

where j ranges from 1 to the number of middle-layer neurons.

Thus,

$$\delta_\theta = (0.6812 - 0.5869) * (0.24244) = 0.02286$$

and

$$\delta_{yj} = \begin{bmatrix} 0.2489 \times (0.02286 \times 0.1) \\ 0.2458 \times (0.02286 \times 0.2) \\ 0.2362 \times (0.02286 \times 0.3) \end{bmatrix} = \begin{bmatrix} 5.689 \times 10^{-4} \\ 1.123 \times 10^{-3} \\ 1.621 \times 10^{-3} \end{bmatrix}$$

Step 7: Having obtained the error matrices, the weight matrices need to be updated. This is achieved by the use of the following equations:

$$\Delta W = \beta \delta_o y^T$$

and

$$\Delta V = \beta \delta_y I^T$$

where β is defined as the learning constant and it is between 0 and 1. The closer the value of β to 0, the better the training and the more robust is the network to oscillations due to local minima problems. The disadvantages of using a low value of β is increased training time. The value of β chosen for this example is 0.5. Hence, the incremental weight matrices are:

$$\Delta W = 0.5 \times 0.02286 \times \begin{bmatrix} 0.5324 & 0.5645 & 0.6168 \end{bmatrix}$$
$$= \begin{bmatrix} 6.085 \times 10^{-3} & 6.452 \times 10^{-3} & 7.05 \times 10^{-3} \end{bmatrix}$$

$$\Delta V = 0.5 \times \begin{bmatrix} 5.689 \times 10^{-4} \\ 1.123 \times 10^{-3} \\ 1.621 \times 10^{-3} \end{bmatrix} \times \begin{bmatrix} 0.433 & 0.866 \end{bmatrix}$$

$$= \begin{bmatrix} 1.231 \times 10^{-4} & 2.46 \times 10^{-4} \\ 2.43 \times 10^{-4} & 4.86 \times 10^{-4} \\ 3.5 \times 10^{-4} & 7.01 \times 10^{-4} \end{bmatrix}$$

Having calculated the incremental weights for **W** and **V**, the weight matrices need to be updated.

$$V^{n+1} = V^n + \Delta V$$

$$W^{n+1} = W^n + \Delta W$$

Thus, the new weights are calculated as:

$$V = \begin{bmatrix} 0.1 & 0.1 \\ 0.2 & 0.2 \\ 0.3 & 0.4 \end{bmatrix} \times \begin{bmatrix} 1.231 \times 10^{-4} & 2.46 \times 10^{-4} \\ 2.43 \times 10^{-4} & 4.86 \times 10^{-4} \\ 3.5 \times 10^{-4} & 7.01 \times 10^{-4} \end{bmatrix}$$

$$= \begin{bmatrix} 0.1001231 & 0.100246 \\ 0.200243 & 0.200486 \\ 0.300350 & 0.400701 \end{bmatrix}$$

and

$$W = \begin{bmatrix} 0.1 & 0.2 & 0.3 \end{bmatrix} + \begin{bmatrix} 0.6085 \times 10^{-3} & 0.64522 \times 10^{-3} & 7.05 \times 10^{-3} \end{bmatrix}$$

$$= \begin{bmatrix} 0.1060805 & 0.20645 & 0.30705 \end{bmatrix}$$

Step 8: Having obtained the new weight matrices, it must be ascertained if the network has been trained to identify the first input pattern. To check if the network has been trained to identify the first pattern, steps 2 through 4 are repeated with the weights computed above as the weight matrices. The output from the middle-layer neurons is computed as:

$$\begin{bmatrix} 0.1001231 & 0.100246 \\ 0.2002430 & 0.200486 \\ 0.3003500 & 0.400701 \end{bmatrix} \times \begin{bmatrix} 0.433 \\ 0.866 \end{bmatrix} = \begin{bmatrix} 0.130166 \\ 0.260326 \\ 0.477058 \end{bmatrix}$$

The transfer function when applied to this matrix yields the output from the middle-layer neurons:

$$y = \begin{bmatrix} \dfrac{1}{1+e^{-0.130166}} \\ \dfrac{1}{1+e^{-0.260326}} \\ \dfrac{1}{1+e^{-0.477058}} \end{bmatrix} = \begin{bmatrix} 0.532618 \\ 0.564716 \\ 0.617053 \end{bmatrix}$$

The resulting matrix y is then multiplied with the W matrix to yield:

$$\begin{bmatrix} 0.1060805 & 0.20645 & 0.30705 \end{bmatrix} \begin{bmatrix} 0.5324 \\ 0.5645 \\ 0.6168 \end{bmatrix} = \begin{bmatrix} 0.362552 \end{bmatrix}$$

Applying the transfer function to this matrix gives:

$$o = \frac{1}{1+e^{-0.362552}} = 0.58965$$

It is seen from the output that there is still an error since the desired output is 0.6812. However, there is an improvement in the network's prediction when compared to the output of the previous iteration.

Hence, steps 2 through 8 must be repeated using the converged weights from the previous iteration as the starting weights. The above procedure is repeated until the calculated output matches the desired output with reasonable accuracy.

Step 9: Having trained the network for the first input set ($T_R = 1.3$, $P_R = 2.6$), the network is then trained to identify the second input set ($T_R = 1.5$, $P_R = 2.6$). This is done by repeating steps 2 through 8 by using the converged weights for the previous input set ($T_R = 1.3$, $P_R = 2.6$) as the initial starting weights. This is done for all the input sets in the training data.

Step 10: The network is trained until it is able to identify all four patterns with reasonable accuracy. When this is achieved, the network is said to be trained.

The final matrices obtained are:

$$W = \begin{bmatrix} -5.599281 & 3.989932 & -4.884027 \end{bmatrix}$$

and

$$V = \begin{bmatrix} -0.715912 & -1.188941 \\ -5.167781 & 3.7914790 \\ -1.294071 & -0.917158 \end{bmatrix}$$

Predicting Unknown Data using a Trained BPN

In this section, the BPN developed in the previous section is used to predict unknown output values that are not part of the training set.

It is desired to compute the value of Ψ for $T_R = 1.4$ and $P_R = 2.65$.

The input matrix after normalizing the input (dividing by 3) is:

$$I = \begin{bmatrix} 0.467 \\ 0.883 \end{bmatrix}$$

Multiplication of $\boldsymbol{V}$ and $\boldsymbol{I}$ yields:

$$\begin{bmatrix} -0.715912 & -1.188941 \\ -5.167781 & 3.791479 \\ -1.294071 & -0.917158 \end{bmatrix} \times \begin{bmatrix} 0.467 \\ 0.883 \end{bmatrix} = \begin{bmatrix} -1.384 \\ 0.934 \\ -1.4138 \end{bmatrix}$$

Applying the transfer function on the resulting matrix yields the output from the middle layer.

$$\begin{bmatrix} 0.2000 \\ 0.7178 \\ 0.1956 \end{bmatrix}$$

Multiplication of the above matrix with W yields:

$$\begin{bmatrix} -5.599281 & 3.989932 & -4.884027 \end{bmatrix} \times \begin{bmatrix} 0.2000 \\ 0.7178 \\ 0.1956 \end{bmatrix} = 0.7888$$

The transfer function when applied to this output provides an output with a magnitude of 0.6875. Recall that the output has been normalized by dividing by 5 during training. As a result, the network's prediction of the output is 3.4378. Note that linear interpolation for the desired values T_R and P_R yields $\psi_R = 3.45875$.

The discrepancy between the network's prediction and the result from the linear interpolation is approximately 0.6%, which implies that the network has been successfully trained and was able to solve the problem for the new data set that it had not seen before.

Example 8.3: Training a Recurrent Network

Recurrent networks are similar to back-propagation network with the important exception of the presence of feedback connections between layers. Figure 8.20 shows the structure of a simple recurrent network. There are three input neurons, two middle-layer neurons, and one output layer neuron. The input layer has neuron labeled "R". This is the recurrent neuron which is connected by weights to the middle layer neuron. The input to the recurrent neuron in the input layer is the output from the middle layer neuron. Similarly, there is one middle-layer neuron in the middle layer which is

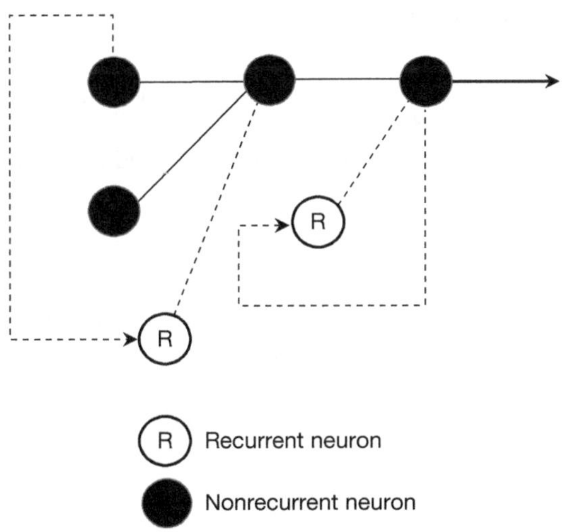

FIGURE 8.20 Structure of the recurrent network used in the numerical example (Doraisamy, 1998).

connected to the output layer neuron. Thus, the input to the recurrent neuron "R" in the middle layer is the output from the output layer. The number of recurrent neurons in the input layer is equal to the number of nonrecurrent middle-layer neurons.

Likewise, the number of middle-layer recurrent neurons is equal to the number of output layer neurons. The above discussion suggests that the output at any time step depends on the output at the previous time step. This system of feedback connections gives the recurrent network the important advantage of solving highly non-monotonous functions.

Training a Recurrent Network

Consider the following problem with the input-output specifications as a function of time.

Time	Input #1	Input #2	Output
1	0	1	0.5
2	0	1	1.0

Step 1: The first step is to initialize the weight. In a recurrent network, there are four weight matrices:

Nonrecurrent weights

- $W^{in\text{-}mid}$: the weight matrix for the connections between the input and the middle layer.
- $W^{mid\text{-}out}$: the weight matrix between the middle and the output layer.

Recurrent weights

- $W^{mid\text{-}mid}$: the weight matrix between the middle-layer neurons to itself.
- $W^{out\text{-}out}$: the weight matrix for the connections between the output layer neurons to itself.

Assume all weights to be initialized to 0.1 and the output from the neurons in the middle layer at $t = 0$ is initialized to be 0.3.

Step 2: The output from the middle layer is computed by the following equation:

$$I_j^{mid}(t) = \sum_{i=1}^{NINP} W_{ij}^{in-mid} y_i^{in}(t) + \sum_{k=1}^{NMID} W_{kj}^{mid-mid} y_k^{mid}(t-1)$$

where *NINP* is the number of nonrecurrent input layer neurons, and *NMID* is the number of nonrecurrent middle layer neurons. In this example, *NINP* is 2 and *NMID* is 1. Thus,

$$I_j^{mid}(t=1) = [(0.1 \times 1) + (0.1 \times 0)] + (0.1 \times 0.3) = 0.13$$

and

$$f(I) = \frac{1}{1 + e^{-0.13}} = 0.5324$$

The activity of the output layer neurons at this time ($t = 1$) is:

$$I_j^{out}(t) = \sum_{i=1}^{NMID} W_{ij}^{mid-out} y_i^{mid}(t) + \sum_{k=1}^{NOUT} W_{kj}^{out-out} y_k^{out}(t-1)$$

where $NOUT$ is the number of nonrecurrent output layer neurons.

Thus,

$$I_j^{out}(t = 1) = (0.1 \times 0.5234) + (0.1 \times 0.3) = 0.08324$$

and

$$f(I) = \frac{1}{1 + e^{-0.08324}} = 0.5208$$

The response of the middle and output layer neurons are then computed for the next time step in exactly the same manner; the only difference is that the value of the response at $t = 0$ is replaced by that of $t = 1$.

Thus,

$$I_j^{mid}(t = 2) = [(0.1 \times 1) + (0.1 \times 0)] + (0.1 \times 0.5324) = 0.15324$$

and

$$I_j^{out}(t = 2) = (0.1 \times 0.5382) + (0.1 \times 0.5208) = 0.1059$$

Applying the transfer function to the above computed activities,

$$f(I_1^{mid}) = 0.5382$$
$$f(I_1^{out}) = 0.5264$$

Step 3: The error needs to be computed in order to update the weights. The errors are always computed backwards in a recurrent network.

Hence, the error at time $t = 2$ for the output layer is:

$$\begin{aligned} E_j^{out}(t = 2) &= \left[y_j^{desired}(t = 2) - y_j^{computed}(t = 2)\right] \frac{d}{dI}\left\{f\left[I_j^{out}(t = 2)\right]\right\} \\ &= (1 - 0.5264) \times [0.5264 \times (1 - 0.5264)] = 0.118 \end{aligned}$$

j in the above equation ranges from 1 to the number of output layer neurons. The error from the middle layer is:

$$\begin{aligned} E_j^{mid}(t = 2) &= \left(\sum_{i=1}^{NOUT} W_{ij}^{mid-out} E_i^{out}(t = 2)\right) \frac{d}{dI}\left\{f\left[I_j^{mid}(t = 2)\right]\right\} \\ &= 0.1 \times 0.118 \times 0.5382 \times (1 - 0.5382) = 2.934 \times 10^{-3} \end{aligned}$$

where j varies from 1 to the $NMID$.

For the previous time step ($t = 1$), the errors are computed using the following equations:

$$E_j^{out}(t) = \left(\varepsilon_j(t) + \sum_{i=1}^{NOUT} W_{ij}^{out-out} E_i^{out}(t+1)\right)\frac{d}{dI}\left\{f\left[I_j^{out}(t+1)\right]\right\}$$

where

$$\varepsilon_j(t) = y_j^{desired}(t) - y_j^{actual}(t)$$

Thus, the error from the output layer neuron at $t = 1$ is:

$$\varepsilon_j(t) = [(0.5 - 0.5264) + 0.1 \times 0.118] \times [0.5208 \times (1 - 0.5208)] = -3.64 \times 10^{-3}$$

Likewise, the error from the middle layer neurons at time $t = 1$ is:

$$\begin{aligned} E_j^{mid}(t=2) &= \left(\sum_{i=1}^{NOUT} W_{ij}^{mid-out} E_i^{out}(t) + \sum_{i=1}^{NMID} W_{ij}^{mid-mid} E_i^{mid}(t+1)\right)\frac{d}{dI}\left\{f\left[I_j^{mid}(t)\right]\right\} \\ &= [(-0.1 \times 3.64 \times 10^{-3}) + (0.1 \times 2.934 \times 10^{-3})] \times 0.5324 \times (1 - 0.5324) \\ &= -1.757 \times 10^{-5} \end{aligned}$$

Step 4: The weight changes are to be computed from the error computed above.

$$\Delta W_{ij}^{in-mid}(t) = \beta E_i^{mid}(t) y_j^{in}(t)$$

$$\Delta W_{ij}^{mid-out}(t) = \beta E_i^{out}(t) y_j^{mid}(t)$$

$$\Delta W_{ij}^{mid-mid}(t) = \beta E_i^{mid}(t) y_j^{mid}(t-1)$$

$$\Delta W_{ij}^{out-out}(t) = \beta E_i^{out}(t) y_j^{out}(t-1)$$

where β is the learning constant and its value varies between 0 and 1. Assume $\beta = 0.5$, hence,

$$\Delta W_{11}^{in-mid}(t=1) = 0.5 \times (-1.757 \times 10^{-5}) \times 1 = -8.785 \times 10^{-6}$$

$$\Delta W_{21}^{in-mid}(t=1) = 0.5 \times (-1.757 \times 10^{-5}) \times 0 = 0$$

$$\Delta W_{11}^{mid-out}(t=1) = 0.5 \times (-3.64 \times 10^{-3}) \times 0.5208 = -9.47 \times 10^{-4}$$

$$\Delta W_{11}^{mid-mid}(t=1) = 0.5 \times (-1.757 \times 10^{-5}) \times 0.3 = -2.635 \times 10^{-6}$$

$$\Delta W_{11}^{out-out}(t=1) = 0.5 \times (-3.64 \times 10^{-3}) \times 0.3 = -5.46 \times 10^{-4}$$

$$\Delta W_{11}^{in-mid}(t=2) = 0.5 \times (2.934 \times 10^{-3}) \times 1 = 1.467 \times 10^{-3}$$

$$\Delta W_{21}^{in-mid}(t=2) = 0$$

$$\Delta W_{11}^{mid-out}(t=2) = 0.5 \times (0.118) \times 0.5264 = 0.03105$$

$$\Delta W_{11}^{mid-mid}(t=2) = 0.5 \times (2.93 \times 10^{-3}) \times 0.5264 = 7.722 \times 10^{-4}$$

$$\Delta W_{11}^{out-out}(t=2) = 0.5 \times (0.118) \times 0.5208 = 0.03072$$

The above procedure is repeated for all input training patterns.

In the above example, there is only one training pattern, as a result the global weight update matrix is equal to the sum of the error at all the time steps.

$$\Delta W_{11}^{in-mid} = -8.785 \times 10^{-6} + 1.467 \times 10^{-3} = 1.458 \times 10^{-3}$$

$$\Delta W_{21}^{in-mid} = 0$$

$$\Delta W_{11}^{mid-out} = -9.47 \times 10^{-4} + 0.3105 = 0.03103$$

$$\Delta W_{11}^{mid-mid} = -2.635 \times 10^{-6} + 7.722 \times 10^{-4} = 7.69 \times 10^{-4}$$

$$\Delta W_{11}^{out-out} = -5.46 \times 10^{-4} + 0.03072 = 0.03017$$

Recall, all weights are initialized to 0.1, thus the new weights after the first pass are:

$$W_{11}^{in-mid} = 0.101458$$

$$W_{21}^{in-mid} = 0.1$$

$$W_{11}^{mid-out} = 0.130103$$

$$W_{11}^{mid-mid} = 0.1007722$$

$$W_{11}^{out-out} = 0.13017$$

The above weights are the starting weights for the next pass. The above steps are repeated until the error between the desired and the calculated output is within the specified convergence criterion for all time steps.

References

Aydinoglu, G., M. Bhat, and T. Ertekin, "Characterization of Partially Sealing Faults from Pressure Transient Data: An Artificial Neural Network Approach." *Proceedings of SPE Eastern Regional Meeting*. Lexington, Kentucky: SPE, 2002.

Doraisamy, H., *Methods of Neuro-simulation for Field Development*. University Park: The Pennsylvania State University, M.S. Thesis, 1998.

Ertekin, T., and Q. Sun, Artificial Neural Network Applications in Reservoir Engineering. In Basile, A., M. Alavi, and S. Curcio, eds., *Artificial Neural Networks in Chemical Engineering* (pp. 123–204). New York City: Nova, 2017.

Ozdemir, I., Q. Sun, and T. Ertekin, "Structuring an Integrated Reservoir Characterization and Field Development Protocol Utilizing Artificial Intelligence." *Proceedings of the 26th ITU Petroleum and Natural Gas Symposium and Exhibition*. Istanbul, Turkey, 2016.

Putcha, V.B., *Integration of Numerical and Machine Learning Protocols for Coupled Reservoir-Wellbore Models: A Study for Gas Lift Optimization*. University Park: The Pennsylvania State University, Ph.D. Thesis, 2017.

Sun, Q., *The Development of an Artificial-Neural-Network-Based Toolbox for Screening and Optimization of Enhanced Oil Recovery Projects*. University Park: The Pennsylvania State University, Ph.D. Thesis, 2017.

Index

Note: Page numbers followed by *f* denote figures; by *t*, tables.

www.ingramcontent.com/pod-product-compliance
Lightning Source LLC
LaVergne TN
LVHW081514300926
849161LV00006B/1168

9781259585630